World Soils Book Series

Series editor

Prof. Alfred E. Hartemink
Department of Soil Science, FD Hole Soils Laboratory
University of Wisconsin–Madison
Madison,
USA

Aims and scope

The *World Soils Book Series* brings together soil information and soil knowledge of a particular country in a concise and reader-friendly way. The books include sections on soil research history, geomorphology, major soil types, soil maps, soil properties, soil classification, soil fertility, land use and vegetation, soil management, and soils and humans.

International Union of Soil Sciences

More information about this series at http://www.springer.com/series/8915

Borut Vrščaj · Blaž Repe
Primož Simončič

The Soils of Slovenia

Springer

Borut Vrščaj
Department of Agricultural Ecology and Natural Resources, Centre for Soil and Environment Research
Agricultural Institute of Slovenia
Ljubljana
Slovenia

Primož Simončič
Department of Forest Ecology
Slovenian Forestry Institute
Ljubljana
Slovenia

Blaž Repe
Geography Department
University of Ljubljana, Faculty of Arts
Ljubljana
Slovenia

ISSN 2211-1255 ISSN 2211-1263 (electronic)
World Soils Book Series
ISBN 978-94-017-8584-6 ISBN 978-94-017-8585-3 (eBook)
DOI 10.1007/978-94-017-8585-3

Library of Congress Control Number: 2017933858

Printed on acid-free paper

This Springer imprint is published by Springer Nature
The registered company is Springer Science+Business Media B.V.
The registered company address is: Van Godewijckstraat 30, 3311 GX Dordrecht, The Netherlands

Preface

Imagine the land of the planet Earth, but without soil. A bare rock and practically nothing on top of it. Thus, nothing that would enable a plant's root to search for water and nutrients, no living space for myriads of organisms, no moles and earthworms, and no tiny bacteria—nothing that makes our shoes dirty. No soil at all. What a desert!

The air, the water, and the soil. With all due respect, it sounds like the Holy Trinity. Without air, we can live only a few minutes. We can survive a few days without water. Without soil and its products, we would die of hunger within a few weeks.

Soil is a space where life meets its counterpart, the mineral component of this planet. In general, people know that soil is need for plants to grow and thereby for food or timber to be produced. The chemical and physical properties of soils are reflected in fertility—a unique, and complex property essential for the production of the necessary quantities of healthy food as well as the growth of woods and other biomass. We enjoy the food, and its diversity and quantity. To be precise, not all of the Earth's inhabitants enjoy it to the same degree. For many people it is a luxury, as they go to sleep hungry each evening.

However, food is not the only way we benefit from soil. Soils perform key ecosystem services and functions that are crucial for the existence and diversity of terrestrial ecosystems. In addition to providing media for plant roots, soils decompose plant residuals, store and release carbon that is then again available for photosynthesis, and enable the circulation of matter, especially nutrients; soils break down and neutralize harmful substances, filter rainfall, and thus supply sources of drinking water. Soils are a unique habitat for a remarkable number of different organisms. They are also a source of raw materials and store the natural and cultural heritage of the past, etc. The list of functions and services is long and varied.

Soil, or the "skin of the Earth," as some of soil scientists refer to it, appears in many forms showing different characteristics, having a range of capacities, and has variable suitability for plant growth or other uses. The differences in forms and soil types basically determine the diversity of our terrestrial ecosystems and the range of diverse natural habitats. It is due to soils that our eyes can enjoy trees, birds, meadows, butterflies, wheat fields, forests, and deer–everything that lives in and above the soil. The biodiversity of our terrestrial ecosystems stems from the diversity of the soils.

We could go on, but let us be brief: no more no less, the soil is the fundament that enables virtually all land-based life.

And if this is true, why does the general public know so little about the soil? What does mankind gain from the soil and how does it serve us? What is its role in terrestrial ecosystems? How do different soils contribute to the cultural landscapes that we like and are proud of, and, on the other hand, how do they contribute to the valuable natural landscapes we are trying to protect? Although we are well aware of the importance of clean air and water, the importance of soils for the life of this planet is not sufficiently recognised. What is happening with the Earth's soils, how fit they are? Why do we know so little about soils? Or better put, why do we treat soils so badly?

To the typical modern human, soils are self-evident and above all an overlooked resource beneath our feet. People are mostly not aware how soils benefit us and what there role is in ecosystems. In general, we do not recognise them as being valuable. We are not aware that they are not an endless resource. Millennia are required for their formation, and it is important how we treat them.

Yes, we, the "soil people," can talk about soil for hours. There is a good chance many of you remember Dr. Dick Arnold saying "Let's talk soils." The more we know about soils the more we are tempted to discuss them. But how successful have we been in passing on this information regarding the importance of soils to politicians, decision-makers, and other nonsoil people who depend on soils, as does everyone else? Not very successful! But we need to be.

We are faced with the extremely rapid growth of the world's population and therefore of the global needs and pressures on soil resources. At the same time, we are becoming convinced of the consequences of climate change and in large-scale land degradations. In order to mitigate pressures and to respond to the threats to soils adequately, it is necessary to raise awareness of the importance and role of soil in civil society and especially within decision-making centres. We need to better advertise the role and importance of soil in order to achieve food security for successful adaptation to and mitigation of climate change, the maintenance of ecosystems, and sustainable development.

It is high time to design effective policies and measures to achieve sustainable management and better protection of our soil resources.

These and many more other reasons convinced the United Nations to declare 2015 the International Year of Soil and December 5, 2015, the World Soil Day at the 68th General Assembly, held on December 20, 2013.

Slovenia is not a large country, but we, Slovenians, share the same planet with other nations; soils globally are as important as soils in Slovenia, and we believe the opposite also holds true.

This book is a part of the World Book Soil Series initiated by the IUSS. Additionally, it is a small contribution to the UN—FAO and Global Soil Partnership initiative, as well as part of raising awareness of the world's soil and soil protection policies.

Ljubljana, Slovenia

Borut Vrščaj
Blaž Repe
Primož Simončič

Acknowledgements

This book could not have been written without the fundamental, crucial information and data on Slovenian soils. In the past decades, many people have been involved in surveying, digging, shovelling, opening soil profiles, and describing the soil properties. The authors would like to take this opportunity to thank them for their work and timeless contribution to Slovenian soil science and soil survey activities.

Our colleagues who have passed away should be acknowledged first of all. Prof. Bogdan Vovk (1902–1988), the first scientist with a Ph.D. in soil sciences in Slovenia was a researcher and full professor of soil science. He determined the fundamentals of Slovenian soil science, pioneered the soil curriculum at the University of Ljubljana, and initiated international collaboration. Dr. Dušan Stepančič (1921–2007), an excellent soil scientist and soil surveyor who was responsible for the soil survey and field research for a long period, contributed to several important soil survey reports and maps. Dr. Jože Sušin (1922–1988), a full professor of soil science, was a researcher and the author of the first dictionary of soil science in the Slovenian language. Prof. Marjan Ažnik (1927–2010) dedicated his efforts to teaching and research activities on the soil–plant relationship, plant nutrition, and fertilization. Special thanks go to Prof. Mirko Leskovšek (1923–2007), a full professor and researcher in the field of soil fertilization and plant nutrition. Numerous long-term field tests were fundamental to a series of practically oriented books on soil fertilization. Dr. Narcis Mršić (1951–1997), a biologist, was the author of the outstanding publication "Animals of Our Soils," a book on Slovenian pedozoology with exhaustive texts and numerous and remarkable drawings of soil organisms that might never be surpassed.

Very warm thanks and special merit go to Prof. Dr. Albin Stritar (1925–1988) for his enthusiasm in teaching the subjects soil science and land use and soil protection. His lectures attracted many students and guided their professional path towards soil science. His "*Pedologija—Kompendij*," is a textbook that numerous students of agriculture have had to read. His outstanding book "*Landscapes, landscape systems and Soil Use and Soil Conservation in Slovenia*," which describes Slovenian soils, catenas, and landscapes, should be mentioned here. Professor Stritar was one of the soil scientists that passionately protected the best agricultural soils in Slovenia against sealing and urbanization.

Among geographers, Prof. Svetozar Ilešič (1907–1985) deserves mention. He is one of the greatest Slovenian geographers of all time and was also the first geographer who fully recognised the true meaning and importance of soils. The late Prof. Ivan Gams (1923–2014) must also not be overlooked. He devoted his life to nearly all aspects of physical geography and karstology.

Several colleagues, soil surveyors, and scientist still dig, auger, and study soils. There is no doubt that Tomaž Prus, M.Sc., deserves special credit for his lifework as a soil surveyor, and the first president of the Slovenian Soil Science Society. In the opinion of several colleagues, he is "the guy" who has studied the largest number of soil profiles now available in Slovenian digital databases. For elaboration of the Soil Map of Slovenia, the majority of credit has to be given to late Janez Rupreht, Marjan Šporar, Dr. Borut Vrščaj, Dr. Nataša Vidic, Marko Zupan M.Sc., and several others. Dr. Franc Lobnik, a retired professor and long-time Head of the

Centre for Soil and Environment Science, played an important role in Slovenian soil science. The credit for ensuring funding of the Digitalization of the Soil Map of Slovenia programme should be attributed to him.

The lines from the manuscript soil maps, often not easy to read, and the numbers from the reports had to be digitized, typed into datasets, verified, arranged, and linked. A large portion of this great work was carried out by Irena Tič, a colleague whose contribution was often overlooked and perhaps even underestimated.

There would not be complete soil information without soil analytical data. A large number of soil samples were carefully analysed by the staff in several soil laboratories. Here, we would like to mention the late Drago Razboršek, Vera Omerza Premru, and Andrejka Hodnik, colleagues who spent almost their entire life's work to the soil samples. Several others such as Vida Pongrac Žnidaršič, Daniel Žlindra, M.Sc., Zalka Ilc, and Dr. Nives Kugonič, who still work with soil analytics, should be mentioned here.

Several colleagues from the Agricultural Institute of Slovenia participated in the content of this book. The authors wish to thank and give credit for their help and contributions. We thank Dr. Tomaž Kralj for his careful reading of the texts related to the Slovenian Soil Classification and the WRB. Marjan Šinkovec and Janez Bergant did an excellent job on the GIS and spatial data on geology. Together with Janez Sušin, they made a significant contribution to various research reports used to write this book. Tone Godeša, M.Sc., has designed the hydraulic-driven and computer-controlled apparatus for measuring soil penetration resistance and shear strength and has patented an excellent device for drilling undisturbed soil profile samples.

The Soils of Slovenia comprises texts, photographs, and tables from a number of publications and research reports. We wish to thank Dr. Helena Grčman and Mr. Marko Zupan, M. Sc., for ther work on soil contamination and the soil classification. The reports written by both of them were used to an important extent within the contamination topic of this book.

For their work in the field of forestry and soil science, great appreciation goes to our colleagues that intertwined these two fields in the 1970s, namely Mr. Janko Kalan, long-time head of the Forest Ecology Department as well as head of the soil laboratory; Mihej Urbančič, hard-core field pedologist and practitioner and the main author of the Forest Soils Atlas; and Dr. Polona Kalan, head of the soil laboratory. Lidija Kodrič, Marjan Pavšar, and others were constantly breaking new ground in soil research starting in 1947. This book has also benefited from contributions including texts, citations, photographs, and comments mostly from forestry colleagues: Dr. Milan Kobal, Dr. Lado Kutnar, Dr. Aleksander Marinšek, Dr. Klemen Eler, Robert Robek, M.Sc., Dr. Urša Vilhar, Andrej Verlič, Daniel Žlindra, Rok Pisek, M.Sc., Franc Perko, M.Sc., and others.

The most important Slovenian soil geographer, who should be given nearly all credit, is the now retired but fully active and tireless researcher of Slovenian and global soils Prof. Franc Lovrenčak. He is a true pioneer in geographic soil research in Slovenia. More than 40 years ago, he elevated soil geography to an equal and integral part of Slovenian geography at the Geography Department of the University of Ljubljana. He is the author of a university textbook (Soil Geography) and an author/co-author of three manuals on laboratory analysis and field research. Professor Ana Vovk Korže has carried out a significant amount of work. Education and raising public awareness are also topics that Dr. Milena Petauer devoted her research to. Dr. Blaž Komac and Dr. Matija Zorn together research mostly processes related to soil erosion and landslides. Some parts of this publication would have been impossible without their contributions. There are other geographers whose research also broadened the knowledge of Slovenia's soils. Professor Borut Belec from the University of Maribor published yet another university textbook. Professor Darko Radinja researched soils on alluvial deposits and in karstic areas. Professor Jurij Kunaver dealt with terminology, soil erosion, and didactic approaches to research. Dr. Barbara Lampič explored the vulnerability of Slovenian Soils.

Several other people directly or indirectly contributed to this book. The authors expressly wish to thank the English language editor of the project, Dean J. DeVos, who successfully strove under difficult circumstances to understand the vagaries of soil science well enough to be an important part of the team, as well as Žan Pižorn, a pilot who took us to the skies to observe and photograph the spatial distribution of Slovenian soils.

Last but not least, it is very true that the book would not exist without the patience of our families, who generously tolerated our absence while it was being written.

Borut Vrščaj
Blaž Repe
Primož Simončič

Contents

About the Authors

Dr. Borut Vrščaj (b. 1960) is an agronomist and soil scientist. In order to find out how the soil enables plants to grow, in 1984 he attended a student research camp, where he first came into touch with pedology and soil surveying. As a university student, he intensively contributed to the research work and teaching at the Center for Soil and the Environment of the University of Ljubljana, where he subsequently graduated with a degree in soil information systems. His completed his studies at the University of Turin with a Ph.D. thesis on soil quality evaluation in urban environments. He is a specialist in soil quality, soil properties, environmental modelling, and soil classification. He has worked in the fields of soil surveys (mapping), soil environmental informatics, and landscape ecology. His international experience is extensive; he was a USDA Cochran Fellow, worked as a visiting scientist at the European Soil Bureau, JRC, Ispra, was a member of the EC ESBN Digital Soil Mapping and the TWG ESDAC—INSPIRE Working Groups, and an FAO DSM expert. For the last eleven years, he has been working at the Agricultural Institute of Slovenia as a researcher and as Head of the Department of Agricultural Ecology and Natural Resources, where he actively participates in national and international research projects. He is active in raising awareness of soils and transferring knowledge on soils and the environment by teaching soil science, soil ecology, land use, and soil protection at the Environmental Protection College in Velenje, University of Maribor and the University of Primorska. He is currently the president of the Slovenian Soil Science Society.

Dr. Blaž Repe (b. 1972) is a geographer and an assistant professor at the Department of Geography, Faculty of Arts, University of Ljubljana. He joined the department as a student and actively participated in the field of cartography. After completing his undergraduate thesis, he worked in the field of soil geography, biogeography, geographical information systems, and digital cartography, on which he also lectures today. He has been invited as a guest lecturer to other universities in Slovakia, Croatia, and Bosnia and Herzegovina. He has participated and written several university and other textbooks and manuals. In the field of soil geography, he is active in digital soil mapping, spatial modelling of soil formation factors, soil degradation, and soil classification. An important part of his work involves educating secondary school teachers and transferring knowledge to lower levels of education. Raising public awareness of the importance of soils is one of his additional soil-related activities. For a substantial part of his career, he has acted as Head or Vice Head of the Department of Geography, Faculty of Arts, University of Ljubljana. He is an active member of the Slovenian Soil Science Society and several other international societies.

Dr. Primož Simončič (b. 1961) is a forestry scientist from Ljubljana, Slovenia, and works as a senior researcher at the Slovenian Forestry Institute (SFI). After completing his degree in Wood Science and Technology, he worked at various institutions. As a designated "Young Researcher," he completed a Master's Degree in Agronomy at the Biotechnical Faculty, University of Ljubljana, in 1984, and a Ph.D. in Forestry at the Biotechnical Faculty, University of Ljubljana, in 1992. Between 1988 and 2012, he worked as a researcher at the

Department of Forest Ecology at the SFI and Head of the Department between 1996 and 2013. The scope of his work includes forest soil quality, forest soil nutrients and tree nutrition, and carbon pools and fluxes in forest ecosystems. He has specifically focused his work on intensive monitoring of forest ecosystems, carbon sequestration, and greenhouse gas reporting. He is currently the director of the SFI. He is a member of several national and international societies and an active member of the Soil Science Society of Slovenia.

List of Figures

List of Tables

1 Introduction

Borut Vrščaj, Blaž Repe, and Primož Simončič

Abstract

This chapter presents soil and soil-related research activities in Slovenia. It discusses early soil surveying and mapping, including the first soil map, which was published in approximately 1873. In general, soil research in Slovenia started at the beginning of the twentieth century and intensified after WWII. This chapter presents the research activities of a number of Slovenian soil scientists, agriculturalists, foresters, and geographers. The institutions that carry out soil research and/or teach soil science are also briefly presented.

1.1 About Soils in Slovenia

Soils in Slovenia are, in the literal sense of the word, a finite resource. The country is hilly and diverse; the extent of plains is limited. The soils are frequently shallow and stony on steep slopes or shallow or sandy on permeable alluvial gravels in lowlands next to hydromorphic alluvial clays and clayey terraces. Soils that rank into the groups Rendzina/Rendzic Leptosols and Leptosols are the most common.

In terms of spatial distribution, the diversity of soils is high. The rough relief of the cold and moist Alpine environments, the hot and dry Pannonian Basin in combination with the waterless Karst features and phenomena make the country very diverse. The landscapes of this relatively small territory, featuring Alpine, Continental, and somewhat transitional Mediterranean and Pannonian climates, are appealing to visit and enjoy every season of the year. But the spatial diversity of the soils also significantly contributes to the Slovenian landscapes; they are interesting; we could say even spectacular in places.

Yet the soils of Slovenian have to feed Slovenians and perform ecosystem functions as well. In comparison with the neighbouring lowlands of Hungary, Croatia, and Austria, our soil resources seem to be very limited. Agricultural production decreased substantially over the last two decades, and positive signs that the food security of the country is improving are still not apparent. Agricultural production has to be increased in spite of the significantly reduced extent of good agricultural soils.

Nowadays, in Slovenia, we are becoming aware that our soil resources were not adequately managed in the past. Too many high-quality soils were sealed in the name of economic development. True, development is very much needed, but do we have to urbanize the best soils for such goal? Or would it be wiser to choose very carefully which soils—and where—we can afford to urbanize and which soils have to be protected and preserved because they are needed? In the future, land use planning in Slovenia should focus more on the information on soils in order to steer development towards a more sustainable approach.

Now, much more than in the past, we need focused information on the quality, quantity, and spatial distribution of soils of Slovenia. This information, focused and purpose-driven, should be available in order to direct agricultural and forestry production towards greater sustainability. The soil organic matter content should be increased

B. Vrščaj (✉)
Department of Agricultural Ecology and Natural Resources, Centre for Soil and Environment Research, Agricultural Institute of Slovenia, Ljubljana, Slovenia
e-mail: Borut.Vrscaj@kis.si

B. Repe
Geography Department, University of Ljubljana, Faculty of Arts, Ljubljana, Slovenia
e-mail: blaz.repe@ff.uni-lj.si

P. Simončič
Department of Forest Ecology, Slovenian Forestry Institute, Ljubljana, Slovenia
e-mail: primoz.simoncic@gozdis.si

B. Vrščaj et al., *The Soils of Slovenia*, World Soils Book Series,
DOI 10.1007/978-94-017-8585-3_1

where possible; conventional tillage practices should be adapted to better mitigate the frequent droughts and nutrient depletion; the use of pesticides needs to be lowered, and the fertility and the productivity of the soils need to be retained or even increased in spite of lowering the fertilizer inputs.

Being shallow and sandy, the soils above the groundwater sources need to be very carefully managed in order to better protect important sources of potable water. This very sensitive issue is frequently in conflict with intensive agricultural production.

Threats to soils are in general not visible enough to convince the decision-makers to invest resources in soil protection. Soil erosion, soil compaction, and soil contamination are the main forms of degradation that we need to struggle against.

The great amount of work in the past yielded a significant volume of soil information, which is now available for designing sustainable agricultural practices on the national level. According to the needs, they are still not sufficient to steer agricultural production, groundwater protection, and forest management on the detailed level of individual plots. In general, we lack information about the fertility; is it decreasing or is it stable? Can it be increased, and if yes, where and how? What is the content of the soil organic matter? How much and where has soil carbon been lost? Where, regarding which soils and tillage practices, can it be increased? Is our soil becoming increasingly poor or too rich in nutrients? Are the soils under the pure pine forest stands becoming more acidic? Is the soil organic matter increasing or not? How severe is soil compaction under heavy machinery in agriculture and forestry? Many questions have to be answered in the near future, and within a scenario of climate change, even more questions are expected to arise.

1.2 About the Soil Research History

The history of soil research in Slovenia started in the nineteenth century when the first rough soil map of Carniola was elaborated. After WWI, soil data was collected mainly for plant nutrition purposes. The collection of general multi-purpose soil information was carried out within the context of land aggregation and land amelioration projects after WWII.

1.2.1 Early Soil Surveying and Mapping

To the best of the author's knowledge, the oldest printed and available soil map dates back to the second half of the nineteenth century, to approximately 1873. The cartography was made by Franz Shollmayr and corrected by Markus Lipold (Fig. 1.1). The lithograph was printed by the Jože Blaznik printing shop in Ljubljana. Being a part of the Austro-Hungarian Empire, the text was in Slovenian and German. The map covers the central part of the Slovenian territory and presents the twelve classes which delineate the areas of fundamentally different soils, such as the Histosols of the Ljubljana Moor, the Eutric Cambisols area of the Vipava valley and Ilirska Bistrica, and the Luvisols of Bela Krajina.

In the twentieth century, medium- to large-scale soil mapping of small selected areas was carried out after WWII in the late fifties and early sixties. Many documents were published in scientific reports. Unfortunately, most of the collected data and maps have been lost or are unavailable. A good example of such mapping is the soil mapping of the Ljubljana Moor (Ljubljansko Barje) conducted by the Agricultural Institute of Slovenia (KIS 1962). The manuscript map (Fig. 1.2) found in the archives was digitized and is available as a digital polygon database. Unfortunately, the mapping methodology was not developed nor was soil classification available. These two main reasons make this information difficult to directly incorporate into the contemporary Digital Soil Map 1:25,000.

In Slovenia, systematic soil mapping on a scale of 1:25,000 began in the year 1963 when Slovenia was a constituent republic of the Federal Republic of Yugoslavia. The mapping process was financed by the authorities of the individual republics. This was initiated on the federal level, while the mapping process, financial resources, and the methodologies slightly varied between republics.

Numerous soil profile data were collected, and several soil maps were published in paper form. In 1981, the initial soil mapping scale in Slovenia was changed from 1:50,000 to 1:25,000. By the end of 1986, about 50% of the Slovenian territory had been mapped partly on a scale of 1:50,000 and partly on a scale of 1:25,000. Only a few map sheets were published on top of old military topographic maps on a scale of 1:50.000; others existed as manuscripts and soil profile data as typescripts. Three soil maps on a newer updated 1:50,000 scale topography were published between 1984 and 1986: sections of Ljubljana, Murska Sobota, and Ptuj. Each of them was accompanied by a booklet containing soil descriptions, information on basic soil properties, representative measured soil profile data, and land use information.

1.2.1.1 Before 2000

Early works in the field of soil research were publications on soil fertility. The engineer Zobec worked on plant nutrition and published a book on agrochemistry, wherein he described soils and explained fertilization in a practical manner (Zobec 1930). He was followed by Prof. Bogdan Vovk (1902–1988), who determined the fundamentals of Slovenian soil science. He was the first scientist with a Ph.D. in soil sciences in Slovenia. Dr. Dušan Stepančič (1921–2007) carried out the soil survey and field research for several years.

Fig. 1.1 The soils of inner Carniola—early soil map dating from approximately 1873

Besides his early work on agricultural soil quality, which was subsequently published as a decree (Geodetska uprava RS [Slovenian Geodetical Survey] 1984), he contributed to several important soil survey reports and maps (Stepančič et al. 1984a, 1985, 1986a, b, 1987a, b). Prof. Dr. Jože Sušin (1922–1988) was a full professor of soil science who contributed significantly to the research on Karst soils (Sušin 1964) and authored the first dictionary of soil science in the Slovenian language (Sušin 1983). Prof. Dr. Marjan Ažnik (1927–2010) dedicated his efforts to teaching and research activities on the soil–plant relationship, plant nutrition, and fertilization. Prof. Dr. Mirko Leskovšek (1923–2007) worked in the field of soil fertilization and plant nutrition. He carried out numerous long-term field tests that were fundamental to a series of practical books on soil fertilization (Leskošek 1993; Leskošek and Mihelič 1998; Lobnik et al. 1992b). Prof. Dr. Albin Stritar (1925–1988) was a teacher who also worked on land use and soil protection. His was an early voice against sealing the best agricultural soils in the mid-eighties. An important part of his work was to define Slovenian soils in relation to catenas and landscapes (Stritar 1990b, 1991). A significant part of the 1:25,000 soil map of Slovenia was mapped by Janez Rupreht and Marjan Šporar. Several soil surveyors, teachers, and scientist are still contributing to soil science in Slovenia. Tomaž Prus, M.Sc., Marko Zupan, M. Sc., and Dr. Helena Grčman work on soil contamination, soil quality, and soil protection. Janez Sušin from the Agricultural Institute of Slovenia is still active in the field of soil fertilization and, more recently, on groundwater protection against contamination with nitrates. We remember Dr. Nataša Vidic for her excellent work on soil age research (Lobnik et al. 1992b; Prus et al. 1986a; Vidic et al. 1991; Vidic and Lobnik 1997). Dr. Franc Lobnik, a retired full professor of soil science, guided soil teaching and soil research for more than two decades. Dr. Tomaž Kralj specialized in soil classification problems. He provided an important contribution to Slovenian Soil Classification in relation to the WRB. Dr. Narcis Mršić (1951–1997), a biologist, dedicated his life to a podo-zoology. His outstanding publication "Animals of Our Soils" might never be surpassed (Mršić et al. 1997b).

In the forestry sector, a small group of researchers formed around their common interest in forest soil research,

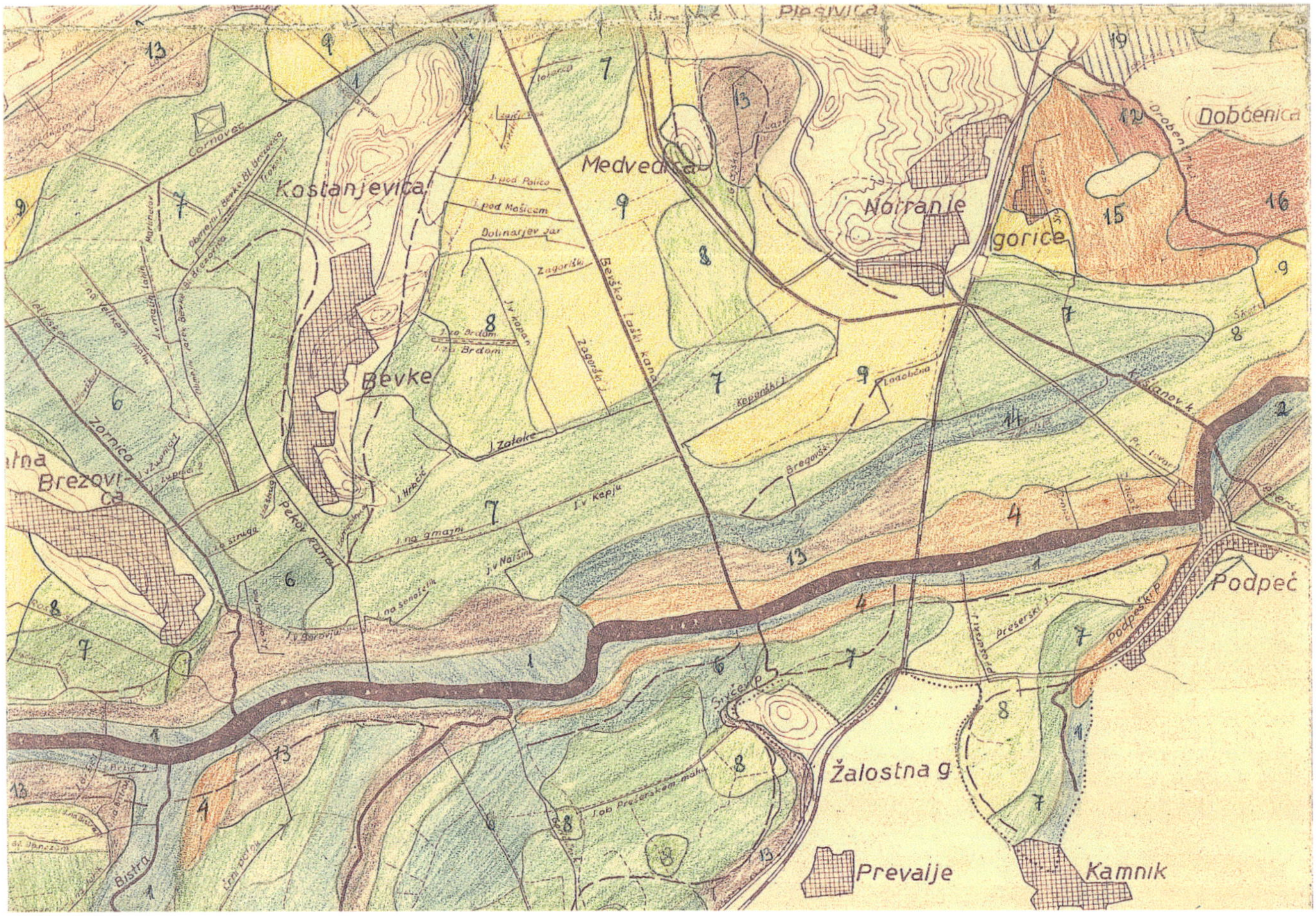

Fig. 1.2 A section of the hand-drawn soil map of Ljubljana moor (Agricultural Institute of Slovenia, 1962)

especially from 1947 onwards, when the Slovenian Forestry Institute was established. With the establishment of the laboratory in 1956, better conditions were created for the development of forest soil science. Marija Kodrič worked for the institute from 1949 to 1961, Marjan Pavšar from 1955 to 1974, Niko Pehani from 1965 to 1969, Janko Kalan from 1975 to 1994, and Mihej Urbančič from 1981 to 2012. Marija Kodrič was a pioneer in the development of organized research on forest soil science in the 1950s. In recent decades, the contributions of Janko Kalan and Mihej Urbančič were especially evident in the fields of forest soil mapping, forest soil monitoring, the development of the laboratory (the development of methods and the introduction and implementation of new analytical procedures), studies on the status of soils under forests and forest stands, the mineral nutrition status of forest trees, etc.

The first contours of Slovenia's soil geography started to form at the beginning of the 1950s. At that time, it was clearly marked by the heritage of pedology (Vovk Korže 2003). Soil geography used pedology's results and findings, and geography in general took responsibility for the educational part on the primary and secondary levels. Primary and secondary school geography curricula provide basic soil knowledge also for those who will not continue with studies in geography, agronomy, forestry, landscape architecture, geology, etc. This basic knowledge includes soil properties and processes, soil profiles and horizons, and domestic and foreign soil types. Also on the university level, soil geography deals with soil classification beyond Slovenia's borders. Geographers are the only ones who publish expert articles about soils from all the continents (Lovrenčak 1976, 1994; Vovk Korže 2003; Vovk Korže and Lovrenčak 2004; Repe 2005a, b, 2006c). Nevertheless, the soil topics dealt with by Slovene geographers are colourful and quite varied. The first true soil geographical research was carried out by Prof. Lovrenčak (1970). He attempted to find the relationships between soils and vegetation and the reasons for their formation, development, and spatial extent. Prof. Lovrenčak, furthermore, endeavoured to find all the possible interactions within the soil–vegetation system, investigated the soils of high altitudes, and karstic areas, and provided regional soil geographical descriptions of Slovenia's different regions. In the 1990s, Prof. Vovk Korže joined him in soil research in the north-eastern part of Slovenia (Vovk Korže 1995).

1.2.1.2 After 2000

A number of young colleagues, i.e. Dr. Polona Kalan (soil chemistry, soil sampling design, and statistics), Dr. Urša Vilhar (water balance), Dr. Milan Kobal (forest soil modelling, soil monitoring), and Daniel Žlindra (soil chemistry, LQ/LA), are involved in emerging forest soil research areas and in integration at the international level in the context of projects and international interest groups (ICP Forests). The phytocenologist Dr. Lado Kutnar, Dr. Aleksander Marinšek, and Dr. Klemen Eler cover the field of forest vegetation and soil properties, site conditions in terrestrial ecosystems. Prof. Dr. Domen Leštan works on soil remediation practices. Dr. Mateja Muršec teaches soil science and is involved in soil research primarily related to soil organic matter (Fig. 1.3).

An important field of research in forestry is the biotic part of the soils under forests, which is covered by Prof. Dr. Hojka Kraigher and Dr. Tine Grebenc. Forest soil compaction studies and forest operation influences are monitored by Robert Robek, M.Sc (Fig. 1.4).

Prof. Ana Vovk Korže focused more on explaining the role of soils in the ecosystem (Vovk Korže 2007), the methodology of soil research (Vovk Korže and Lovrenčak 2004), and permaculture aspects of soils (Vovk Korže 2014). Dr. Repe and Dr. Petauer further extended soil knowledge in geography. Petauer works in the field of soil geography in schools (Petauer 2005) and on raising public awareness of soils (Petauer 2011). Repe deals with the methodology of soil research (Repe 2001), soil degradation (Blaž. Repe 2002), soil classification (Repe 2005a, b, 2006c, 2010b), soil maps, and digital soil mapping (Repe 2006b). He devotes special attention to presenting soils and soil-related issues to the lay public. We should not forget the authors who do not primarily deal with soils. By far the most creative are Komac and Zorn, who research soil erosion and landslides (Zorn and Komac 2008). Kunaver also researched the same topic (Kunaver 1989). Lampič assessed soil vulnerability (Lampič 2002). Many other authors treat soil topics as a smaller part of some broader research on physical geography, regional geography, landscape ecology, biogeography, etc. One of the very rare, but very successful collaborations between geography and pedology is a research study on plot gardening in Ljubljana (Jamnik et al. 2009).

Fig. 1.3 Taking a picture of soil profile; Pohorje (*Photo* B. Vrščaj)

Fig. 1.4 Forest soil research; Pohorje (*Photo* B. Vrščaj)

1.2.2 Soil Research Institutions

1.2.2.1 National Institutes

The Agricultural Institute of Slovenia, Central Laboratory and the Department of Agricultural Ecology and Natural Resources (AIS; www.kis.si) (AIS) carries out soil fertility and plant nutrition research and provides services, builds and maintains soil databases, improves the existing digital soil maps and georeferenced data, and carries out research projects related to the sustainable use of agricultural soil. The AIS soil analysis laboratory participates in ring tests and is accredited by the relevant Slovenian and the French accreditation agencies. In 2005, 7487 soil samples for plant nutrition control were collected and analysed. A significant portion of these samples are currently being georeferenced and stored in a soil information system developed by the AIS. Within recent soil research projects, 100 sets of topsoil data are being collected in the municipality of Ljubljana to assess soil contamination, along with 150 soil profiles from locations across Slovenia that are representative as regards describing the most frequent soil types (Figs. 1.5 and 1.6).

The Slovenian Forestry Institute (SFI; www.gozdis.si) carries out research on soils under forests. It collects forest soil data information and has information on 100 forest soil profiles in digital form, and an additional 200 are to be added within the framework of a BioSoil research project (Simončič 2006). The soil profile data were not integrated into the soil information system. The Agricultural Extension Service of the Chamber of Agriculture and Forestry of Slovenia collects and analyses soils for plant nutrition purposes. Six regional laboratories process a large number of soil samples. The information is not georeferenced.

The above-mentioned two public institutions and the CSES carry out the bulk of field and laboratory research at the national level, as well as within international, mainly EU research projects.

1.2.2.2 Universities—Teaching and Research Institutions

The Center for Soil and Environmental Science of the University of Ljubljana (CSES; www.bf.uni-lj.si/cpvo) is the most important soil research institution at the university. To a lesser extent, the research work linked to the teaching process is conducted at other institutions or university departments:

The Faculty of Agriculture of the University of Maribor, Department of Chemistry, Agrochemistry, and Soil Science (http://fk.uni-mb.si), which teaches soil science. Soil research is carried out to a limited extent within the framework of research projects.

The Geographical Institute of the Pedagogical Faculty of the University of Maribor carries out soil research for educational purposes. The data on the collected 150 soil profiles are not georeferenced (Vovk Korže 2006).

The Department of Geography of the Faculty of Arts of the University of Ljubljana (http://www.ff.uni-lj.si/an/study/departments/department_geography) and the Karst Research Institute (http://kras.zrc-sazu.si/) sample and describe soil within teaching programmes and several research projects. The researchers have examined 200 sets of soil profile data, which are georeferenced but not yet in digital form (Repe 2006).

1.2.2.3 Small and Medium-Sized Enterprises (SME)

The ERICO Environmental Institute (www.erico.si) carries out environmental research projects mainly related to soil contamination assessments from 1999. This company conducts soil environmental research, mainly soil contamination and partly soil quality. Since 2005 small private consultancies have emerged. They conduct research in collaboration with

Fig. 1.5 Field soil research at AIS. The soil drilling device propelled by tractor (*left*) specially designed for sampling undisturbed 20-cm-wide core of soil profile to the 130 cm depth. The apparatus drills through soft soil and hard rock (*right*) (*Photo* B. Vrščaj)

Fig. 1.6 Measuring soil penetration resistance and shear strength (*Photo* B. Vrščaj)

public research institutions or, in the case of smaller studies, on an individual basis. Such research is typically related to soil quality assessments, soil suitability for irrigation, soil fertility, and minor soil contamination studies. Here we present three typical small- and medium-sized enterprises (SME). Envit d.o.o. (www.envit.si), which is a product of the university SME incubator, is focused on the development of remediation techniques regarding contaminated soils and brownfields. ProTellus d.o.o. (www.protellus.si) is a professional company that performs soil mapping, soil contamination studies, and soil quality monitoring activities. Agrarius s. p. (www.agrarius.si) is a small consultancy focused on soil surveying and interpretation, soil environmental assessments, soil suitability studies, and agricultural soil quality evaluation as a contribution to the broader resource studies needed in local or regional planning processes. It is expected that the introduction of new national legislation related to soil quality, soil contamination, and remediation in the future will encourage research in these fields.

1.2.3 Websites

The most comprehensive public soil information at the moment is available via the WEB GIS Website of the Agricultural Institute of Slovenia, Department of Agroecology and Natural Resources, Centre for Soil, and the Environment: www.kis.si/Talni_podatkovno_informacijski_center. A minor part of the datasets is presented in English at: www.kis.si/en/Soil_data_information_system.

2 Regions and Landscapes

Blaž Repe

Abstract

This chapter briefly explains the general and geographical features of Slovenia, a country located in Central Europe bordering Italy, Austria, Hungary, and Croatia. Geographically, it is situated where four large European regions meet—the Alps, the Pannonian Basin, the Dinaric Mountains, and the Mediterranean. This endows Slovenia with extreme natural diversity, probably one of the highest levels in Europe. As a result of this situation, there are five main natural regions in Slovenia: (1) the mountainous Alpine Region—glacially reshaped and transformed, it features the highest altitudes, the steepest slopes, and high rates of precipitation; (2) the hilly, largely forested Pre-Alpine Region, with swift alternations between carbonate and noncarbonate parent materials; (3) the forested and agricultural Karstic–Dinaric Region, with an absence of surface waters and the absolute dominance of soluble and permeable limestone and dolomite; (4) the lowland Sub-Pannonian Region, with its hot continental climate in summer, significantly lower mean annual precipitation, undulating hills, and flat river plains; and, finally, (5) the warmest of them all, the Sub-Mediterranean Region—the coast of the Adriatic Sea.

2.1 Slovenia—Country Information

Slovenia is a Central European country bordering Italy to the west, Austria to the north, Hungary to the north-east, and Croatia to the south-east (Fig. 2.1). It is situated at the crossroads of four large and diverse European regions: the Alps to the north and north-west; the Pannonian Basin to the east; the Balkan Peninsula with the Dinaric Mountains to the south-east, and part of the Mediterranean—the Adriatic Sea—in the south-west. The geomorphological and climatic transitions over a very short distance (apx. 200 km) make the country extremely diverse. This diversity is reflected in the architecture, cuisine, cultural heritage, dialects, and, of course, soils. In spite of being one of the smallest countries in Europe, Slovenia is considered one of the most diverse.

B. Repe (✉)
Geography Department, University of Ljubljana, Faculty of Arts, Ljubljana, Slovenia
e-mail: blaz.repe@ff.uni-lj.si

2.1.1 Basic Country Data

- Official name: Republic of Slovenia
- Area: 20,273 km^2
- Maximum distance: north–south 163 km, west–east 248 km
- Geometric Centre: 46° 07′ 11.8″N, 14° 48′ 55.2″, 64 m.a.s.l. (Spodnja Slivna, near Litija)
- Total length of the borders: 1370 km (with Austria 318 km, Hungary 102 km, Croatia 670 km, Italy 280 km)
- Coastline: 46.6 km on the Adriatic Sea
- Highest peak: Mt. Triglav (2864 m)
- Deepest sea point: 37 m (300 m NW from Cape Madona, close to the town of Piran)
- Average elevation: 556.8 m
- Average inclination: 13.17° (23.4%)
- Mean annual precipitation: 1567 mm
- Mean annual temperature (2001–2010): 9.2 °C
- River network density: 1.32 km/km^2
- Number of known species: 22,000

B. Vrščaj et al., *The Soils of Slovenia*, World Soils Book Series,
DOI 10.1007/978-94-017-8585-3_2

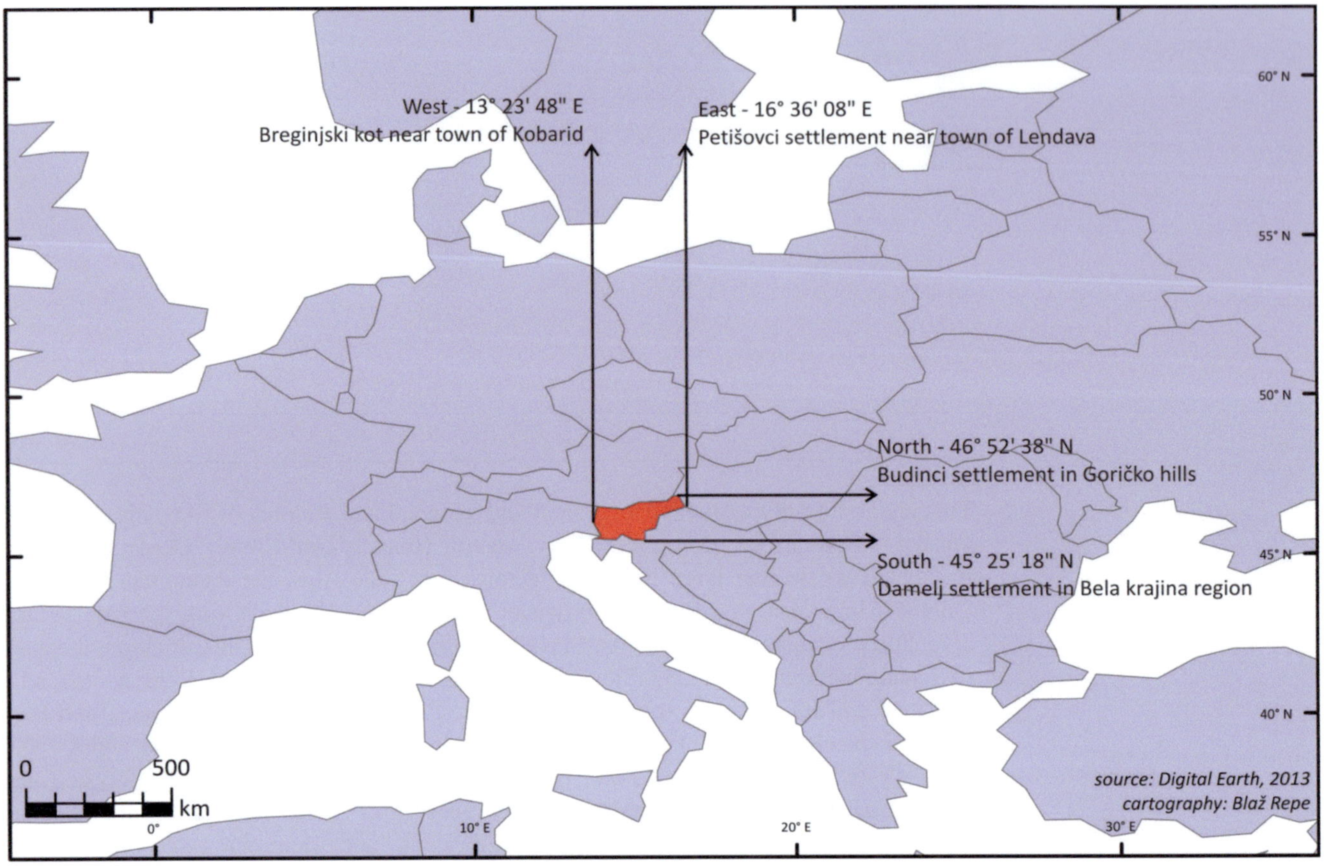

Fig. 2.1 The location of Slovenia in Europe

- Population (2013): 2,025,866
- Population density: 99 persons per km^2
- GDP (2012): €35.466 billion; €17,343 per capita

2.1.2 The Geographical Position

Slovenia lies in Central Europe at the crossroads of the 46° northern latitude and 15° eastern longitude, which is in the direct vicinity of its geometrical centre. Slovenia is 20,273 km^2 and ranks 154th of the world's 249 countries ("List of countries and dependencies by area", 2014). Despite its small size, Slovenia represents a very special natural environment on a European as well as global scale. The natural diversity of Slovenia is extremely high, even exceeding many other much larger countries. It is especially rich in terms of biodiversity. Although Slovenia occupies only 0.00136% of the entire land surface of our planet, it represents a habitat for more than 1% of all the living species on earth and more than 2% of all living land species, including fresh water ecosystems (Mršić et al. 1997a). The large number of different habitats shows extreme diversity (CKFF 2010a, b, c, d). The reasons for this enormous variety can be attributed to the variety of its natural features, which also includes the soils. Slovene soil scientists have identified more than 400 soil type varieties and forms (Ministry of Agriculture and the Environment, 1999). The variety of natural features is determined by the country's geographical position. To the north it borders Austria, to the east Hungary, to the south Croatia, and to the west Italy. Slovenia shares a maritime border with the latter two.

Considering the physical geography of Europe, four large units intersect in the territory of Slovenia: the Alps, the Pannonian Basin, the Dinaric Mountains, and the Mediterranean (Fig. 2.2). This junction position results in enormous variety, with the transitioning and changing of natural (and social) features over extremely short distances. Tectonically and geologically speaking, Slovenia, with all its diversity of mostly sedimentary rock, lies in a tectonically active area where the African tectonic plate sub-ducts under the Eurasian plate. There are nearly all possible varieties of geomorphological features, from highlands to lowlands (mountains, hills, high and low plateaus, broad and narrow valleys, basins, and river and coastal plains). Nevertheless, the topography is mainly hilly, horizontally and vertically very

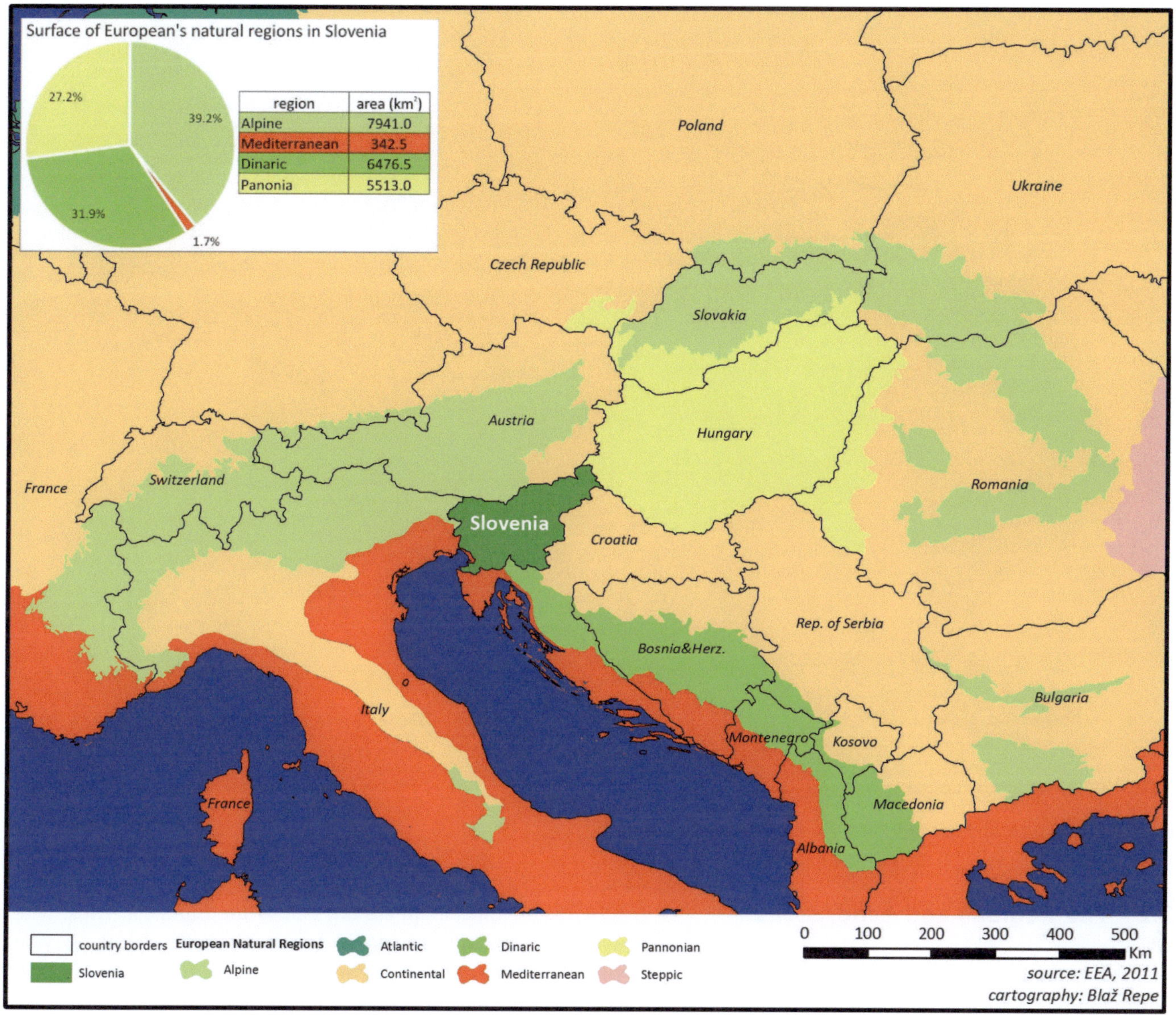

Fig. 2.2 The position of Slovenia within the natural macroregions of Europe

variable over short distances. A substantial part of Slovenia is karstic, with one of the highest karstic areas in Europe, due to the abundance of limestone and dolomite. Entering English via German, the term "karst" originates in Slovenia as a descriptor for the geomorphology of the Karst region, which comprises the south-western part of the Republic of Slovenia.

Climatically Slovenia is influenced by western air circulation, which brings moist oceanic air masses and consequently precipitation. There are also other strong influences: cold Alpine influences from the north, warm Mediterranean influences from the south-west, and dry continental influences from the east. Sufficient quantities of precipitation make Slovenia humid, rich in water and therefore the soils largely have udic properties. On the other hand, 1/3 of the country is karstic and without any surface waters. Forests are also one of the dominant land covers of Slovenia, covering nearly 60% of the country, thus ranking third in Europe behind Finland and Sweden (Gale 2011). Slovenia hosts around 24,000 recognised living species, which live from the highest altitudes down to the deepest, dark corners of the karstic underground. The total number of species is estimated at between 50,000 and 120,000. A large portion of this estimation is accounted for by the soil organisms (Mršić et al. 1997a). According to the total land area, Slovenia has a disproportionally high biodiversity, comparable to tropical regions (Ogrin and Plut 2009, p. 127).

Fig. 2.3 Natural (ecological) regions of Slovenia (Natek 2002)

2.2 The Regions and Landscapes

In order to present the general characteristics and regions of Slovenia, the ecological-landscape method has been chosen from the many efforts to define the natural regions (Natek 2002). Considering topography, lithology, and climate, three large units can be distinguished: the Alpine, Pre-Alpine, and Karstic-Dinaric. Adding in the climatic factors, two more can be extracted: the Sub-Mediterranean and the Sub-Pannonian. The prefix "Sub-" in the latter two units places a stress on the transitory, marginality, and less prominent features that are otherwise typical of European natural macroregions. There is also another region, the Slovenian Sea or the Gulf of Trieste area. Typically, it is not included in general research or is treated as an integral part of the Sub-Mediterranean region (Fig. 2.3).

2.2.1 The Alpine Region

The Alpine region extends over the north-western and the northern parts of the country. The Slovene Alps belong to the extreme south-eastern end of the European Alps, which are already substantially lower when they reach Slovenia. Nevertheless, the topography has a clearly recognizable mountainous appearance and it reaches the highest elevations of Slovenia (Fig. 2.4). There is a prevalence of permeable and soluble limestone and dolomite with an extensive area of high mountain karst. The bottoms of intermediate valleys are covered with glaciofluvial deposits, mostly carbonate gravel and sand. The mountains, glacially reshaped and transformed during the Pleistocene Era, have very steep, east-west oriented slopes. Flat or plain land is scarce. The harsh mountainous weather has a short vegetal growing period. Precipitation, although extreme in quantities (Slovenian Environment Agency, 2010) falls mostly as snow. In contrast, soil conditions, although generally udic, can appear xeric-like due to the extreme slopes. Forest is one of the dominant features of this region and its coverage (70%) is significantly above the Slovenian average. The upper forest line appears between 1700 and 1900 m.a.s.l. (Lovrenčak 2007).

Soils are mainly shallow and belong to different types of Leptosols (Rendzic, Nudilithic, Eutric, Hyperskeletic, etc.).

Fig. 2.4 The Julian Alps (*Photo* B. Repe)

Agriculturally, the region is sparsely inhabited and limited to the narrow bottoms of the valleys and mostly oriented to dairy cattle farming. The same areas are also occupied by transport infrastructure and industry. Therefore, some areas are facing a serious lack of space, which significantly affects soils (urban sprawl).

2.2.2 The Pre-Alpine Region

The Pre-Alpine Region is a 10–30 km wide belt of lower mountains, high hills, and intermediate basins and valleys, with an average height of 560 m.a.s.l. It is a very diverse and heterogeneous region, characterized by swift alternation between carbonate and noncarbonate parent materials. In most cases, higher areas and summits are made of limestone and dolomite, while less steep lower areas and foothills are made of siliceous sedimentary rocks, mostly sandstones and shale. In the eastern part, some magmatic and metamorphic rocks can be found (the Pohorje Mountains). Basins and valleys are filled with thick deposits, mostly gravel and sand, rarely with silt and clay. In spite of the lower altitudes, the surface is very diverse and dynamic, with slopes still very steep and valleys often gorge-like. The climate can be described as a moderately continental or moderately warm and humid climate with warm summers (Ogrin 1996). The slopes are well-drained, with an udic regime, but the valleys and basins are often aquic. On a noncarbonate parent material, the river network is very well developed and dense. The streams and rivers are very torrential and cause annual flooding. The natural vegetation is forest, mostly of European beech (*Fagus sylvatica*) associations. These forests frequently have a protective function, due to the steep slopes and heavy erosion processes. For the past 60 years, people have constantly been migrating to the cities, leaving a once cultural landscape to degradation and reforestation (Fig. 2.5).

Soils are extremely diverse, as all known Slovene types can be found here. In higher areas, there are still Leptosols

Fig. 2.5 Pre- Alpine hills of western Slovenia (*Photo* B. Repe)

(Rendzic, Eutric), with hydromorphic types in lower areas (Fluvisols, Gleysols, Stagnosols), however Cambisols prevail (Dystric, Chromic, Eutric). To an even greater extent than in the Alpine region, the soils in the valleys and basins are seriously degraded by urban sprawl.

2.2.3 The Karstic-Dinaric Region

This region lies over a substantial part of southern Slovenia. It is a part of the larger Dinaric mountain system that extends further in the south-easterly direction to the Balkan Peninsula (Gams 2003). The main feature of this region is the absolute dominance of soluble and permeable limestone and dolomite and consequently the karstic characteristics of the topography, hydrology, soils, and vegetation. There are some fluvial deposits on larger karst poljes. The topography is extremely rough (dolines and collapsed dolines) and diverse, with alternating hilly mountains, plateaus, karstic poljes, and plains. Mountain ridges follow a typical NW-SE (Dinaric) direction. Water corrodes the limestone in distinct karst-related features (caves, dolines, poljes, etc.). The climate is typically udic, moderately continental, with higher rates of precipitation (over 2000 mm/year). Hydrology is distinctly karstic with no surface running or still water. Nearly all the water runs underground through permeable and cracked limestone (Fig. 2.6). The only surface streams run over fine loamy deposits on Karst poljes (Fig. 2.7). The area is particularly suitable for forest growth. The European beech (*Fagus sylvatica*) and silver fir (*Abies alba*) forests, extending as far south as Macedonia, are the largest contiguous forested area in Europe. With forest cover over 80%, this region ranks highest in this respect in Slovenia.

Soils are directly parent material dependant. Different types of Leptosols (Eutric, Rendzic, etc.), Cambisols (Chromic), Luvisol (Haplic) tightly intertwine mosaically. The physical and chemical soil properties are favourable, but the varying depth, the presence of surface rocks, and mostly the lack of water makes this region very unsuitable for arable

Fig. 2.6 Chromic Cambisol—the corrosion of limestone at the lithic contact (*Photo* B. Repe)

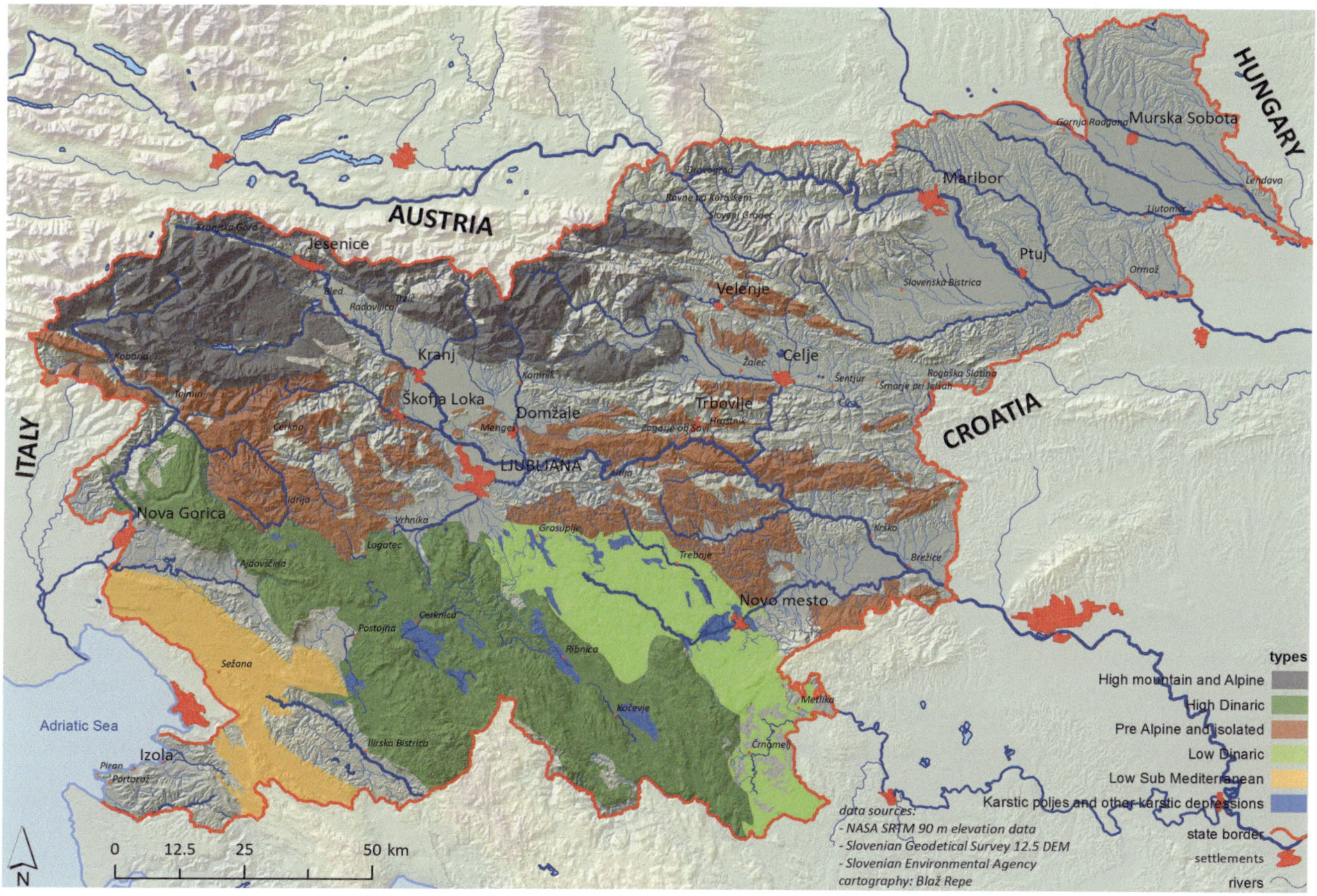

Fig. 2.7 The Karst and hydrographic network

Fig. 2.8 Vineyard terraces of Haloze in the Sub-Pannonian region, eastern Slovenia (*Photo* B. Repe)

land farming. All karstic areas are characterized by very sparse population densities.

2.2.4 The Sub-Pannonian Region

The Sub-Pannonian Region occupies the entire north-eastern and eastern parts of Slovenia, on the most western edges of the Pannonian Basin. It is a topographical and climatic transition into a typical, levelled, and continental European inland. The geological structure is mainly of soft tertiary sediments, which are only partially solidified (marl, mudstone, sandstone, etc.). The topography is a mixture of alternations between low hills, basins, and river plains. The plains are thickly covered with mostly noncarbonate clayey, silty, or sandy material; the floodplains are wide. The altitudes are low, on average apx. 250 m.a.s.l. Openness to the Pannonian Basin characterizes the region with a moderately continental climate, with the most typical continental distribution of precipitation and average quantities below 1000 mm. The soil water regime is generally udic, but there is a certain duality. Due to the topography, the largest areas of hydromorphic soils can be found here, but on the other hand, droughts are also the most frequent in this part of the country. Due to the levelled, even concave surface and impermeable parent material, groundwater is shallow beneath the surface and the river network is very dense. Rivers are typically lowland, making meandering loops, and regularly flooding. The natural vegetation in the lowlands is hygrophyle forest of English oak (*Quercus robur*), black alder (*Alnus glutionsa*), and willows (*Salix alba*), while the low hills are covered with an acidophilic association of European beech and sweet chestnut (*Castanea sativa*). Forest cover is well below the Slovene average (30%). The natural features and conditions favour an agricultural orientation of the region (60%) (Fig. 2.8).

The soils are young, mostly under the direct influence of water (Fluvisols, Gleysols, Stagnosols). Where water drains normally, the soils are influenced by lithology and the most frequent types are Cambisols (Dystric) and Umbrisols.

2.2.5 The Sub-Mediterranean Region

This region encompasses the quite lower, south-western part of Slovenia, which opens onto the Adriatic Sea.

Fig. 2.9 A narrow strip of alluvial flat surface with saltpans on the Slovenian coast (*Photo* B. Repe)

Geologically, the region can be divided into two entirely different parts. The first one is built of flysch and is mainly low, flat, or undulating with small-scale low hills; the second part is of limestone and is represented by higher hills and a karstic plateau. The average temperatures are above 0° throughout the entire year. The distribution of precipitation shows a transitional pattern between the Mediterranean and continental regimes, and generally does not meet the criteria for xeric. The cold, strong, and gusting Bora wind (Slovenian: *burja*) is very common during the winter months, which can cause some soil wind erosion. Along the Soča River, valley maritime influences reach far north, deep into the Alps (Ogrin 1996). Impermeable flysch allows surface running water with a dense river network, while the hydrology on the limestone is typically karstic. The vegetation, although still thermophile and drought tolerant, is not evergreen, but deciduous of downy oak (*Quercus pubescens*), manna ash (*Fraxinus ornus*), and hop hornbeam (*Ostrya carpinifolia*).

The dominant soil types on limestone are Leptosols (Eutric, Rendzic, etc.) and Cambisols (Chromic, Eutric, etc.). The Karst is known for *Terra Rossa*. Eutric Cambisols on flysch are of great importance and are known as Slovenia's best agricultural soil. Marshes with Gleysols commonly appear on the bottoms of the flysch valleys. Fluvisols form along rivers and on wider and mostly gravely floodplains. Directly on the coast, a limited area of salt-affected soils (Solonchaks) can be found.

The abundance of preclusive activities (tourism, transport, industry, and settlements) leads to land use conflicts at the expense of the natural environment and soil degradation.

This region includes the sea, i.e. the most north-eastern part of the Adriatic Sea or approximately one-third of the Gulf of Trieste (180 km^2) (Radinja 1990), and the 46.6 km of the coast between Italy and Croatia. By far the largest areas of sub-aqueous soils can be found in this region (Repe and Pristovšek 2011) (Fig. 2.9).

3 Factors of Soil Formation

Blaž Repe, Primož Simončič, and Borut Vrščaj

Abstract

The parent material is considered the most important of Slovenia's soil formation factors. Chemically, only partially changed parent material heavily influences soil properties. 93% of the Slovenian territory is composed of sedimentary rocks, mainly limestone and dolomite, and only 4% belongs to the metamorphic group, while 3% is of magmatic origin. After the parent material, topography is certainly deemed to be Slovenia's next most important and prominent natural soil formation factor. Very young structural relief predominates, which has a direct impact on the soil water and micro- and mezzo-climatic conditions. The influence of climate on Slovenian soils is more general; it is mostly considered to be a constant and does not have an important interregional distinguishing aspect. Due to its importance, water as a soil formation factor deserves a special place in the Slovenian Soil Classification, since on the first level of classification soil is divided on the basis of hydromorphism. Forests thrive in Slovenia due to the natural conditions. Most of the Slovenian territory is covered by forests (66%), which ranks Slovenia as the third most forested country in Europe, while agriculture occupies only 27.8% of the country. Time is reflected mostly in the very young parent material briefly exposed to soil formation factors; the land is therefore covered with young and mostly shallow soils.

3.1 Factors of Soil Formation

Dokuchaiev in Russia and independently Hilgard in America discovered that a given parent material may form different soils depending on the environmental conditions, particularly climate and vegetation. This was redefined later by adding additional independent factors to the following soil formation equation (Jenny 1941):

$$S = f(cl, o, r, p, t, \ldots)$$

This formula was adapted to accommodate the variety of soil formation factors found in Slovenia (Prus 2000; Stritar 1991):

$$S = f(p, w, r, cl, o, m)t$$

where soil (S) is an function (f) of the parent material (p), water (w), relief /topography (r), climate (cl), organisms (o), and man (m), all working in time frame (t). One factor can predominate locally, while others can be barely noticeable in the short term. Due to the enormous natural (and social) variability, Slovenia's soil formation factors, soil genesis, and soil processes are very specific.

B. Repe (✉)
Geography Department, University of Ljubljana, Faculty of Arts, Ljubljana, Slovenia
e-mail: blaz.repe@ff.uni-lj.si

P. Simončič
Department of Forest Ecology, Slovenian Forestry Institute, Ljubljana, Slovenia
e-mail: primoz.simoncic@gozdis.si

B. Vrščaj
Department of Agricultural Ecology and Natural Resources, Centre for Soil and Environment Research, Agricultural Institute of Slovenia, Ljubljana, Slovenia
e-mail: Borut.Vrscaj@kis.si

B. Vrščaj et al., *The Soils of Slovenia*, World Soils Book Series,
DOI 10.1007/978-94-017-8585-3_3

3.1.1 Parent Material

Parent material is considered the most important of Slovenia's soil formation factors. The moderate and humid climate results in relatively slow soil genesis, especially as regards the weathering of parent material and minerals. Chemically unaltered or only partially changed, the mechanically disintegrated smaller pieces (sandy, silty) of parent material enter the pedosphere, where they heavily influence the soil properties with their own. Because half of the country is karstic and composed of permeable and soluble limestone and dolomite, there is also a direct chemical influence on the soils and water. The parent material also serves as the most important factor for distinguishing among different soil types. In addition to this, the parent material also significantly affects other soil formation factors (Stritar 1991; Lovrenčak 1994a; Prus 2000; Repe 2006a; Kočevar and Vidic 2003).

Geologically speaking (Plenčar et al. 2009), Slovenia is a transitional land, situated at the tectonically active contact point of Eurasian and African tectonic plates. In particular, it lies at the junction of the Alpine and Dinaric mountains and the Pannonian Basin. At first glance, there is little variability in Slovenia's lithological structure: 93% of the surface is made of sedimentary rocks, only 4% belong to the metamorphic group, and 3% are of magmatic origin (Fig. 3.1). The reason for this one-sided lithological situation can be found in the geological history. Throughout most of history, the majority of the Slovene territory had been under the sea. Particularly, until the end of the Cretaceous period, approximately 65 million years ago, occasionally and only some parts had managed to become mainland. At that time, the Alps started to rise due to the collision between the aforementioned tectonic plates. The area north-east of the Pohorje massif belongs to the Eurasian and the remainder to the Adriatic tectonic plate, which is a part of the African tectonic plate. Many earthquakes have demonstrated that the junction in between, called the Periadriatic Seam (or fault), is still very active. Despite the seeming lack of variability at first glance, since the tectonic and the geological history of Slovenia is very diverse, the lithological structure also varies to a great extent, as will be explained below (Verbič 1998). A very important consequence of this is also the extremely diverse and variable soil cover.

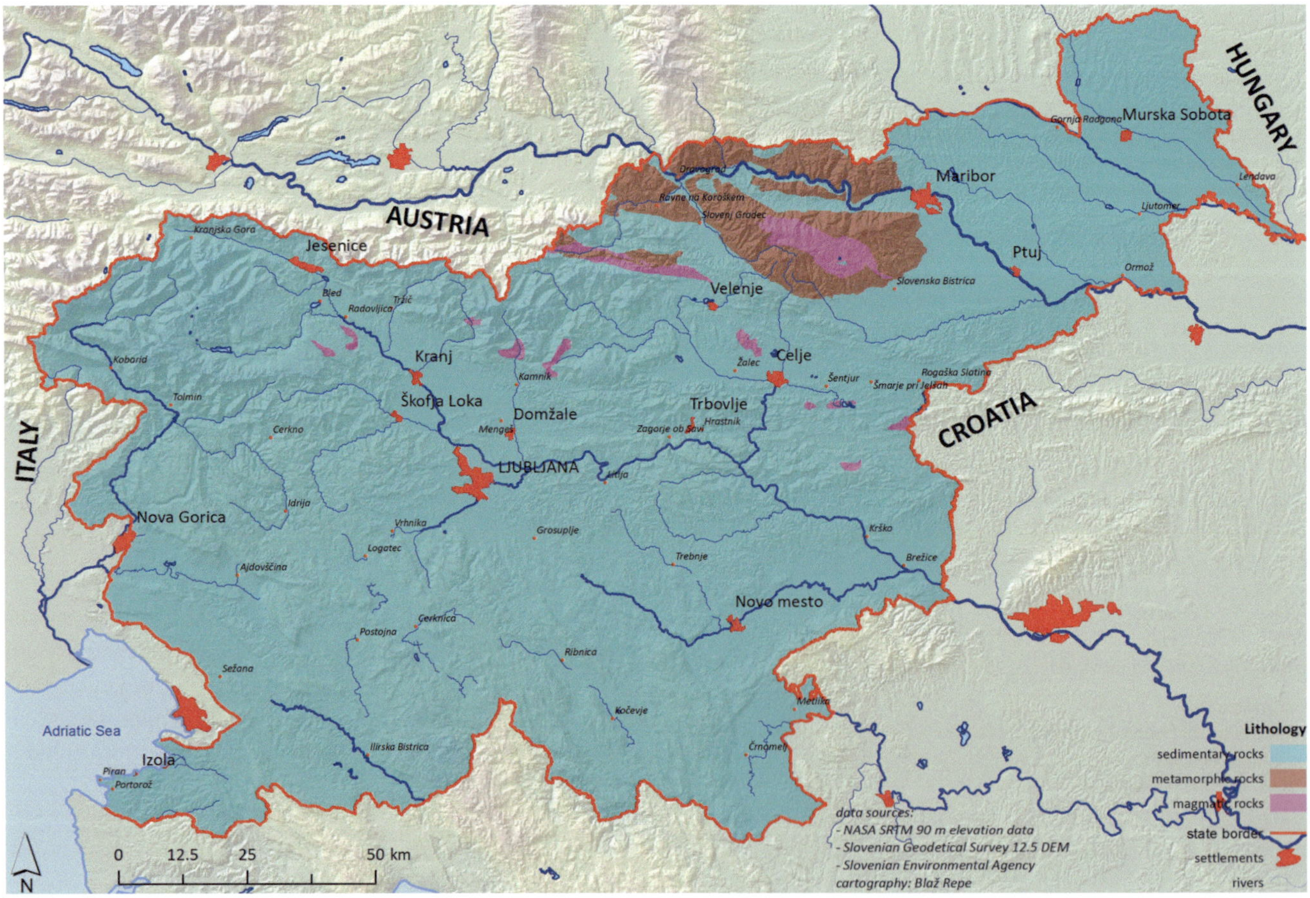

Fig. 3.1 The lithology of Slovenia

The oldest rocks are Devonian limestones between 360 and 410 million years old. The metamorphic rocks from the Pohorje massif are much older, but later metamorphosis has erased all the datable evidence. From the Carboniferous period to the Permian period, the territory of Slovenia became mainland with vast marshes and swamps. Sediments forming slates, sandstones, and limestones of the western pre-Alpine mountains deposited. Throughout nearly the entire Mesozoic period, the territory of Slovenia was again under the sea. Vast and thick layers of limestone deposited. They are the foundation of the major part of today's Slovenian Alpine and Dinaric mountains. Generally, this very calm period was periodically interrupted by temporary volcanism, resulting in that some magmatic rocks (keratophyre, porphyry, and tuff) can be found in the Slovenian pre-Alpine mountains. Thick layers of flysch started to deposit at the beginning of the Cenozoic (Palaeocene and Eocene). The Pannonia Sea remained in the east and the north-east, where sandstones and mudstones deposited. In the Miocene period, the Alpine orogeny reached its climax. At the same time, layers from the Pannonian Sea also started folding. In the Pliocene, the karst began to form and some of the surface descended and all of Slovenia's largest basins were formed. In the Pleistocene, the climate cooled and much of the territory was covered with ice and snow. On the valley slopes, the rivers running below the retreating glaciers formed a system of terraces. Meanwhile, the basins were filled with coarse and fine partly fluvial, partly glacial sediments (Plenčar et al. 2009). In the central and western areas, the sediment is mostly coarse, carbonate gravel, or sand. Sometimes, especially on the oldest terraces, the gravel is consolidated into conglomerate rocks. In the eastern and north-eastern areas, the material is finer, mostly siliceous sand, silt, or clay. In the highlands, screes frequently occur beneath extremely steep slopes. Rivers form fans when entering lowlands from gorges or narrow valleys (Kočevar and Vidic 1998; Verbič 1998).

3.1.1.1 Geological Units and Frequent Parent Material

Table 3.1 presents the geological units of Slovenia. Three by surface most important units in terms of area are Triassic to Lower Jurassic carbonate rocks (18.5%), Quaternary terrestrial deposits (13.45%), and the unit where limestone, dolomite, and nonclastic siliceous sedimentary rocks interlace (10.6%).

According to the Geological Map of Slovenia, the most important lithological groups are clastic sediments (36%), carbonate rocks (54.6%), metamorphic rocks (4.2%), magmatic rocks (1.7%), mudstone (2.3%), and pyroclastic rocks (1.3%). Table 3.2 presents individual lithological units by area (M. Šinkovec after the 1:1 M Geological Map of Slovenia, GeoZS, 2013) (Fig. 3.2).

3.1.2 Topography

Topography is certainly Slovenia's most important and prominent natural feature. It is a direct reflection of the past tectonic events and geological history. Therefore, structural relief (folds, faults, thrusts, etc.) predominates. It provides the general image of the county and is woven into all other landscape elements, natural, and social. Slovenia's relief as we know it today is very young. It was formed in the Tertiary period with the uplift of the Alps and the Dinaric mountains. Quaternary glaciers and the Holocene river network gave the final contours. Topography as a factor of soil formation influences our soil cover indirectly. It has a direct and crucial impact on soil water conditions (quantity, movement direction, the effect on soil processes, etc.) and micro- and mezzo-climatic conditions (solar radiation, temperatures, the quantity and distribution of precipitation, wind, etc.).

3.1.2.1 Landforms

The topography of Slovenia can be described either morphologically or genetically. Morphologically hilly (low and high) topography prevails, although all landforms from low- to highlands are present. On a global scale, the absolute altitudes are considerably low. Therefore, the morphological landforms are also presented by the use of relative elevations and inclinations, since both highly correlate (Fig. 3.5) (Gabrovec and Hrvatin 1998): flat lands: up to 300 m.a.s.l. (alluvial plains, bottoms of basins, and karst poljes); low hills: up to 300 m above the bottom of the local valleys; high hills: between 300 and 1000 m above the bottom of the local valleys; mountains: above the upper forest line, between 1600 and 1900 m; low plateaus: generally flat surfaces, with low inclinations, between 300 and 700 m.a.s.l.; and high plateaus: generally flat surfaces, with variable inclinations, over 700 m.a.s.l.

Automatic GIS landform classification (Peucker and Douglas 1975) of the Slovene digital elevation model (12.5 × 12.5 m resolution) reveals the prevalence of different types of slopes and other topographies with vertically very variable surfaces (57.2%) (Table 3.3, Fig. 3.6).

Genetically, there are four types of relief in Slovenia (Gabrovec and Hrvatin 1998): fluvial, glacial, karstic, and coastal (abrasional). The fluvial type, with its surface river network, occupies most of the Slovene territory. It is further

Table 3.1 Parent material classification and basic statistics

Era	Geological units of Slovenia	Area (%)
Quaternary	Holocene fluvial deposits	7.6
	Quaternary terrestrial deposits	13.5
	Recent marine deposits	0.1
	Pliocene to Holocene terrestrial deposits and residuals	2.2
Tertiary	Neogene predominantly fine-clastic deposits	7.0
	Miocene lithothamnium limestone and marlstone	1.9
	Palaeogene and Neogene coarse-clastic deposits	2.9
	Palaeogene fine-clastic deposits	1.1
	Middle Pliocene basaltic tuff and basalt	0.0
	Middle Oligocene volcaniclastic rocks	1.3
	Middle Oligocene igneous rocks	1.0
Mesozoic	Cretaceous platform carbonate rocks	10.6
	Jurassic to Lower Cretaceous basin deposits	0.5
	Jurassic platform carbonate rocks	7.9
	Triassic to Lower Jurassic carbonate rocks	18.2
	Middle to Upper Triassic turbidites and other basin deposits	4.4
	Middle Triassic volcanic and volcanoclastic rocks	0.6
	Upper Cretaceous to Middle Palaeogene flysch and other basin deposits	4.8
	Upper Cretaceous to Middle Palaeogene carbonate rocks	1.5
	Cretaceous flysch and other basin deposits	3.4
Palaeozoic	Serpentinite	0.0
	Palaeozoic carbonate rocks	0.4
	Middle Permian terrestrial red beds	1.2
	Permian igneous rocks	0.1
	Post-Variscan molasse	3.6
	Variscan flysch	0.1
	Lower Palaeozoic igneous and metemorphic rocks	1.2
	Precambrian to lower Palaeozoic metamorphic complex	3.0

M. Šinkovec after the 1:1 M Geological Map of Slovenia, GeoZS, 2013

divided into destructive (denudation and erosion) and accumulative sub-types. Mechanical weathering and water erosion created systems of river valleys and the ridges in between. Five million m^3 of different types of weathered materials is released and transported every year (Kolbezen 1998). Water freely drains and automorphic soil types prevail. Gradually or abruptly, the waters lose their transportation power and the material is deposited into the accumulative sub-type of relief. Hydromorphic properties with saturated water conditions prevail. This process was especially active and abundant in the Pleistocene age, when rivers carried material from beneath the glaciers and carved riverbeds into their own accumulations while creating systems of terraces. Streams from the hills form numerous alluvial fans on the transitions between flat land and slopes (Natek and Natek 2008). In the hilly areas of the pre-Alpine hills, valleys often take the form of a gorge or dell with literally no space for reasonable agriculture. The accumulation sub-type is the most widespread in the north-eastern lowlands of the sub-Pannonian region. Conditions are not favourable for agriculture due to destructive floods, the high level of underground water, the shallow, young, and weakly developed soils, the acidity of the soils in the higher and older terraces, etc. (Vidic 1998). Nevertheless, these are one of Slovenia's most important agricultural areas due to the flat and even surface that enables intensive use of farming mechanization (Repe 2006b).

Our highest altitudes belong to the glacial type of genetic relief (Gams 1998). The Pleistocene mountain-type ice-cover valley glaciers completely reshaped the highlands. The former V-shaped river valleys were transformed into typical U-shaped valleys and broader, levelled bottoms (Fig. 3.8) are much more suitable for agriculture and habitation (Šifrer 1998; Natek 2007). Highlands, on the other

Table 3.2 Lithological units of Slovenia by area

Lithological group	Lithological units	Area (%)
Clastic sediments	Clastic sediment + conglomerate + diamictite	7.6
Clastic sediments	Clastic sediment + conglomerate + diamictite + organic rich sediment	13.5
Clastic sediments	Clastic sediment + impure carbonate sediment + sandstone + conglomerate + coal	7.0
Clastic sediments	Conglomerate + sandstone + impure carbonate sedimentary rock	2.9
Carbonate rocks	Dolomite + limestone + clastic sedimentary rock + impure carbonate sedimentary rock + pyroclastic rock	18.2
Metamorphic rocks	Foliated metamorphic rock + amphibolite + eclogite + marble + quartzite	3.0
Metamorphic rocks	Foliated metamorphic rock + basalt + marble + quartzite	1.2
Magmatic rocks	Granodiorite + dacite + tonalite + andesite + quartz diorite	1.0
Carbonate rocks	Impure carbonate sedimentary rock + mudstone + limestone + sandstone + nonclastic siliceous sedimentary rock	3.4
Carbonate rocks	Impure carbonate sedimentary rock + sandstone + mudstone + breccia	4.8
Carbonate rocks	Limestone + clastic sedimentary rock + impure limestone + dolomite + nonclastic siliceous sedimentary rock	4.4
Carbonate rocks	Limestone + coal	1.5
Carbonate rocks	Limestone + dolomite + breccia	8.2
Carbonate rocks	Limestone + dolomite + nonclastic siliceous sedimentary rock	10.6
Carbonate rocks	Limestone + impure carbonate sedimentary rock + mudstone + nonclastic siliceous sedimentary rock	0.5
Carbonate rocks	Limestone + impure limestone + sand + mud	1.9
Carbonate rocks	Mud + impure carbonate sedimentary rock + clastic sedimentary rock + coal	1.1
Mudstone	Mudstone + residual material	2.2
Mudstone	Mudstone + sand	0.1
Pyroclastic rocks	Pyroclastic rock + basalt	0.0
Pyroclastic rocks	Pyroclastic rock + tuffite + impure carbonate sedimentary rock	1.3
Clastic sediments	Sandstone + shale + conglomerate	1.2
Metamorphic rocks	Serpentinite	0.0
Clastic sediments	Shale + sandstone + conglomerate + limestone	3.6
Clastic sediments	Shale + sandstone + diamictite + limestone	0.1
Magmatic rocks	Syenogranite + granodiorite	0.1
Magmatic rocks	Trachyte + rhyolite + basalt + pyroclastic rock	0.6

hand, have extreme, even vertical inclinations (the average is 30.5°) that hinder soil genesis and soil can hardly form (Lovrenčak 2007). Soil cover is very sparse and appears only in cracks in the parent material (Surina 2005). Beneath the practically vertical slopes, especially in the less resistant dolomite, large screes form that even further obstruct soil formation (Repe and Blatnik 2013). Large moraines and glacier lakes remained after the retreat of the glaciers.

With less than 46.6 km of Adriatic coast in the south-west, there is very little coastal topography in Slovenia (Fig. 3.9). The coast, evenly built of flysch, is young and originated in the Pleistocene (Ogorelec et al. 1983; Šifrer 1965). Abrupt cliffs, up to 80 m high, alternate with small bays and low sea plains encompassing the river discharges. Although narrow, the sea plains enable habitation and agriculture on medium-developed soils (Eutric Cambisols and Fluvisols), formed on the fine fluvial material. In spots, the intrusion of the saline and brackish waters causes natural soil salinization (Repe and Pristovšek 2011).

One of the main characteristics of Slovenia is the karstic topography (44% of the territory) (Habič 1992) with an extreme small-scale variety of surface features and the absence of surface waters. The term itself (Slovenian: *Kras*) originates from Slovenia's Karst region in the hinterland of Trieste, between the Gulf of Trieste and the Vipava River valley. The contact between soil cover and parent material is completely irregular in shape. Therefore, small-scale and so-called mosaic distribution of soil depths and types with

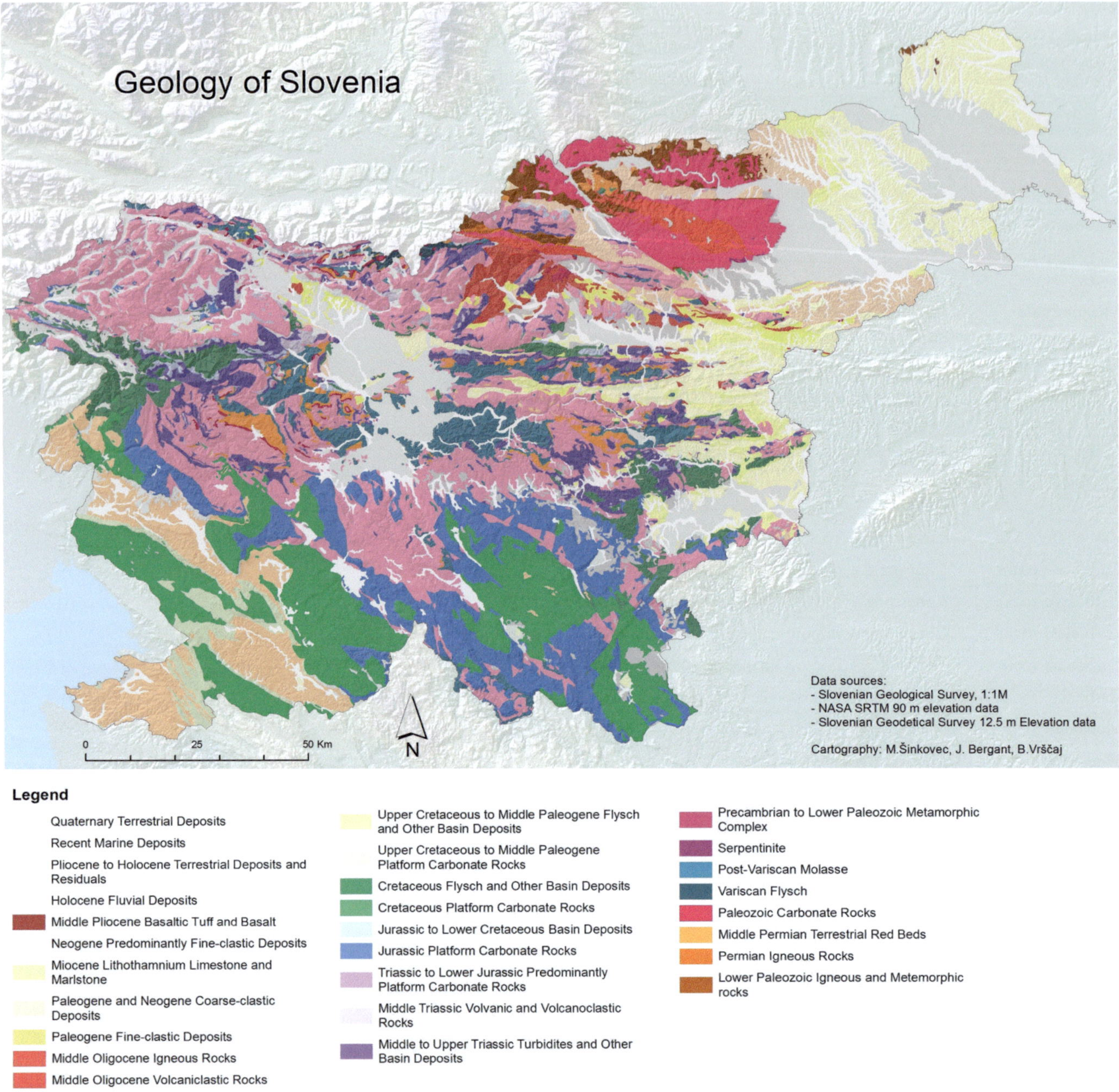

Fig. 3.2 Parent material in Slovenia (M. Šinkovec after the 1:1 M Geological Map of Slovenia, GeoZS, 2013)

protruding boulders developed. This is the most unsuitable area for any agricultural activities, which remains to any significant degree only on larger karstic poljes. In the past, self-sufficient microfields were artificially created on the bottom of dolines (Gams 1974) (Fig. 3.10).

3.1.2.2 Morphometric Characteristics of Slovene Topography

As determined by previous research, 75% of all soil formation factors are related or correlate to topographic characteristics (Gerrard et al. 1992) that condition water discharge and the ratio between erosion and accumulation. GIS analysis of the digital elevation model provides all sorts of variables that describe the aforementioned connections and correlations (Podobnikar 2010). With its young, hilly, diverse, steep, and variable surface, Slovenia's topography is among the most important soil formation factors and its parameters very accurately predict the soil properties (Repe 2010a). All the morphometric parameters are derived from the digital elevation model (digital terrain model, 25 m resolution) for the mainland part of Slovenia, and in order to present them comprehensibly, the country was further

Fig. 3.3 Slope inclination of Slovenia

divided into 13 landscape-ecological region types (Kokalj et al. 2004; Kokalj and Oštir 2006) (Fig. 3.4).

Different authors (Gams 1998; Perko 2001) quote different mean elevations of Slovenia (from 533 to 580 m.a.s.l.). The calculation from the digital elevation model is 556.5 m.a.s.l., which is nearly twice the European mean (300 m; "List of All Countries in Europe|Bridgat.com", 2011). Nevertheless, relatively lower absolute altitudes prevail, since less than 10% of the territory is below 200 m, 11% is over 1000 m, and the main part (33.1%) lies between 200 and 400 m.a.s.l.

Average inclination is rounded up to 13° according to Perko (2001) and 14.1° from the digital elevation model. Not even these quite high averages reveal the actual diversity of Slovenia's topographic features. The actual flat land area is small, only 7.6% (Natek and Natek 2008). Inclinations that enable all the variety of modern agriculture (up to 3°) stretch over 1/5 of Slovenia's territory (19.2%). They are mostly along flood plains in north-eastern Slovenia, in the basins of pre-Alpine Slovenia, and on larger karstic poljes. But on slightly less than half of the country (47.7%), the slopes are too steep (above 13°) for either habitation or any agricultural activities (Fig. 3.3). Only with certain favourable conditions (a southern exposure) is extensive and small-scale stockbreeding possible. Together with shallow soils (less than 40 cm), these are the main reasons why Slovenia is dominated by increasing forest cover (63%; (EUROSTAT 2010)) (Table 3.4).

There are two main mountain ridging directions: the W–E or the Alpine ridging and the NW–SE or the Dinaric ridging. Nevertheless, the aspects are generally very evenly distributed with a slight increase in the southern exposure (14.19%). Taking all this into account, on average Slovenia receives 4013 MJ/m^2 of solar radiation annually (Gabrovec 1996) (Fig. 3.7).

Considering the regional distribution of topographic soil formation factors, the differences are considerable and revealing. They are diametrically opposed between units at higher altitudes and/or inclinations and lowland or flatter units (suitable for modern mechanized farming). The latter (the broader valleys in the highlands, hills and karst; the inner mountain basins; the plains and broader valleys in the low hills of central Slovenia; karstic poljes and lowlands; the low karst of Bela Krajina; the broader valleys and plains of sub-Mediterranean Slovenia) represent only a minor part of the country. They are characterized by low elevations, inclinations, terrain ruggedness, flat or concave features, and a

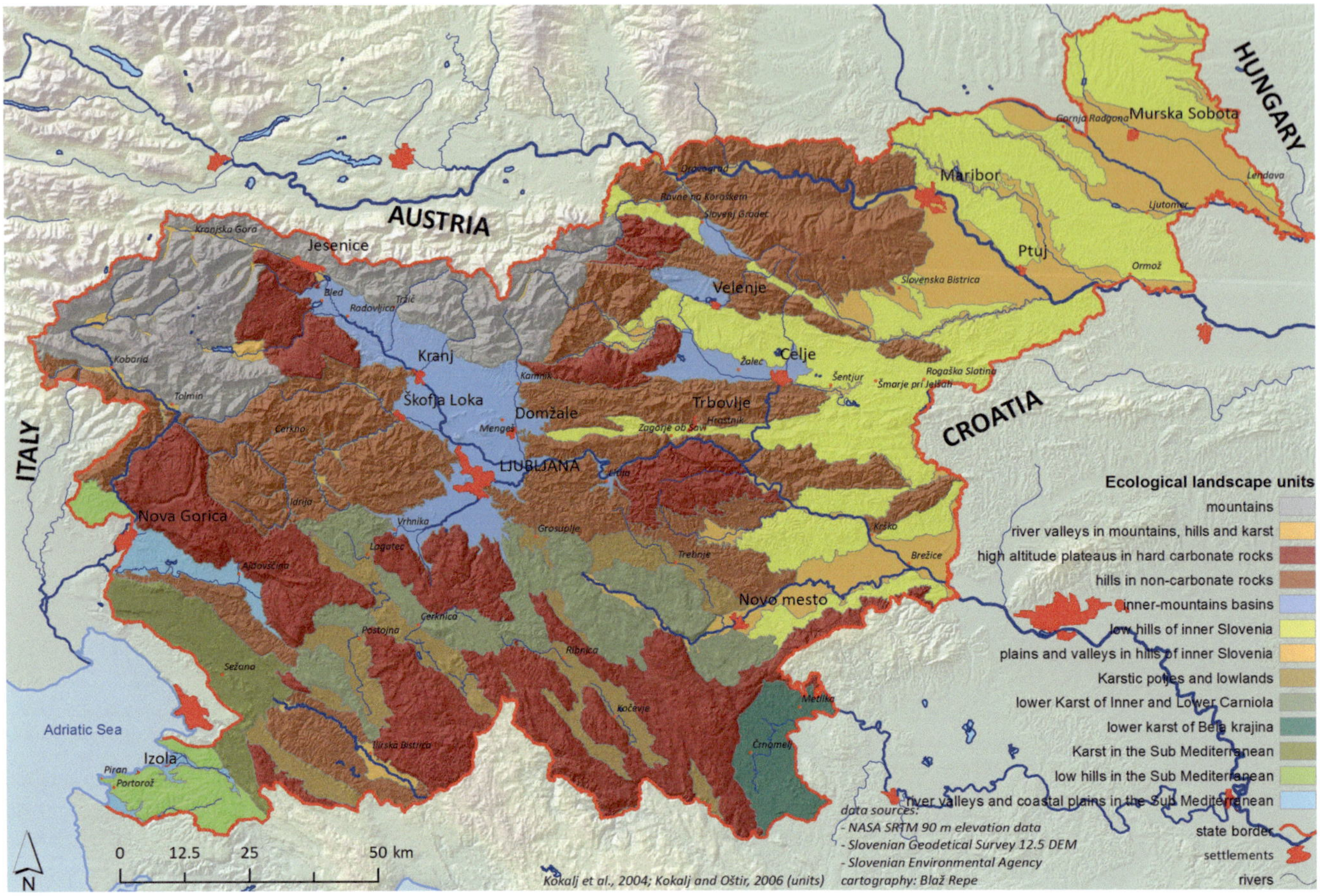

Fig. 3.4 Morphological and genetic relief types

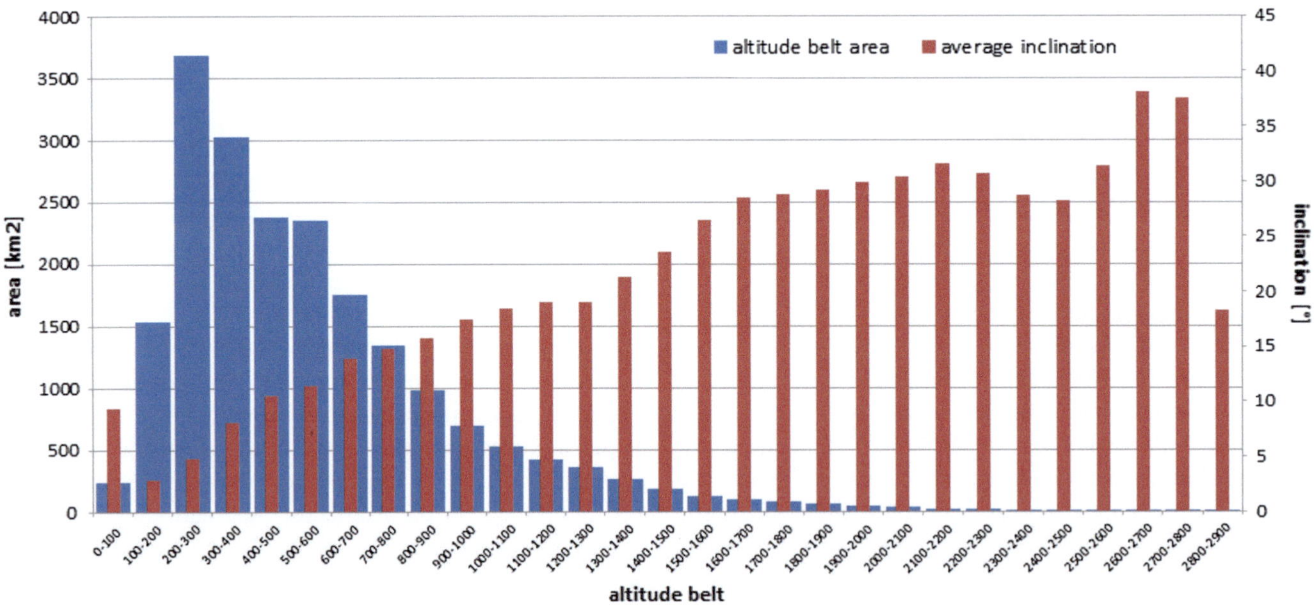

Fig. 3.5 Altitude zonation histogram with attributive average inclination

Table 3.3 Morphological landforms in the total area of Slovenia (Gams 1998)

Morphological landform unit	Area (km^2)	%
Plains/flat lands	3446.4	17.0
Low hills	4865.4	24.0
High hills	6325.1	31.2
Mountains	1702.9	8.4
Low plateaus	1500.2	7.4
High plateaus	2432.7	12.0

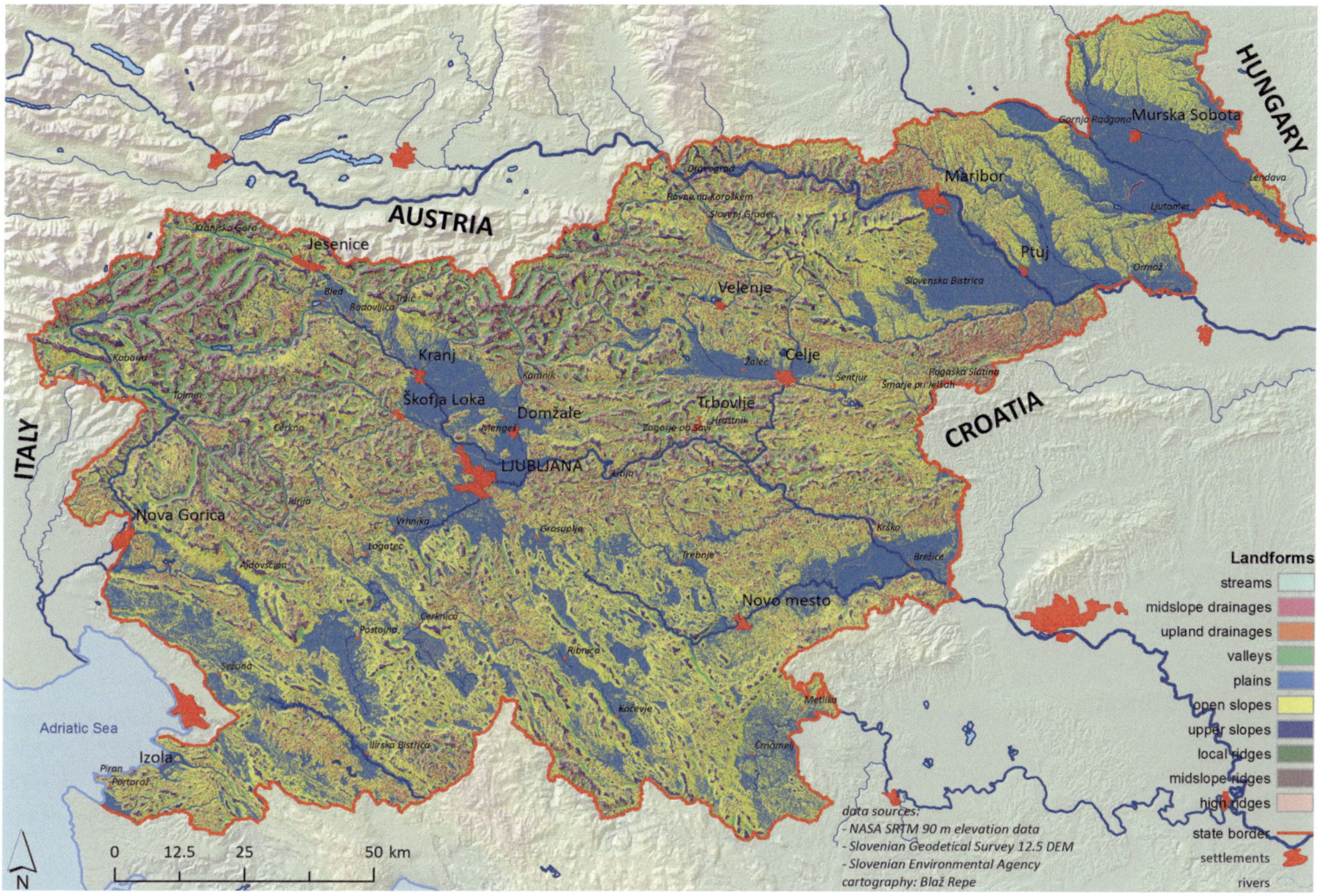

Fig. 3.6 Automatically generated landforms, 25 × 25 m resolution (Peucker and Douglas 1975)

position near the groundwater level. There is a constant inflow of waters into these areas that brings all sorts of weathered materials. The soils are therefore considerably thicker, which all together make these areas much more suitable for farming, especially for modern and mechanized agriculture. In such units, areas under forest cover still exceed all other land use categories, but they are below the national average. The share of agricultural categories is nearly comparable to the forested categories (Figs. 3.11 and 3.12).

On the other hand, in the majority of the country (highlands; high plateaus and hills on carbonate parent material; hills, mostly on noncarbonate parent material; the low hills of central Slovenia; the low karst of Inner and Lower Carniola; the karst of Trieste and Komen and the karst of Podgorje; the low hills of sub-Mediterranean Slovenia) the topographic variables are above average, very variable, and unfavourable. In addition to this, the drainage and discharge of (soil) waters are rather quick; consequently, automorphic soil types with soil depth less than 40 cm, surface rockiness, and stoniness are also important obstructive factors. Farming is limited to extensive, sometimes free-range stockbreeding, vine, or fruit growing on man-made terraces, or is locally entirely self-supporting and still manual. 3.5% of the agricultural area is not of any commercial importance, not even on a regional level. Natural and actual forest vegetation extends over 70% of the units (Repe 2008).

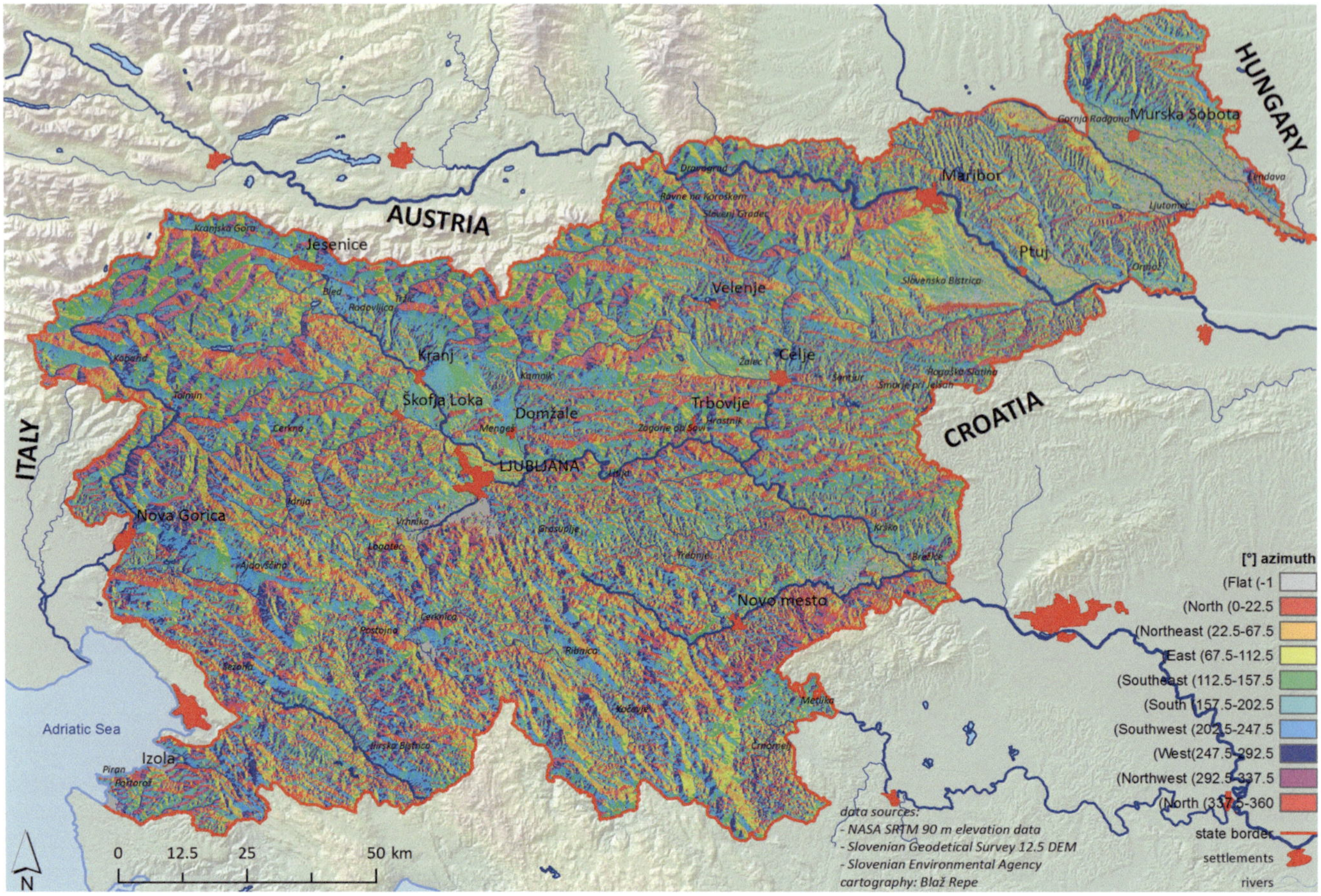

Fig. 3.7 The aspect of slopes

3.1.3 Climate

Mutual interactions between climate and other landscape systems were established in the relatively early stages of soil research. In the larger landscape units, climate generally plays a leading role among other environmental factors (Stritar 1990a). Regarding soil genesis, at 20,000 km^2 Slovenia is definitely not a large landscape unit. The influence of climate on Slovenian soils and soil genesis is more general. It is mostly regarded as a constant and does not have an important interregional distinguishing aspect. The influences of climate, therefore, reflect in general soil genesis processes, such as the rate of parent material weathering and humification, argilogenesis, and braunification, the leaching of nutrients and fine particles, and gleying (Škorić 1986a; Lovrenčak 1994b; Repe 2004a).

Among all environmental factors, the climate of Slovenia is especially determined by the confluence of four large European units: the Alps and the Dinaric mountains, the Pannonian Basin, and the Mediterranean. This position is directly reflected in Slovenia's many climatic divisions (Melik 1935; Furlan 1960; Ilešič 1970; Gams 1972a, 1996; Ogrin 1996). With the exception of the mountainous part, Slovenia (along with most of Europe) is dominated by a warm and humid temperate climate (Ogrin 2004). If we consider the Koppen climatic classification, three main climatic types can be distinguished: the temperate warm and humid climate, with hot summers (Cfa); the temperate warm and humid climate, with warm summers (Cfb); and the mountainous climate (H). However, all three are expressed in atypical form compared to the genuine aforementioned types (with the prefix "sub-" before the names of Slovenia's climatic types and even regions). In more detail, nine sub-types can be extracted according to the distribution of precipitation, the average temperatures of the coldest and the warmest months, and the ratio of October and April temperatures (Ogrin 1996). There are coastal and inland temperate Mediterranean climate in the west; the Mountainous climate of the upper (the highest parts of the Slovene Alps) and lower (the Pohorje massif and the Karstic-Dinaric mountains and plateaus) mountain ranges; western, southern (sub-Alpine), and the central and eastern (sub-Pannonian) temperate continental climate over the rest of Slovenia (Fig. 3.13).

Fig. 3.8 The "U"-shaped valley in Jezersko. Reshaped and reformed by the glaciers in the Pleistocene (*Photo* B. Repe)

Fig. 3.9 Cliffs on Slovenia's coast (*Photo* B. Repe)

A comparison of the generalized map of the World FAO WRB classification (FAO 2006a) and the climatic division of Slovenia reveals the following (Repe 2007a). Macroclimatic features are not an important parameter for the inter-differentiation of Slovenia's soils. Macro- and mezzo-climatic (continental) features determine the general predominance of Cambisols, with the exception of the Alpine region, where Leptosols prevail due to the harsh mountainous climate.

3.1.3.1 Solar Radiation

Solar radiation is the primary energy source for all physical, chemical, and biological processes occurring on the earth (Chang 1968; Cutforth 2013). The amount of solar radiation varies seasonally, regionally, and locally. Due to the more frequent convection at orographic barriers, its amounts are altitude dependent. Radiation decreases with increasing altitudes. Not surprisingly, the highest amount of solar radiation occurs in the summer months, from July until August. In addition, the southern parts of the country receive more energy than the northern ones. As a consequence of convection, there is steep gradient of solar radiation in the zone between 700 and 800 m.a.s.l. The formation of fog and low-level clouds decreases the amount of solar radiation in

Fig. 3.10 Man-made dolines. People filled dolines with soil that was scraped and collected from the surrounding areas in order to create some flat, protected from winds, and stone-free areas where they could grow some crops for domestic use (*Photo* B. Repe)

the coastal areas in the spring and in the inner basins and valleys in the colder part of the year. In the winter months, the highest amounts are registered at higher locations and altitudes. In spring, the amount of radiation decreases with altitude more rapidly in the lowlands than in the hilly parts. The maximum amount of solar radiation occurs on the coast, between 2100 and 2400 h/year, and the minimum is in the Alps, between 1600 and 1900 h/year (Dolinar 2006). There is also an evident difference between southern and northern exposures of slopes. The amount of quasi-global radiation of the inclined surfaces with southern sun exposures can be more than 20% higher compared to horizontal surfaces; on sunless slopes, the difference can be up to 70% lower. On average for the entire Slovenia, northern slopes receive 9% less and southern slopes 4% more solar radiation then exposed horizontal locations (Kastelec et al. 2007). The sub-Mediterranean region has the highest average energy of quasi-global radiation (4375 MJ/m^2), the mountain region has the lowest (3713 MJ/m^2), and other regions vary more or less around the average, which is 4000 MJ/m^2. There are very large differences in the annual energy of quasi-global radiation within short distances. One consequence being that the variability between climate types is smaller (1:1.178) than the variability within an area of a particular climate type (1:1.281) (Gabrovec 1996) (Fig. 3.14).

3.1.3.2 Temperatures and Altitude Belts

The distribution of temperatures in Slovenia is directly dependant on the amount of solar energy, topographic features. The distance from the sea, and land use effect actual temperature values (Ogrin and Plut 2009). The warmest places are in the coastal region and its immediate hinterland, with average annual temperatures above 12 °C (January 4 °C and July 22 °C). Due to the continental influences, the north-eastern part of the country follows with averages above 10 °C (January −1 °C and in July 20 °C). The averages in the central part are between 8 and 10 °C (January 0 °C and July 20 °C). The Pohorje massif and high karstic plateaus in

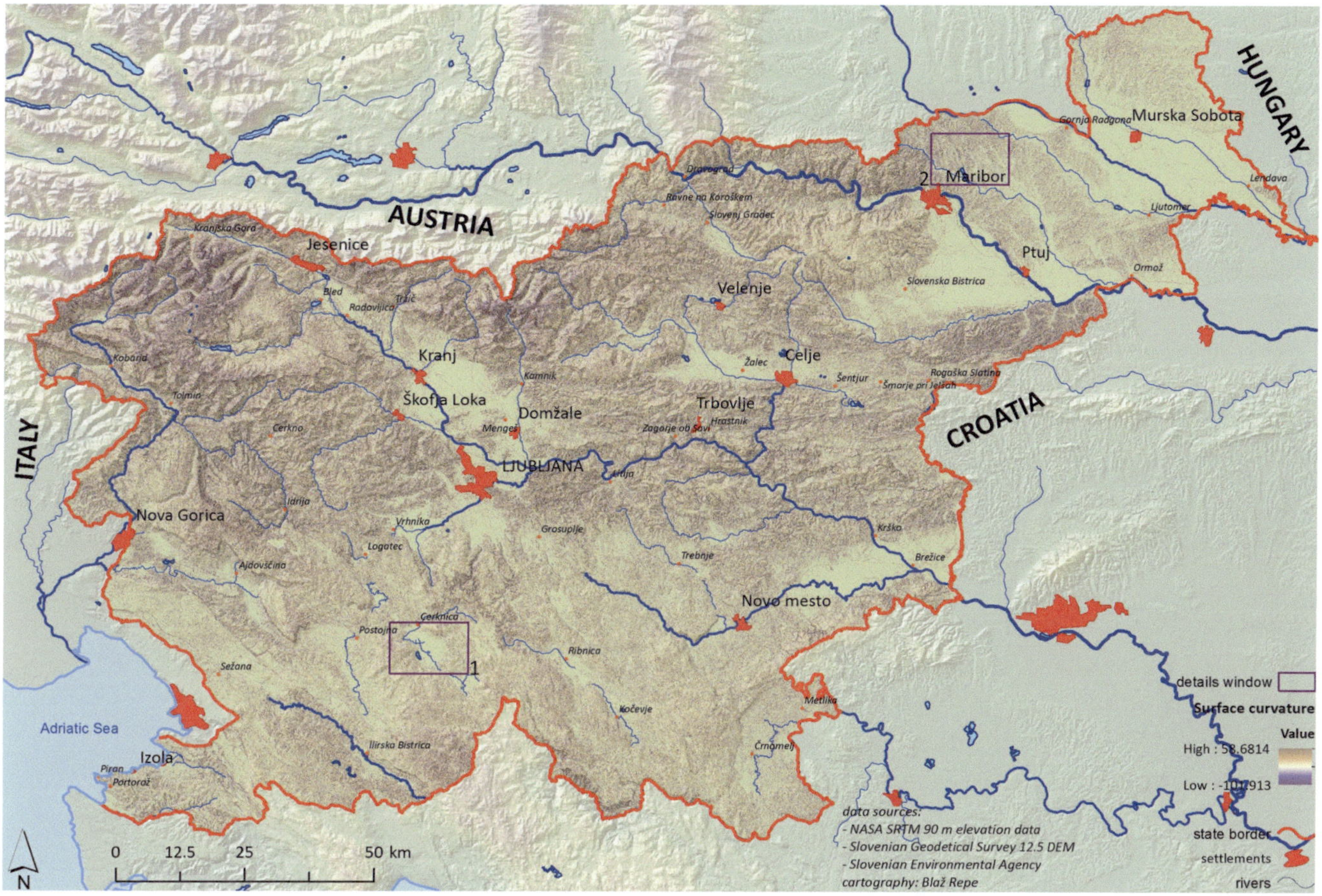

Fig. 3.11 Slope curvature of Slovenia

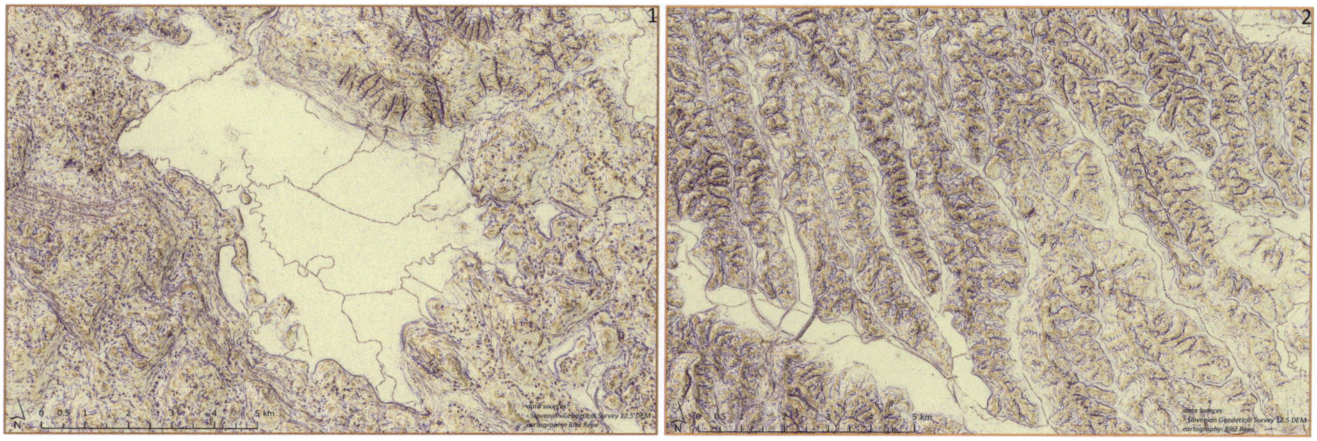

Fig. 3.12 The curvature of slopes. The area of Cerknica Lake (*left*) and Slovenjske Gorice (*right*). There is an extreme microvariability in Slovenia's topography. This is clearly visible on the carbonate karstic surface (*left*), as it is on siliceous rocks (*right*). The exceptions are karstic poljes, filled with fine sediments (*left*) and more or less narrow valleys in hilly parts of the country (*right*)

the west and south face a drop in the annual average to 6 °C or less. In the highest part of the Alps, the annual average is below 0 °C (January −8 °C and July 6 °C) (ARSO 2006) (Fig. 3.15).

Among topographic features, elevation above sea level is the most important since with its increase there is generally an evident decrease in temperature. The average annual temperature drops 0.53 °C for every 100 m.a.s.l. higher

Table 3.4 Areas of the inclination units in the total area of Slovenia

Inclination unit (°)	Area (km^2)	%
0–3.0	3891.3	19.2
3.1–7.0	2724.1	13.4
7.1–13.0	4013.2	19.8
13.1–21.0	4309.2	21.3
21.1–33.0	3782.1	18.7
33.1–55.0	1479.7	7.3
55.1–90	73.0	0.4

Fig. 3.13 Climatic divisions of Slovenia

altitude. Due to the relatively low elevations, altitude zones (belts) are not very clearly expressed and not very diverse. In some places, the variability is much greater within one belt itself due to the variability of other factors than due to the temperature change (Repe and Blatnik 2013). Nevertheless, Gams (1972b) divided Slovenia into 7 climatic altitude zones: (1) the bottoms of basins, valleys, and karstic depressions, which are marked by the occurrence of temperature inversion. Although climatically not ideal, a thick soil cover and level surface concentrate almost all agricultural activity in this zone. (2) There is a thermal zone 100–300 m.a.s.l. above the bottoms and inversions. (3) In the hilly zone (up to 800–850 m.a.s.l.), meagre agriculture and settlements are limited to southern, sun-exposed locations. (4) The lower mountainous zone (up to 1200 m.a.s.l.) with shallow soils only allows solitary farms that live on forestry and/or free-range stockbreeding. (5) The higher mountainous zone (up to 1550/1600 m.a.s.l. or 1800 m.a.s.l. in the NE parts of Slovenia) cannot sustain cultivation, except some high mountain pastures or meadows. (6) The

Fig. 3.14 Solar radiation in Slovenia

sub-Alpine zone (up to 1700/1900 m.a.s.l.) reaches the upper forest line. Soils are poorly developed and very shallow (less than 10 cm). (7) The Alpine zone (up to 1200 m.a.s.l.) reaches the highest elevations.

3.1.3.3 Precipitation

According to the amount of precipitation, Slovenia is above the worldwide average (730 mm) (Linacre and Geerts 1999) and the European average (820 mm) (ARSO 2006). In the period between 1971 and 2000, the average annual precipitation in Slovenia amounted to 1579 mm. About 717 mm of this total amount of precipitated water evaporated, and 862 mm can be attributed to run-off (Dolinar et al. 2008). Slovenia lies on the cyclonal trail following the western air circulation from the Atlantic Ocean and the Mediterranean over Slovenian territory. The mountain barriers of the Alps and the high Karstic-Dinaric plateaus obstruct the aforementioned cyclonal trail and force the moist air to ascend, leading to abundant orographic precipitation. The quantity of precipitation gradually declines towards the east, reaching the lowest values in the far north-east (ARSO n.d.; Gartner 1999) (Fig. 3.16).

On the coast, in front of the orographic barriers, the annual precipitation fluctuates between 1100 and 1200 mm. The maximum is reached in the Alps (3600 mm) and the Dinaric mountains (2500 mm). Most of the central part receives amounts around the average, between 1000 and 1750 mm. In the most continental NE of the country, Slovenia's minimal amounts do not exceed 900 mm. The wettest parts thus receive four times more precipitation then the driest parts (Hočevar and Petkovšek 1988; Kajfež Bogataj 2007). In the coldest part of the year, everywhere in Slovenia, some precipitation can fall as snow. In the mountainous regions, it falls mainly as snow, producing snow cover a couple of metres thick (ARSO 2005). In the spring months, thawing of snow represents an important water source in soils, the river network, groundwater, and consequently for vegetation and agriculture.

Precipitation is not evenly distributed over the year, but there are not really any extremely over-saturated or distinctly dry seasons. In the western part of Slovenia, most of the precipitation falls in the autumn. The wettest month is November and the driest is February. The maximum shifts in the summer months towards the east and with increasing

Fig. 3.15 Mean annual temperatures in Slovenia

continental climatic influences. In spite this maximum, there is still a deficit of water in the summer. This summer precipitation falls in the form of showers and storms with strong winds and hail. In the past 30 years, a change in distribution has been noticed. The autumn maximum is becoming more and more distinct, while other months receive less precipitation (ARSO 2005).

The ratio between precipitation and evapotranspiration is favourable, and Slovenia is regarded as humid. In addition, soil moisture regimes can generally be considered to be udic. Only the highest altitudes of the Alps feature a perudic regime. Due to the flat or concave topography and impermeable parent material, the agricultural north-east features an aquic soil moisture regime, with the largest areas of hydromorphic, waterlogged soils (Fluvisols, Gleysols, Stagnosols). Nevertheless, drought is an increasingly frequent phenomenon and can be attributed to insufficient choice of farming methods or crops, inexpedient or excessive use of water, etc., coinciding with drier seasons (Sušnik and Kurnik 2003; Ogrin 2012; Kikec 2014).

3.1.3.4 The Winds

The wind conditions in Slovenia are influenced by the diverse topography, location in the medium and moderate geographic latitudes, and by the Alps. The winds are mostly westerly orientated, due to the general air circulation in this part of Europe. Compared to the rest of the Europe, Slovenia is not a very windy country since it is sheltered by the Alps (wind shadow) (Bertalanič 2007) (Fig. 3.17).

The most common are south-westerly or north-westerly winds, but the directions can vary locally considerably due to the topography or other obstructions (forest cover, settlements, etc.). In the sub-Mediterranean region, there is a very typical strong and gusty Bora wind. It forms after a cold weather front passes through, particularly in the colder part of the year. The Bora is not a steady wind, but it blows in gusts that can reach speeds over 100 km/h and cause considerate problems and damage (fracturing or uprooting of trees, soil erosion, etc.). Very strong winds also blow in the mountainous areas, especially upon changes in the weather. Elsewhere, strong winds develop during thunderstorms.

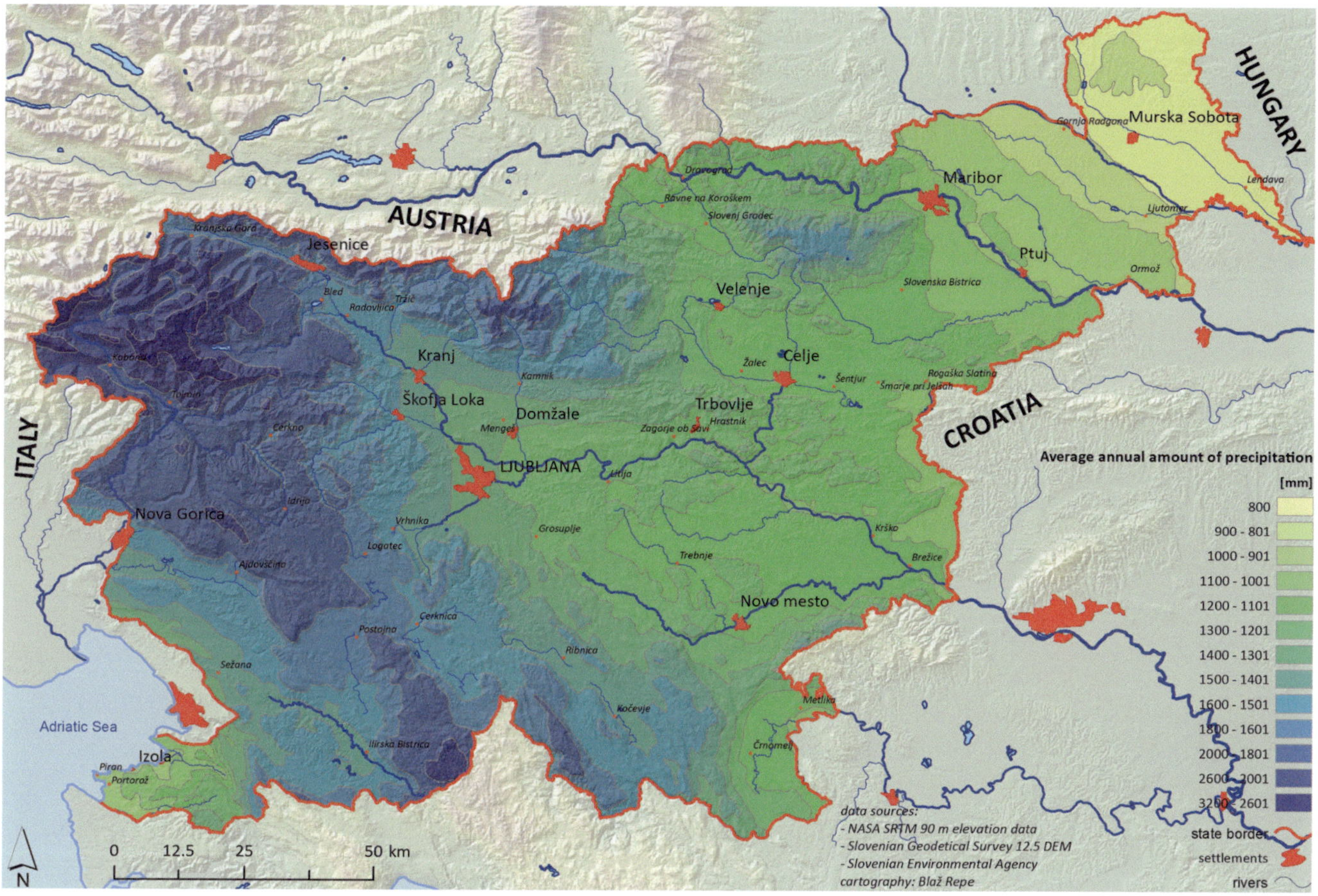

Fig. 3.16 Mean annual precipitation

Otherwise, local winds appear due to the diverse topography and temperature differences and have a typical daily variation (ARSO 2005). Winds blow uphill and from the sea to the mainland during the day, and downhill and from the mainland towards the sea during the night (Pagon 2006).

3.1.4 Hydrology

Due to its importance as a soil formation factor, water deserves a special place in Slovene soil classification. On the first level of classification, soil is divided into four major groups, automorphic, hydromorphic, sub-aqueous, and halomorphic on the basis of water (Prus 2000; Repe 2004b). As is the case with topography, there is also a clear and double hydrological duality. The first duality is related to the parent material. There is the karstic surface of permeable carbonate rocks, with neither surface, stagnant, nor running water (with rare exceptions). On its way into the corroded and cavernous rocks, precipitated water only traverses the pedosphere. Then, there is the rest of the country, consisting mainly of noncarbonated and siliceous rocks with a normal, surface river network. Water eventually makes contact with an impermeable obstacle. If this impediment lies near the surface, hydromorphic conditions can develop.

The source of the second duality is topography. Water is attracted from hilly or mountainous towards lower parts and the local erosion basis. On its way, water carries weathered material and detached soil particles. Soils are shallower and automorphic. Waters with denuded and eroded material run into the valleys, intermountain basins, karstic poljes, lowlands, and plains, where soils are considerably deeper, and again, hydromorphic conditions can frequently occur. Since Slovenia is by far more elevated, hilly, and diverse, there is little hydromorphic soil, only 13.4% (Ministry of Agriculture and the Environment 1999; Repe 2006b).

3.1.4.1 Water Balance

The hydrological conditions of Slovenia are a direct consequence of mostly three factors: climate, above all precipitation, parent material, and topography. Again, Slovenia is a humid country, with above average precipitation evenly distributed over the entire year. At the same time, the

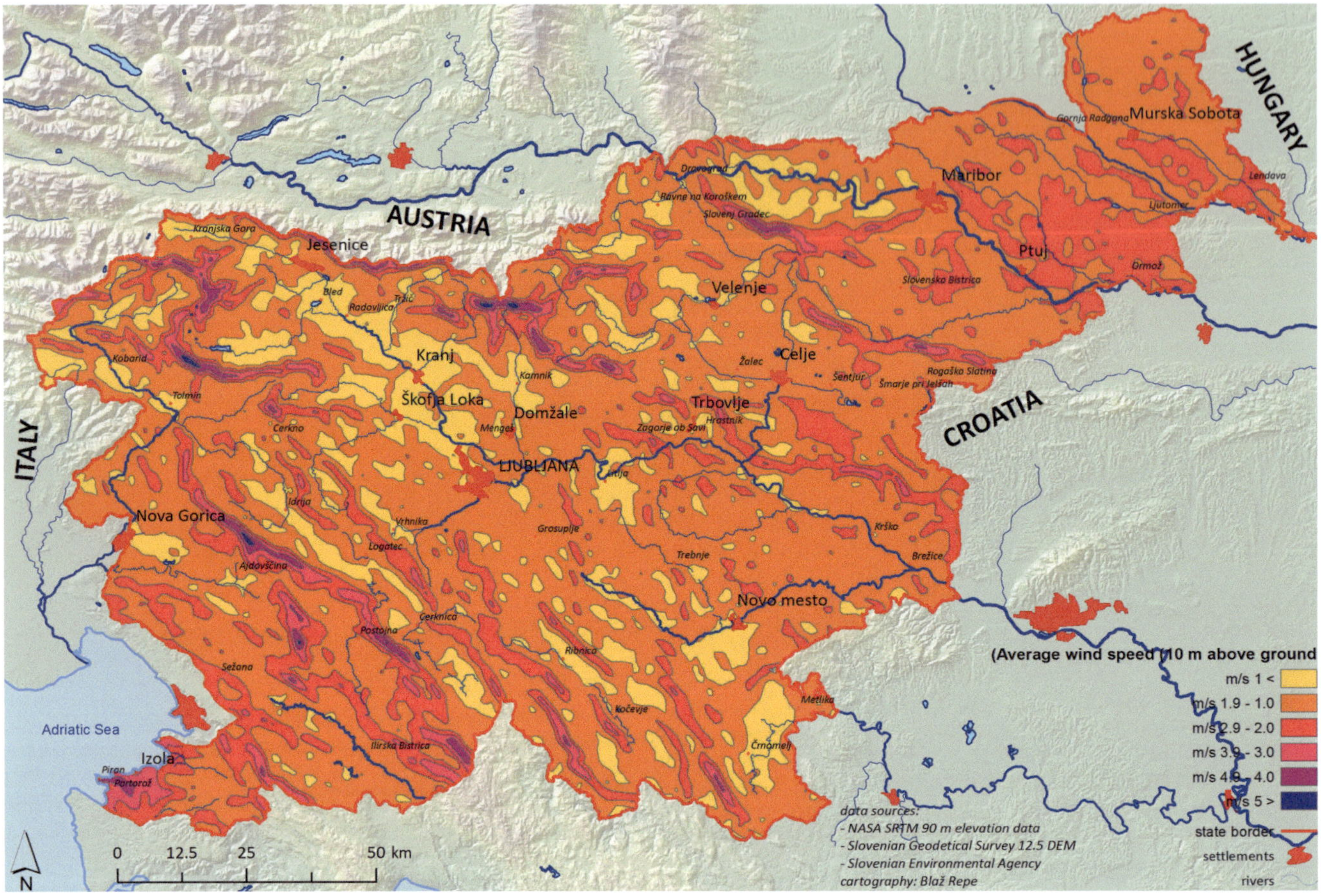

Fig. 3.17 Winds in Slovenia

temperature conditions are such that evapotranspiration does not exceed precipitation. Consequently, besides being humid, Slovenia is a country very rich in water resources. There are all forms of land water, both surface and underground, as well as the sea, but only to a small extent. There are all together around 28,000 km of different watercourses, of which one half never run dry; 1300 lakes with an area over 1 ha; and 46.6 km of coastline. However, since Slovenia is a very small country, its water catchment areas are small and consequently the dimensions and proportions of our water bodies are also small/short. Internal run-off from Slovenian territory not including transitional waters is estimated at 590 m^3/s. The specific run-off is therefore 29 $l/s/km^2$. The entire run-off of Slovenia is 1000 m^3/s. Every second, 400 m^3 of water crosses the borders from our neighbouring countries, but almost entirely from Austria (the Drava and Mura). Slovenia contributes less than 1% to the total European run-off (Bat et al. 2003) (Fig. 3.18).

A simplified equation of the water balance basically explains the general availability of water and also the availability of soil moisture:

$$\text{Precipitation} = \text{run-off} + \text{evaporation}$$

Kolbezen and Pristov (1998) Slovenia receives an annual average of 1579 mm of precipitation (32 km^3), which is unevenly distributed over the entire territory (see Sect. 3.1.3). Evaporation, which is the least spatially variable parameter in the equation (a simplified equation of the water balance basically explains the general availability of water and also the availability of soil moisture), depends only on the altitude and land use type. There is an annual average of 717 mm of evaporated water (13 km^3). The evaporation generally increases from the southern parts (458 mm/year) of Slovenia towards to the north-west and north-east (879 mm/year). The highest values (over 850 mm/year) are in the forested areas of the north-east and in the coastal areas. The evaporation in the intermediate areas is varied and fragmented due to the very diverse land use. The areas with the minimal values of evaporation (below 600 mm/year) are mostly hilly and mountainous.

As regards the run-off, Slovenia exhibits extreme diversity. The average amounts to 862 mm/year (18.5 km^3). The maximum run-off is on the Alpine-Dinaric barrier (over 2000 mm/year) and decreases towards the SW (300–600 mm/year on the coast) and the NE. The lowest level (below 300 mm/year) can be found in Prekmurje, where in

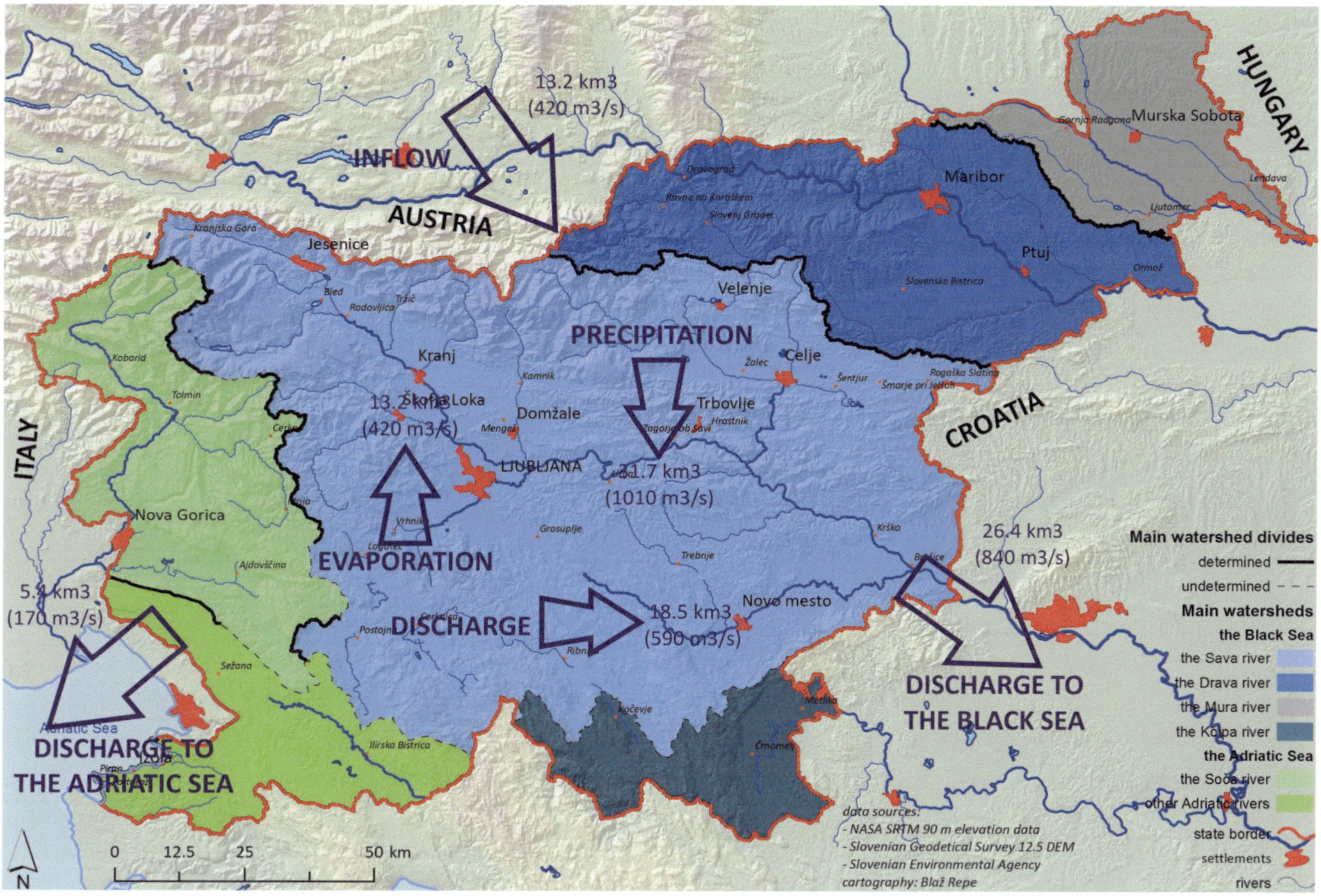

Fig. 3.18 Slovenia's watershed water balance (Kolbezen and Pristov 1998)

certain years evaporation can exceed precipitation (Frantar 2007). The specific run-offs are high, and the average is above 30 l/s/km^2 (Bat et al. 2003).

Overall, Slovenia can be described as wet, since its run-off coefficient is above 55%, which is regarded as well above the global average (36%) (Frantar 2007). Nevertheless, the run-off coefficient has decreased in the last decade. The amount of precipitation has remained at the previous level, but the run-off has decreased and can be attributed to land use changes, especially heavy and widespread reforestation (Kolbezen and Pristov 1998).

3.1.4.2 Surface Inland Waters

Surface inland waters include running (rivers, brooks, torrents, and periodically dry streams) and stagnant (lakes) waters. With a positive water balance, the excess of water eventually starts to form a river network. Slovenia's river network is dense with numerous watercourses, over 31,500 (ARSO n.d.), but due to the smallness of its territory, they are short and over half of them occasionally run dry. According to length, the longest rivers with accompanying watershed areas are the Sava (225 km; 10,872 km^2) and the Drava (142 km; 3489 km^2); other important rivers are the Kolpa, Savinja, Soča, Mura, Krka, and Sotla (Natek and Natek 2008). According to the available digital data (ARSO n.d.), there are nearly 29,000 km of streams, which gives Slovenia a very high river network density of 1.4 km/km^2. This density is very asymmetrically distributed over the country. There are areas with a density over 15 km/km^2 and vast areas totally void of any surface streams. This difference is not climatic, but is attributed to permeable carbonate rocks (limestone, dolomite), where water runs sub-surface within the rocks. It sometimes reaches the surface in the form of karstic springs upon contact with impermeable, noncarbonated rocks.

At first glance, it might seem unusual, but due to the high Karstic-Dinaric barrier, most of Slovenia's waters (80%) runs into the distant Black Sea and only 20% is discharged into the much closer Adriatic Sea (Fig. 3.19). More than in length, Slovenia's watercourses are large in terms of their water quantity. The discharges of Slovenia's largest rivers (the Drava and Sava) exceed 300 m^3/s. In Slovenia, river run-off regimes depend on the amount of precipitation, the type of precipitation (rain and/or snow) and temperatures (the time of the snowmelt). Slovenia's rivers have four types of run-off regimes: those fed only by rain (coastal areas),

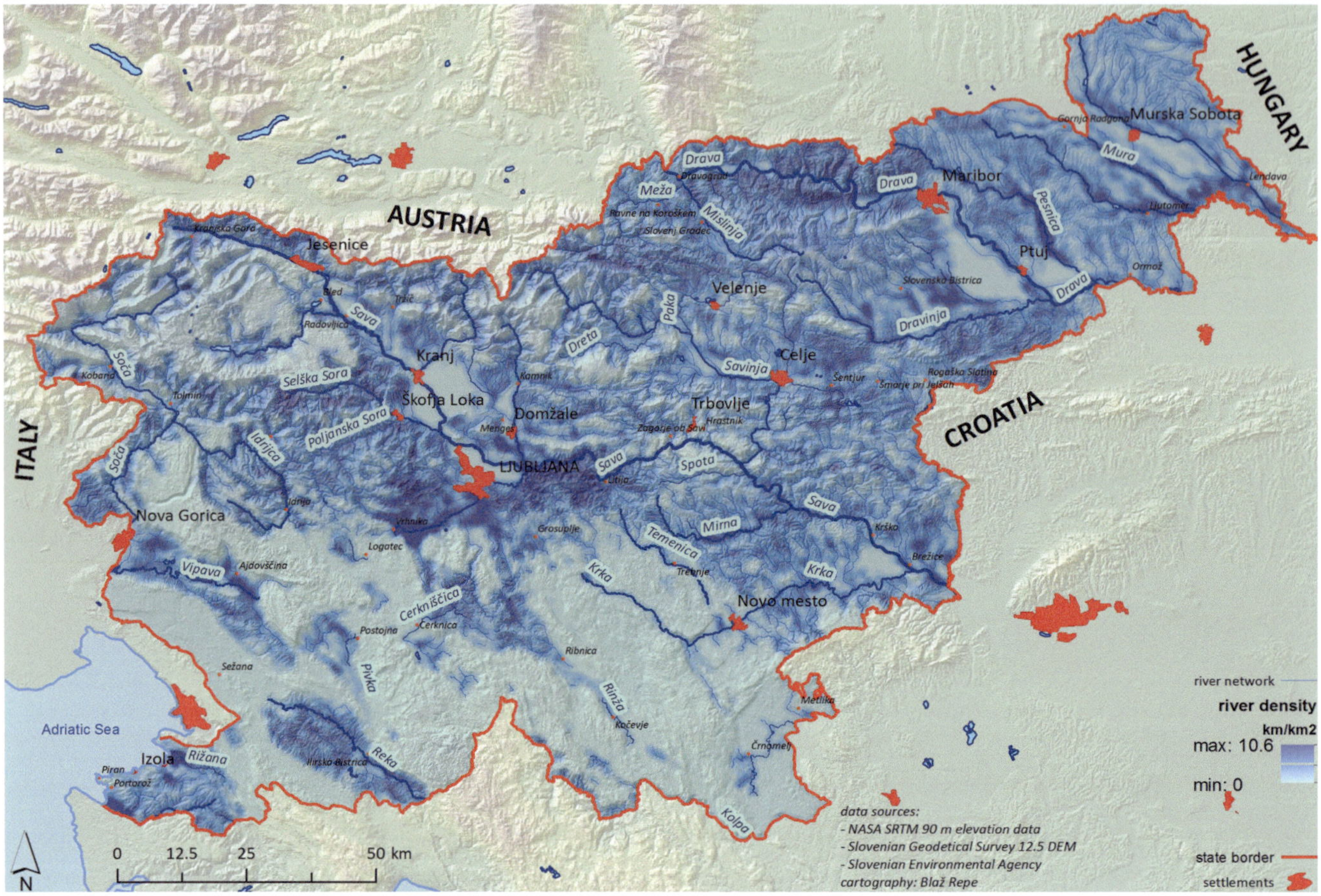

Fig. 3.19 The rivers of Slovenia and their density (discharge)

only by snow (the Drava and the Mura), and the combinations (nival-rain originating in the Slovenian Alps and rain-nival, originating in the north-east). The karst plays an important role, as that is where part of the water is absorbed, retained in the porous and cavernous carbonate rock, and then slowly discharged (Kolbezen 1998; Frantar and Hrvatin 2005).

Slovenia's watercourses are very torrential, with sometimes enormous differences between minimal and maximal discharge amounts (with differences between 10 and 5000%) (Bat et al. 2003) and have extreme erosive and transport power. Annually, around five million m^3 of material is discharged in Slovenia (Kolbezen 1998). In almost all lowlands, this material and an excessive volume of water cannot be properly discharged through the riverbeds, which leads to frequent floods. The estimations of flooded areas vary considerably. Severe floods threaten around 70,000 ha of land, 20,000 ha are very frequently flooded (karstic poljes, the Ljubljana Moor, along the sub-Pannonia rivers) (EWN-Si 2001; Bat et al. 2003).

Over 1300 stagnant water bodies in Slovenia qualify as lakes, with a total area of 69.93 km^2 or 0.3% (Remec-Rekar and Bat 2003). The largest are of glacial-tectonic origin (Bohinj and Bled) and karstic (Cerknica). Nearly every stagnant water has some sub-aqueous soils, at least in their off-shore zone, but only one has been subject to any serious research (Repe and Pristovšek 2011).

3.1.4.3 Underground Waters and Wetlands

The spatial distribution of the underground water is conditioned by the geologic structure and its hydrological characteristics (porosity, permeability). Most of Slovenia is made of very permeable, sedimentary rocks with intergrain porosity (19.8%), solid rock inner-crack porosity (14.2%), and karstic crack porosity (33.2%) (Prestor et al. 2002; Uhan and Krajnc 2003). Aquifers with inter-grain porosity are relatively shallow sandy and gravel alluvial accumulations in tectonic basins, but they represent a substantial dynamic reserve of Slovenia's underground waters (36.8%). The largest one extends along the Sava River (Kranjc-Kušlan 1995; Uhan and Krajnc 2003). The level of groundwater is mostly close to the surface, and reaches and influences the pedosphere (gleying).

Similar conditions also exist in the wetlands. According to the data (VGI 2000; EWN-Si 2001; Beltram 2003), Slovenia is less than 5% wetlands and 61% of these are man-made. As regards soils, the coastal wetlands enjoy a very special place, as tiny, but still the largest, areas of saline and sub-aqueous soil occur there (Repe and Pristovšek 2011; Povše and Repe 2013). In Slovenia, there are considerable interventions in the hydrosphere, especially in wetlands (the loss of 100,000 ha from 1772 to 1990 (Beltram 1996), which (negatively) affect soils (the shrinking of organic matter and peat, reduced biodiversity, etc.).

3.1.4.4 Soil Water and Topography

A topographic wetness index (TWI) has been used as an approximation of the soil water content and its distribution over the entire territory (Kirkby 1978; Moore et al. 1993; Western et al. 1999).

$$\mathrm{TWI} = \ln\left(\frac{A_s}{\tan\beta}\right) \quad (3.1)$$

Kirkby (1978) The average for the entire Slovenia is 13.5, but TWI varies considerably. The histogram shows the majority of TWI values from 10 to 14. This indicates the prevalence of hilly topography and, in terms of soil wetness, transitional areas where water freely drains towards lowlands with higher TWI values. The distribution also directly reflects the ratio between areas of auto- and hydromorphic soils (6.3:1.0) (Repe 2007a). Regionally, TWI values are well distributed across all ecological landscape regions (see Sect. 3.2.2). Units with higher elevation a.s.l., steeper slopes, and a higher altitude above the channel network express higher TWI values (Table 3.5, Figs. 3.20, and 3.21).

3.1.5 Vegetation

3.1.5.1 Forest Vegetation and Soils Under Forests

The territory of Slovenia is one of a few areas where natural conditions enable forests to thrive. 58.5% of the land is covered with forests (Source: SFS Report 2013), which ranks Slovenia as the third most forested country in Europe (Fig. 3.22). Forests are thus an essential feature and constituent part of the environment and their protective and social importance is steadily increasing. The occurrence and structure of forests are therefore to a great extent shaped by the climate, parent rocks, soils, and relief. The macroclimatic types are influenced by the Mediterranean Sea, the Alps, and the Pannonian Plain. Most of Slovenia's forests are located within the natural environment of beech, which—together with the beech-oak and fir-beech forests—cover about 70% of the nation's forestland. In Slovenian forests, 71 tree species can be found in total, of which 10 coniferous and 61 deciduous.

The heterogeneous ecological conditions result in diverse forest vegetation patterns. Based on studies of the phyto-geographic regions (Wraber 1969; Košir 1979, 1994; Zupančič et al. 1987; Zupančič and Žagar 1995), and on different geographic features such as geology, soil, relief, precipitation, temperature, potential vegetation, climate, and phenology, the country was divided into eco-regions and sub-regions as shown in (Fig. 3.23) (Kutnar et al. 2002) and described in the following section. The syntaxonomy of forest vegetation is in accordance with Robič and Accetto (2001), while the characteristics of the forest vegetation types are based upon Marinček and Čarni (2002).

3.1.5.2 Alpine Eco-Region (1)

The Alpine eco-region (1) consists of the Julian Alps, the Kamnik-Savinja Alps, and the Karavanke mountain chain. In the western part of this region, at altitudes of 900 m.a.s.l. up to the upper forest line at approximately 1500 m.a.s.l., beach to beech of different associations grow (*Anemone trifoliae-Fagetum*, *Homogyno sylvestris-Fagetum*, *Ranunculo platanifoliae-Fagetum*) (Fig. 3.24). Many beech and fir-beech forests have been converted to essentially pure monoculture spruce. Some scattered potentially natural sites of primary spruce forest can be found. In the east, above

Table 3.5 Histogram of the topographic soil wetness index

TWI	Area (km^2)	%
Less than 8	26.9	0.1
8–10	127.5	0.6
10–12	7523.9	37.1
12–14	6660.9	32.9
14–16	2362.8	11.7
16–18	1683.9	8.3
18–20	975.7	4.8
20–22	518.7	2.6
22–24	262.5	1.3
24–26	106.2	0.5
More than 26	23.7	0.1
Sum	20,272.7	100.0

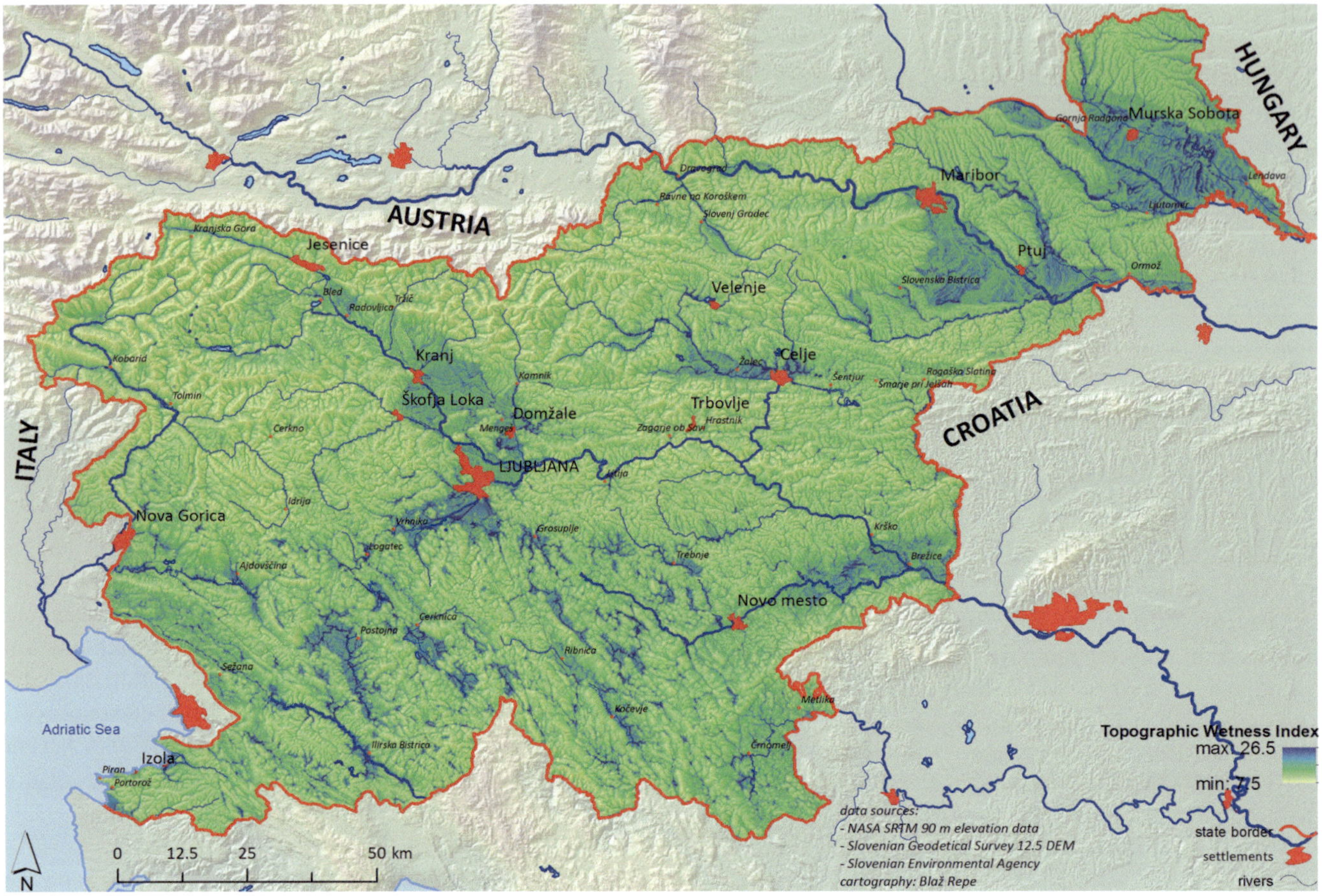

Fig. 3.20 Slovenia's topographic soil wetness index

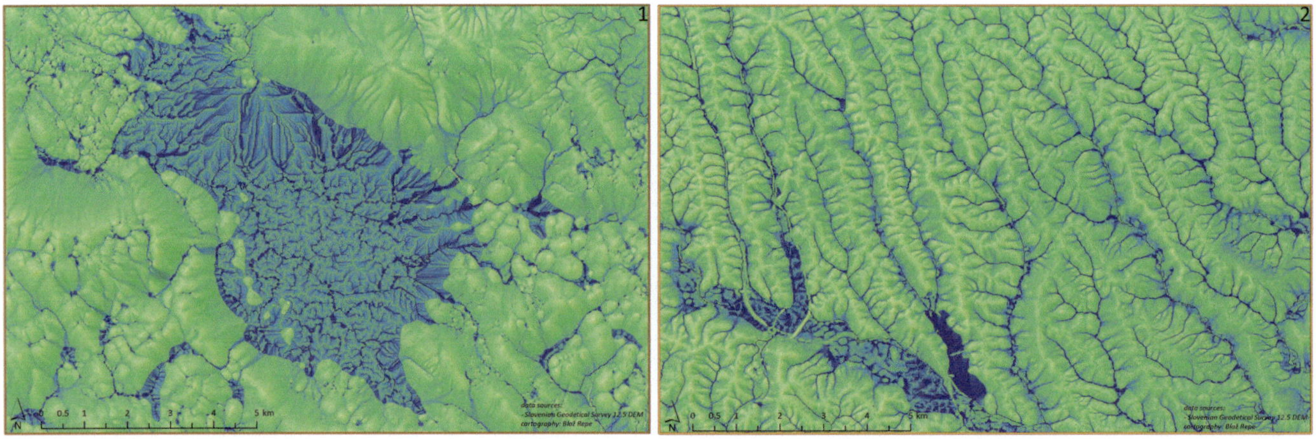

Fig. 3.21 Topographic soil wetness index: the area of Cerknica Lake (*left*) and Slovenjske Gorice (*right*). The microstructure of the surface (much as with curvature) is reflected in soil water conditions. They vary over very short distances. All this is reflected in a very diverse soil cover, especially in the karstic areas

1000 m.a.s.l., secondary spruce forests of *Avenello flexuosae-Piceetum* extend over a large area, on noncalcareous parent material of potentially natural sites of acidophilic beech forests. At lower altitudes, acidophilic beech forests of *Castaneo sativae-Fagetum* grow. In the Alps and Karavanke, the upper tree line reaches 1700 m.a.s.l. The woody

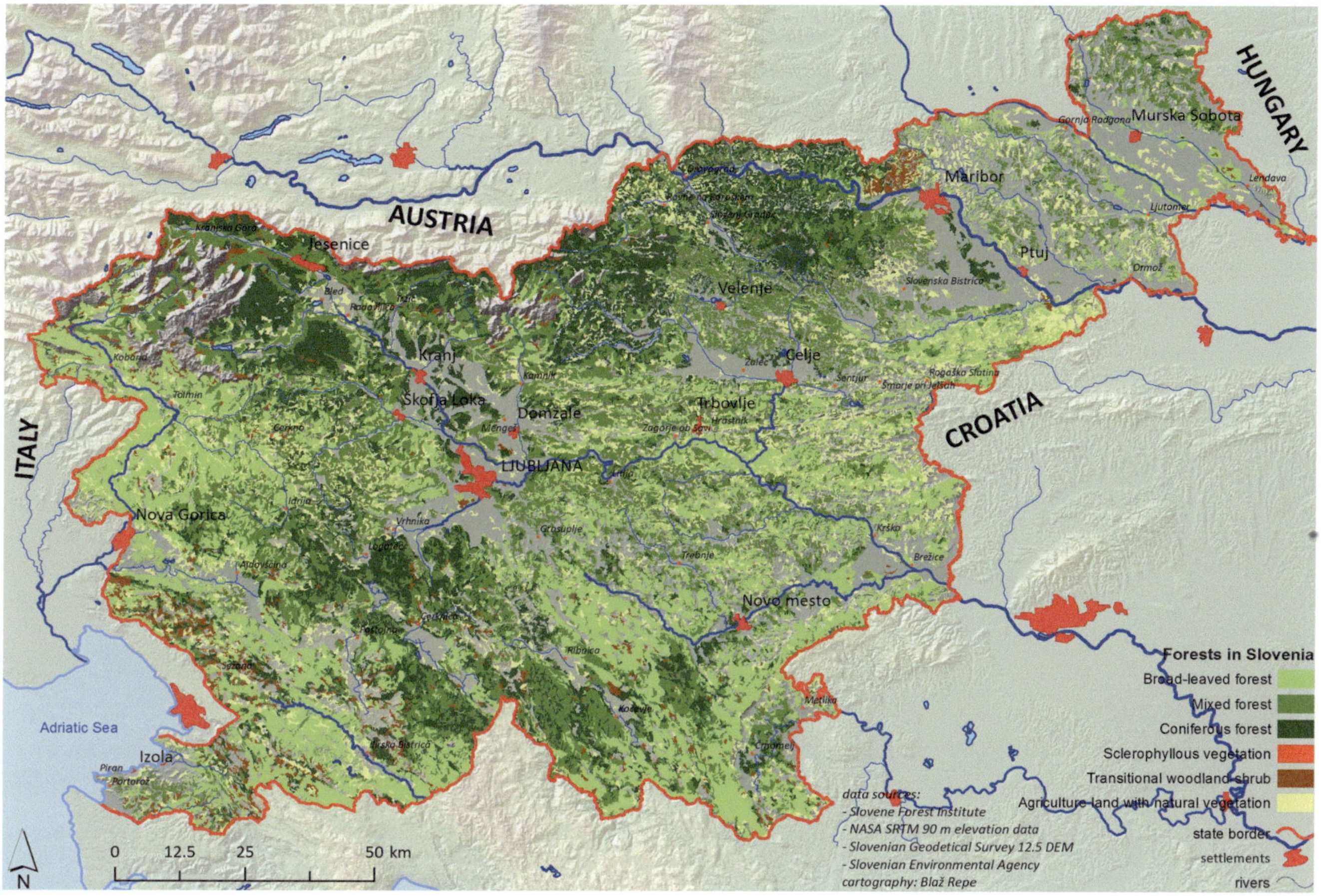

Fig. 3.22 Forests in Slovenia

vegetation that can be found is dwarf-mountain-pine shrubbery (*Rhodothamno-Rhododendretum hirsuti* association), mixed with single larch (*Larix decidua*) trees.

3.1.5.3 Soils Under Forests

Dominant soil units are Rendzic Leptosols formed on limestone, dolomites, calcaric moraines, and alluvial deposits; Dystric Cambisols on silicate clastic sedimentary rocks. Other significant soil units are Lithic Leptosols and Eutric Cambisols on calcareous, Dystric and Umbric Leptosols on noncalcareous parent material. Special soil units are Folic Histosols on limestone, dolomites; Fibric Histosols on raised bog, and Podzols on moraines, chert (Figs. 3.25, 3.26 and 3.27).

3.1.5.4 Pohorje Eco-Region (2)

The Pohorje eco-region includes the crystalline Pohorje massif and Kozjak mountains. In the lowland and hilly areas, mixed forests of beech, sweet chestnut, and sessile oak (*Castaneo sativae-Fagetum*) grow. In the northern part of this region, on shady and very steep slopes at altitudes from 400 to 1000 m.a.s.l., *Galio rotundifolii-Abietetum* forests predominate. The parent materials consist of noncarbonate tonalite, gneiss, mica schist, amphibolite, and others. Predominantly at altitudes between 900 and 1300 m.a.s.l., the so-called acidophilic alti-mountain beech of the *Luzulo albidae-Fagetum* association occurs. The geological material consists of moderately acidic noncalcareous stones. These forests have always been subject to significant human impact, such as clearing for meadows and livestock grazing. In the Pohorje massif range, beech was locally exterminated in the period of glassworks. All these activities resulted in the complete predominance of Norway spruce (*Picea abies*), which has an evident influence on the sites. The secondary spruce forests were classified as the *Luzulo sylvaticae-Piceetum* association. The *Cardamini savensi-Fagetum* is a zonal beech association of the upper parts of the Pohorje mountain range. It grows on silicate metamorphic and igneous rocks between 800 and 1300 m.a.s.l. In these forests, European silver fir (*Abies alba*) and Norway spruce are mixed with beech.

Soil Under Forests

Dominant soil unit: Dystric Cambisols on acid crystalline and volcanic rocks. Other significant soil units are Dystric

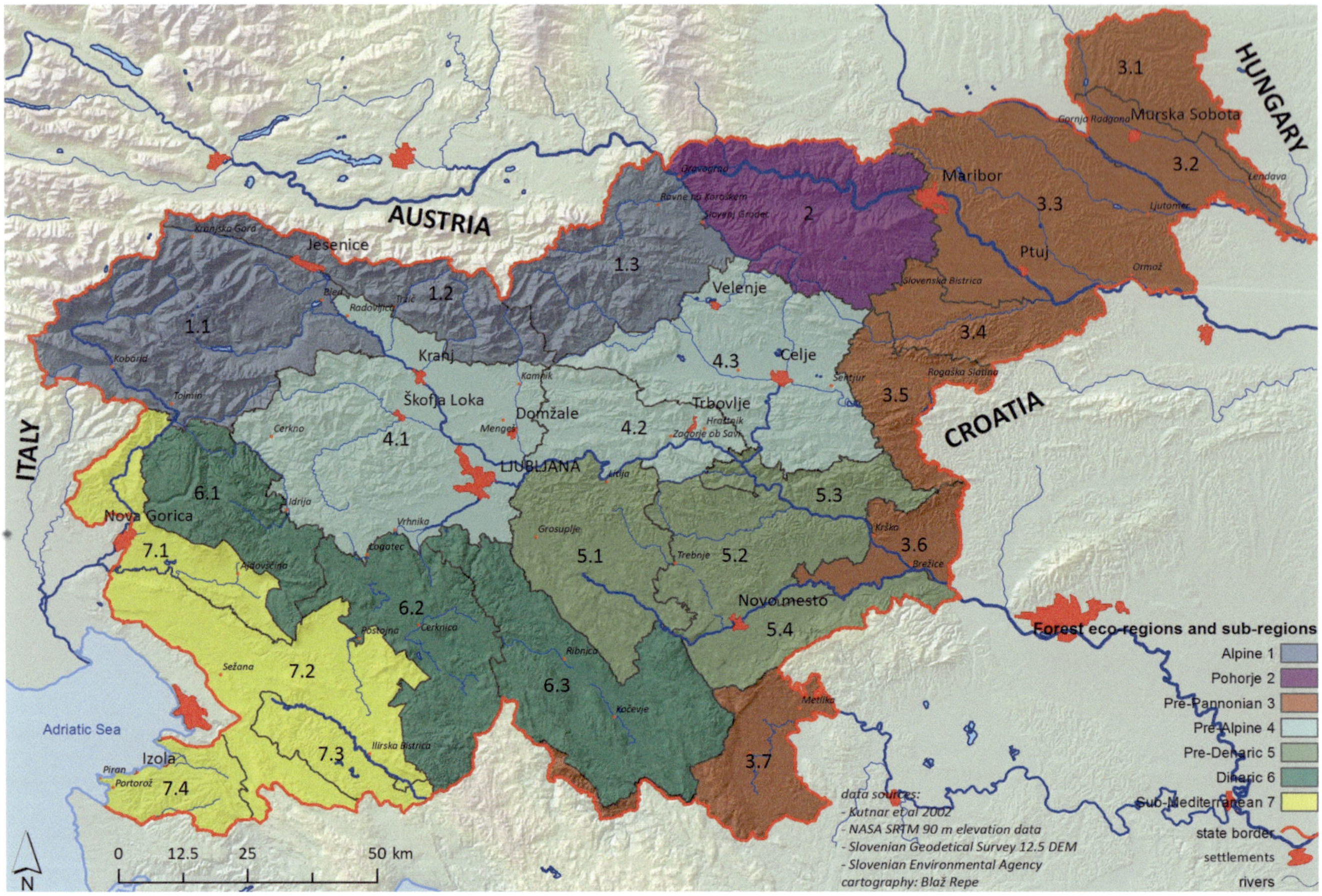

Fig. 3.23 Map of eco-regions and sub-regions in Slovenia

Fig. 3.24 An Anemono trifoliae-Fagetum forest at the Gozd Martuljek site in the Julian Alps, north-western Slovenia (*Photo* M. Kobal)

Fig. 3.25 A Leptic Rendzic Phaeozem, soil profile (*left*); organic layer (*right*) at the Gozd Martuljek research site (*Photo* M. Kobal). A Folic Leptosol, as a more shallow soil, and an Endoleptic Phaeozem, a deeper soil (WRB 2006) at the Gozd Martuljek research site, situated at an altitude of 1133 m.a.s.l. in north-western Slovenia (Julian Alps). On the *left* side is a photograph of a whole soil profile, on the *right* side is a photograph of the organic layers that are characteristic of "well-preserved" soils under forests. Sustainable forestry and close-to-nature forest management principles are the basis in Slovenia for better forest soil protection (permanent tree cover) and consequently a better-preserved forest floor

Fig. 3.26 An Adenostylo glabrae-Piceetum forest at Fužinske planine, the Pokljuka plateau in the Julian Alps (*Photo* M. Kobal)

and Umbric Leptosols, Humic Cambisols; on glaciofluvial deposits Dystric and Eutric Cambisols, Fluvisols. Special soil units are Fibric Histosols on raised bogs, Eutric Cambisols on marl (Figs. 3.28 and 3.29).

3.1.5.5 Pre-Pannonian Eco-Region (3)

The pre-Pannonian eco-region extends from the south-east to the north-east of Slovenia. Forests cover only a small part of this region. Forests of hornbeam (*Carpinus betulus*) and

Fig. 3.27 A Histosol profile at Fužinske planine (*left*); an organic layer (*right*) (*Photo* M. Kobal). An Endoleptic Sapric Folic Histosol (Eutric, Skeletic) profile (WRB 2006) at Fužinske planine (altitude 1500 m.a.s.l.) near the Pokljuka plateau and above Bohinj Lake in NW Slovenia. The soil layer is above carbonate bedrock, the process of the accumulation of organic material is rather expressed, under 50 cm and organic soils mixed with the parent substrate prevail (a higher proportion of stoniness; 50–80 cm +)

Fig. 3.28 A secondary spruce monoculture stand on a potential beech site of Luzulo albidae-Fagetum at Komisija on the Pohorje massif (*Photo* M. Kobal)

oaks (*Quercus* sp.), which would be characteristic of lowlands and the hilly parts of Slovenia, were mainly cleared and the land has primarily been used for agriculture. On small areas along watercourses, associations of white willow (*Salicetum albae*), forests of black alder (*Alnetum glutinosae*), and pedunculate oak-ash forests (*Querco roboris-Ulmetum laevis*) occur. In the plains, which are influenced by high underground water and temporary flooding, oak-hornbeam forests (*Querco roboris-Carpinetum*) can also be found. These riparian forests are threatened since the marshes and wetlands were intensively drained in the past. In many places, they have been converted to wet meadows.

Fig. 3.29 Cambisol profile at the Komisija site on Pohorje (*left*), soil organic layer (*right*) (*Photo* M. Kobal). Halpic Cambisol (Humic, Hyperdystric, Skeletic, Chromic) soils (WRB 2006) at the Komisija site on the Pohorje massif in N Slovenia at an altitude of 1318 m.a.s.l.; soils at Komisija overlie noncarbonated bedrock, such conditions are typical of a smaller proportion of forest area; soils are deep, the pH value ($pHCaCl_2$) is relatively low, between 3.1 and 4.6, partly as consequence of spruce monoculture stands at a certain altitude range and the nature of the bedrock

In the northern part of this region, on the lower altitudes of hills, *Pruno padi-Carpinetum betuli* occur, while at higher altitudes, zonal beech forests (*Vicio oroboidi-Fagetum*) grow. On slopes from 300 to 700 m.a.s.l., moderately acidophilic beech forests with admixed sweet chestnut and sessile oak (*Castaneo sativae-Fagetum*) grow. They thrive on very different noncalcareous parent material. On Luvisols and degraded soils, forests of *Galio rotundifoli-Pinetum sylvestris* prevail.

Soils Under Forests

Dominant soil units are Dystric Cambisols, Dystric Planosols on consolidated and unconsolidated silicate clastic sedimentary parent material, Eutric Gleysols, Eutric Fluvisols on alluvial sediments. *Other significant soil units* are Dystric Leptosols; Eutric Cambisols, Planosols, Gleysols, Fluvisols; Stagnic, Gleyic, Haplic Luvisols. *Special soil units* are hydromeliorated Planosols, Gleysols, Fluvisols (Figs. 3.30, 3.31, 3.32, and 3.33).

3.1.5.6 Pre-Alpine Eco-Region (4)

The pre-Alpine eco-region is located between the Alpine region (1) in the north, and the Dinaric (6) and pre-Dinaric regions (5) in the south. On these plain areas and gentle slopes, azonal forest associations of *Vaccinio-Pinetum sylvestris* grow. This is a pioneer association occurring mostly on poor sites. Much more common in this area are acidophilic beech forests of *Blechno-Fagetum* associations and *Castaneo sativae-Fagetum* associations, which are growing on Dystric Cambisol on various noncalcareous geological formations. In many cases, forests of *Castaneo sativae-Fagetum* have been changed either to spruce (*P. abies*) monoculture forests or to coppiced forests with a mix of sweet chestnut (*Castanea sativa*), Scots pine (*Pinus sylvestris*), and sessile oak (*Quercus petraea*). In this region, thermophilous beech forests (*Ostryo carpinifoliae-Fagetum*) cover large areas of the southern exposed sites from lowlands to an altitude of around 1000 m.a.s.l. with a relatively high amount of precipitation. In the mountain belt, on dolomite and limestone of various geological formations, beech forests of *Hacquetio epipactis-Fagetum* and *Lamio orvalae-Fagetum* are also very common.

Soils Under Forests

Dominant soil units are Eutric Cambisols on limestone, dolomite, calcaric moraines and alluvial deposits; Dystric Cambisols on silicate clastic sedimentary rocks. *Other significant soil units* are Rendzic, Dystric and Umbric Leptosols; Eutric and Dystric Fluvisols; Humic Gleysols. *Special soil unit is* Terric Histosol (Figs. 3.34 and 3.35).

3.1.5.7 Pre-Dinaric Eco-Region (5)

The pre-Dinaric eco-region extends from the central to the eastern part of the country. Here, beech forest of *Hacquetio*

Fig. 3.30 A Polysticho setiferi-Abietetum forest at the Gruškovje site in the pre-Pannonian eco-region in north-eastern Slovenia (*Photo* M. Kobal)

Fig. 3.31 A Cambisol profile at Gruškovje (*left*) and the soil organic layer (*right*) (*Photo* M. Kobal). An Endoleptic Cambisol (Humic, Eutric, Ruptic; WRB 2006) profile at Gruškovje in north-eastern Slovenia, Haloze hills near the border with Croatia. More than 1-m-deep soils overlie marl. The microlocation (foot slope, the presence of surface water soil erosion processes, upper soil layers) has colluvial characteristics

epipactis-Fagetum and *Lamio orvalae-Fagetum* associations are very common, along with a variety of other forest associations. On gently configured hills and flatlands, at altitudes from 150 to 350 m.a.s.l. there are some remnants of hornbeam forests (*Abio albae-Carpinetum betuli*). However, they have mostly been converted into agricultural and urban areas. On noncalcareous parent material, *Castaneo sativae-Fagetum* forests grow. The majority of these moderately acidophilic beech forests have been changed either to monoculture spruce or to coppiced forest. On Dystric

Fig. 3.32 A forest stand at the Bunčani site, along the riverbank of the Mura River, north-eastern Slovenia (*Photo* M. Kobal)

Fig. 3.33 A Fluvisol profile at Bunčani, topsoil (*Photo* T. Kralj). A Mollic Fluvisol (Calcaric Humic Hypereutric Episiltic; WRB 2006) profile in Bunčani, right next to the Mura River in north-eastern Slovenia (see Fig. 6.9, p. 175). The soil is under the influence of fluvial deposition, with a high exchangeable base saturation proportion, the soils are deep with a mollic organic horizon

Cambisol, acidophilic hornbeam forests of *Vaccinio myrtilli-Carpinetum betuli* occur. Like the majority of the forests in this hilly region, they have been intensively utilized due to their location close to settlements and easy access, as well as favourable land configuration (Figs. 3.36 and 3.37).

On noncalcareous parent material, *Castaneo sativae-Fagetum* forests grow. The majority of these moderately acidophilus beech forest have been changed either to spruce monoculture or to coppiced forest (Figs. 3.38 and 3.39).

On a Dystric Cambisol, acidophilus hornbeam forests of *Vaccinio myrtilli-Carpinetum betuli* occur. Like the majority of the forests in this hilly region, they have been intensively utilized due to their proximity to settlements and easy access, as well as favourable land configuration (Figs. 3.40 and 3.41).

Fig. 3.34 An Arunco-Fagetum forest at the Lubnik site near the city of Škofja Loka, the pre-Alpine eco-region (*Photo* M. Kobal)

Fig. 3.35 A Cambisol profile at Lubnik (*left*), soil organic layer (*right*) (*Photo* M. Kobal). A Haplic Cambisol (Calcaric, Sodic, Hyperhumic, Hypereutric, Laxic, Skeletic, Episiltic; WRB 2006) profile at Lubnik, Gorenjska region, close to the city of Škofja Loka, at an altitude of 530 m.a.s.l. The bedrock is dolomite, colluvial processes take place. In the photographs, dense roots of different thicknesses, both in the organic and mineral part of the soil, are clearly visible

3.1.5.8 Dinaric Eco-Region (6)

The Dinaric ridge separates the littoral and continental regions. It is characterized by a mixed fir-beech forest (*Omphalodo-Fagetum*). It covers parts of the high Karst plateaus of southern Slovenia and extends to Croatia. It grows mostly at altitudes from 700 to 1200 m.a.s.l. The parent material is limestone with sporadic dolomite. In the lower part of the mountain belt, between 600 and 900 m.a.s.l., beech forests of *Lamio orvalae-Fagetum* occur. From lowlands up to altitudes of around 600 m.a.s.l., sub-montane beech forests of *Hacquetio epipactis-Fagetum* can be found (Figs. 3.42 and 3.43).

Soils Under Forests

Dominant soil units are Rendzic Leptosols on limestone and dolomite. Associated or other significant soil units are Eutric

Fig. 3.36 A Cardamini savensi-Fagetum forest on Gorjanci Mountain, south-eastern Slovenia (*Photo* M. Kobal)

Fig. 3.37 A Luvisol profile at Trdinov Vrh (*left*), soil organic layer (*right*) (*Photo* M. Kobal). A Haplic Luvisol (Humic Endoeutric Silitic; WRB 2006) profile at Trdinov Vrh, in the Gorjanci flat hills in south-eastern Slovenia, 767 m.a.s.l. The bedrock is dolomitized limestone. In soil, eluvial and illuvial pedogenic processes are present, which can also be observed from the analytical results for soil texture and the exchangeable base saturation (*V*) value. In the organic horizon, a litter layer is present, mostly beech leaves (Ol; *right figure*) and a thin of layer; there is no Oh layer, which indicates fast organic matter decay processes

and Chromic Cambisols, Haplic and Chromic Luvisols on limestone and dolomite. Special soil units are soils in virgin forest remnants.

3.1.5.9 Sub-mediterranean Region (7)

In the sub-Mediterranean eco-region, limestone and flysch are the characteristic parent materials. The most common type of forest vegetation is *Ostryo carpinifoliae-Quercetum pubescentis*. In the past, many sites of these forests were converted to pastures. In the second half of the nineteenth and at the beginning of the twentieth centuries, many of them were planted with Austrian pine trees (*Pinus nigra*). After WWII, many pastures were abandoned and have now been gradually overgrown by forests of *Ostryo carpinifoliae-*

Fig. 3.38 A Hedero-Fagetum forest near the Hinje site in Suha Krajina area near the city of Novo Mesto (*Photo* M. Kobal)

Fig. 3.39 A Luvisol profile at the Hinje site (*left*), soil organic layer (*right*) (*Photo* M. Kobal). An Epileptic Luvisol (Humic, Eutric, Clayic; WRB 2006) profile at the Hinje site (395 m.a.s.l.), in the Suha Krajina area in south-eastern Slovenia. The bedrock is limestone. The process of ion leaching from upper to lower layers is present, the *V* value increases with depth also in deeper layers where soils are not in contact with the bedrock

Quercetum pubescentis associations. The mountainous parts of the sub-Mediterranean region are covered by beech forests of *Seslerio autumnalis-Fagetum*. This occurs up to 1100 m.a. s.l. in the littoral mountain region and grows on the coldest sites of the sub-Mediterranean region. On the edge of its area, many of these forests have been changed into secondary forests of *Seslerio autumnalis-Ostryetum carpinifolia* associations. On medium deep to deep Dystric Cambisol on acidic Flysch formations in the Brkini sub-region, the association *Castaneo sativae-Fagetum* is very common.

Soils Under Forests

Dominant soil units are Chromic Cambisols on limestone, dolomites; Eutric Cambisols on flysch, marl. *Other significant soil units* are Rendzic Leptosols, Eutric Cambisols, Chromic Luvisols on limestone, dolomites; Dystric

Fig. 3.40 An acidophilus beech forest of a Blechno-Fagetum association at the Jelša site in the central part of Slovenia (*Photo* M. Kobal)

Fig. 3.41 A Cambisol (Dystric) profile at the Jelša site (*left*), soil organic layer (*right*) (*Photo* M. Kobal). An Endoleptic Folic Cambisol (Dystric) (WRB 2006), profile at the Jelša site, east of the capital city of Ljubljana, 355 m.a.s.l., on the upper part of the slope. Less developed forms of raw humus can be found in the organic horizon

Cambisols on flysch. *Special soil units are* Calcaric Regosols, Calcaric Cambisols on flysch, marl (Figs. 3.44, 3.45, 3.46 and 3.47).

3.1.5.10 Nonforest Vegetation

In Slovenia, nonforest vegetation is present either naturally, where forest growth is precluded by the climatic or soil factors or, predominantly, is man-made in cases where forests have been cleared for the agricultural production or the settlements. Naturally, the nonforest vegetation in central Europe is present above the treeline, in very dry or excessively wet soil, in steep terrains with high erosion (rocky walls, cliffs, screes) or in extremely nutrient limited conditions (bogs).

Natural Nonforest Vegetation

Most of the natural nonforest vegetation in Slovenia is present above the tree line (1700–1800 m.a.s.l.). Alpine

Fig. 3.42 A Dinaric fir-beech forest of an Omphalodo-Fagetum association in the Snežnik area, southern Slovenia (*Photo* M. Kobal)

Fig. 3.43 A Phaeozem profile in Snežnik, profile (*left*) and soil organic layer (*right*) (*Photo* M. Kobal). An Epileptic Phaeozem (Episiltic) profile (WRB 2006) close to Snežnik mountain (1796 m.a.s.l.), at an altitude of 1208 m.a.s.l. the bedrock is limestone, the surface stoniness and rockiness are relatively high (20–30%, Figs. 3.28 and 3.29), also in other horizons. Due to rocks and surface stoniness, we can observe fir roots on the floor which end in small sinkholes and "soil pockets"

vegetation comprises different (sub)-Alpine grassland types in lower Alpine belts and sparse rocky and scree vegetation in the upper belts up to the top of the highest mountains. Alpine grasslands differ with respect to the soil conditions, being either acidophilic (graminoid-dominated *Caricetea curvulae* class) or basophilous to neutrophilous (species-rich *Seslerietea albicantis* class). Where soil is nutrient-rich (e.g. mountain pastures) eutrophic *Poo alpinae-Trisetetalia* vegetation is present. The vegetation in the upper mountain belt (in Slovenia, almost exclusively on carbonate rocks) becomes increasingly sparse and the soil shallow and less developed. Depending on the firmness of the substrate, either vegetation of firm rock outcrops (*Caricetea firmae*) or more dynamic scree vegetation (*Thlaspietea rotundifoliae*) can be found. Small areas of the Alpine belt are also characterized by snow-bed vegetation (*Arabidetalia coeruleae*,

Fig. 3.44 A Seslerio autumnalis-Quercetum petraeae forest at the Križ site, south-western Slovenia (*Photo* M. Kobal)

Fig. 3.45 A Cambisol profile at Križ (*left*) and the soil organic layer (*right*) (*Photo* M. Kobal). An Epileptic Cambisol (Humic, Eutric, Clayic, Rhodic; Terra Rossa) profile (WRB 2006) at Križ, on limestone bedrock in the Karst area in south-western Slovenia close to Italy. Terra Rossa, "Ilovka", typical of the Karst in the Sežana Karst area; the soil has a high proportion of skeleton in the Brz horizon; the pH value ($CaCl_2$) increases rapidly with depth, from 4.6 to almost 7 within only 10–15 cm

Salicetalia herbaceae) that is adapted to extremely short vegetation periods, wet soil, and heavy snowpack.

The second major group of natural nonforest vegetation in Slovenia can be found in areas too wet or too nutrient-poor for forest growth. This includes different herbaceous wetland types such as marshes, transition mires, bogs, and fens. Marshes are located in areas with less permeable soil or close to open (often standing) water and are characterized by different rather nutrient-demanding reeds, grasses, and sedges (e.g. *Phragmites australis*, *Phalaris arundinacea*, *Schoenoplectus lacustris*, *Typha* spp., and *Carex* spp.). Wet areas in the coastal region are combined with the saline conditions; saline marshy vegetation has developed there, characterized by halophytes or salt-tolerant species such as *Juncus maritimus*, *Juncus acutus*, *Salicornia europaea*, *Salsola* spp., *Spartina juncea*, *P. australis*. Peatlands (mires)

Fig. 3.46 An Ornithogalo pyrenaici-Fagetum forest at the Padež site on flysch in south-western Slovenia (*Photo* M. Kobal)

Fig. 3.47 A Cambisol profile at the Padež site (*left*) and the soil organic layer (*right*) (*Photo* M. Kobal). A Haplic Cambisol (Humic, Eutric, Endoskeletic) profile (WRB 2006) at the Padež site in the Brkini area (altitude about 530 m.a.s.l.), south-western Slovenia. Eutric Cambisol, deep soils on flysch bedrock, with noticeable colluvial processes, is present on the site; the humus is classified as a mull formation

in Slovenia are rare, small in size, and endangered. Of them, groundwater-fed fens are somehow more widespread, also located in lowlands but always in very small patches, often only as fragments in wet forests. This includes vegetation from the *Caricetalia davalianae* and *Caricetalia fuscae* orders. Bogs with active peat producing moss vegetation (class *Oxycocco-Sphagnetea*) are at their marginal distribution in Slovenia and are located only in the higher altitudes in the Alpine region. The bogs are often surrounded or combined with transition mires (*Scheuchzeretalia*), peaty spruce forests (*Piceo-Sphagnetum flexuosi*), and dwarf-pine shrublands (*Sphagno-Pinetum mugi*).

Secondary Nonforest Vegetation

Due to the natural conditions (rough terrain, shallow soil), permanent grasslands are the most widespread agricultural type of land use in Slovenia. They are used either for mowing or grazing and could be more or less intensified (eutrophic). Many types of grassland are relatively low-productive, but high in diversity, and are a target of different protection measures due to their possible intensification, on the one hand, or abandonment and afforestation with secondary forest, on the other. The driest grasslands in Slovenia, bordering on the sparse vegetation of screes, rocks, and sands, belong to the *Xero-Bormion* alliance in inland Slovenia and *Scorzoneretalia* in the sub-Mediterranean part. Both of these are bound to carbonate bedrock. On slightly deeper soils or in wetter conditions (Rendzinas, Cambisols, steeper flysch terrains), semi-dry grasslands can be found that are either acidic (*Nardetalia*) or calcareous (*Brometalia erecti*), both characterized by their species diversity and threat of abandonment.

The most productive grasslands are of the *Arrhenatheretalia* alliance and are located in deeper Eutric and Dystric Cambisol on flat or moderately steep terrain. They are often intensified and cut 3–6 times per year or continuously grazed, which results in lower species richness and a larger proportion of the most productive grasses and legumes. Grasslands on wetter soils (periodically also flooded) belong to the Molinietalia order in nutrient-poor (oligotrophic) conditions and *Calthion* and *Filipendulion* alliances in nutrient-richer conditions. Many of these grasslands were intensified by drainage and fertilization and are increasingly protected by nature conservation and agri-environmental measures. Many *Molinietalia* grasslands in Slovenia have developed in former fens and bogs (Ljubljansko Barje) that were drained and later regularly mown for fodder or bedding for farm animals. Many other, mostly minor but widespread nonforest vegetation types can be found in Slovenia. Many of them are related to a specific disturbance type such as ploughing and soil cultivation, trampling, and river gravel erosion and accumulation. Disturbance-related (ruderal) vegetation types can be found on arable lands (weed vegetation), in urban environments, brownfields, and along streams. These vegetation types are frequently infested with invasive plant species, which are increasingly changing the ecosystem properties and functioning.

3.1.6 Land Use and Land Cover

Most of the Slovenian territory is covered by forests (66%), while agriculture occupies 27.8% of the country. Built-up and traffic areas together occupy 3.9% of the land. Table 3.6 shows the main land use/land cover categories. The main land use change since WWII has been the increase in the share of forest and the decrease in agricultural land. In 1933, forests covered 35% of the country (Kotar 2009). Fast industrialization after the war led to land abandonment. In approximately 80 years, the size of the forests has practically doubled.

High fragmentation is typical of Slovenian land use, resulting from the topographic conditions and historical legislation, including the inheritance law of the Austro-Hungarian Empire. The land was divided between children, and the result was very long and narrow fields. This pattern is especially characteristic in lowlands on gravel deposits covered by Dystric or Eutric Cambisols or Leptosols.

The agricultural land use dataset shows a similar picture. The precise agricultural land use database (LU5), regularly updated by the interpretation of airborne digital ortho-photoimagery at a scale of 1:5.000 (MKGP 2014), gives detailed information on the structure of the agricultural land, semi-natural, and natural land cover/land use classes. Nonagricultural land uses—i.e. forests, urban, and waters—are represented by only one class (Fig. 3.48).

Within the agricultural land use classes (in total 33%), permanent meadows and pastures predominate (17%),

Table 3.6 Land cover of Slovenia (Statistical Office of the Republic of Slovenia 2005)

Land cover	%	km^2
Forests	66	13,380
Agricultural	27.8	5,636
Bare rock	1.6	324
Surface waters	0.7	142
Built-up	2.8	568
Traffic	1.1	223
Total	100	20,273

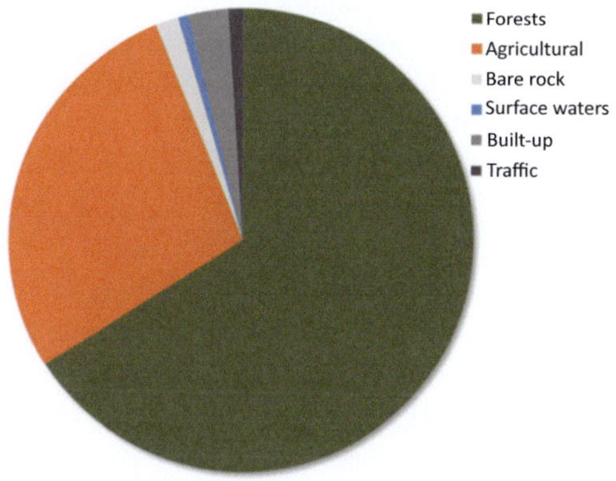

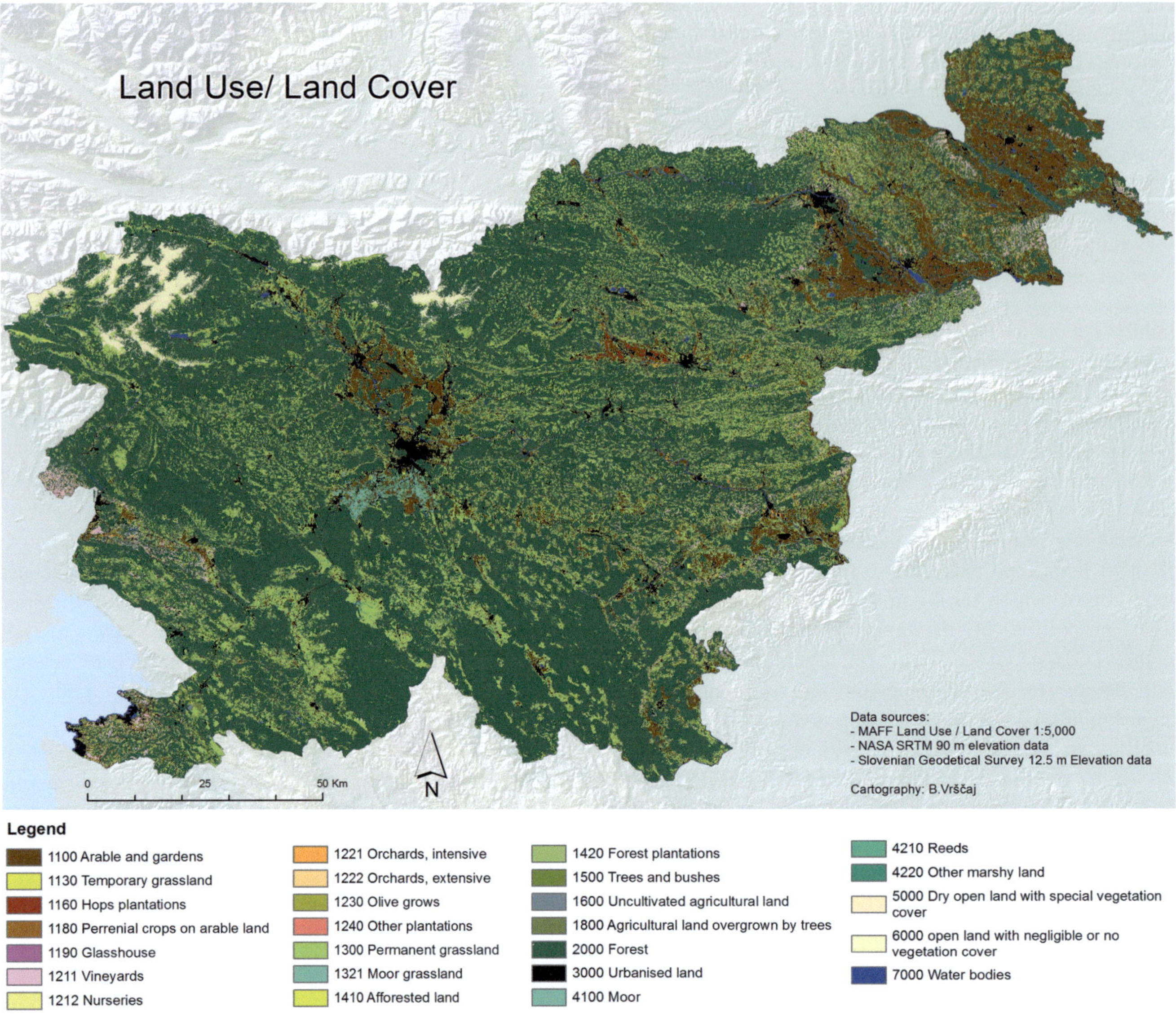

Fig. 3.48 Forest, urban, and agricultural land uses and water bodies in Slovenia (MKGP/MAFF 2014)

followed by arable (9%). The high share of classes such as abandoned land (1.63%), trees and shrub (1.35%) classes indicate ongoing land abandonment processes. Both classes can be considered as a transitional stage to the climax vegetation—forests. Highly productive agricultural land uses such as hops plantations, intensive orchards, vineyards, and greenhouse production, occupy a minor part of agricultural land (Table 3.7). The table reflects the fact that agriculture contributes only 2.1% to the GDP of Slovenia.

The agricultural land is very fragmented (Fig. 3.49). According to the statistics form 2013, in Slovenia there are 72,400 farms. The average size of a farm is 6.6 ha of agricultural land, which is 0.2 ha larger than in 2010. On average, the number of farms has decreased by 2% every year. The average size of a farm slightly increased from 6.4 ha in 2010 to 6.6 ha in 2013 (Kmetijski inštitut Slovenije and Ministrstvo za kmetijstvo in okolje, 2014) (Figs. 3.50 and 3.51).

The high fragmentation presented in Fig. 3.51 is a result of inheritance law, while the irregular pattern of interlacing small forest, grassland, and arable land and households is a result of both the hilly relief and the law (Figs. 3.52 and 3.53).

3.1.7 Time

Considering time as a factor of soil formation actually entails considering the time of other factors of soil formation activities, i.e. those that influence soil formation, soil development, and soil properties. Very young parent

Table 3.7 Agricultural land use data (MKGP/MAFF 2014)

Land use/land cover	ha	%	%
Arable land	183,158	9.03	33.09
Hop fields	1824	0.09	
Permanent crops on arable land	335	0.02	
Glasshouses	140	0.01	
Vineyards	19,297	0.95	
Nurseries	43	0.00	
Orchards, intensive	4088	0.20	
Orchards,	26,111	1.29	
Olive grows	1964	0.10	
Other permanent crops	469	0.02	
Permanent meadows and pastures	346,441	17.08	
Bog meadows	6015	0.30	
Land abandoned	33,165	1.63	
Forest plantations	341	0.02	
Trees and shrubs	27,342	1.35	
Untillaged agricultural land	10,123	0.50	
Agricultural land with forest trees	10,405	0.51	
Forest	1,202,826	59.30	59.30
Built and related areas	109,105	5.38	5.38
Moor	45	0.00	1.57
Reeds	66	0.00	
Other marshy land	1270	0.06	
Dry open land with special vegetation cover	17,129	0.84	
Open land with or without infertile vegetation cover	13,301	0.66	
Water	13,468	0.66	0.66
Total	2,028,471	100.00	100.00

Fig. 3.49 Fragmented agricultural land in Goričko, north-eastern Slovenia (*Photo* B. Vrščaj)

Fig. 3.50 Fragmented agricultural land in north-eastern Slovenia. The soils are Eutric Cambisols formed on Pleistocene deposits (*Photo* B. Vrščaj)

Fig. 3.51 Fragmented agricultural land in north-eastern Slovenia. The soils are Dystric Cambisols formed on silicate fluvial gravel [(GURS 2011), eTla Web GIS]

material briefly exposed to soil formation factors is covered with young, shallow, and poorly developed soils. Recent, young soils can therefore be found on Holocene alluvial deposits, alluvial fans, and screes. Older rocks, being exposed for a longer period, have accumulated more weathered parent material and collected more organic matter, etc. These soils are more developed and extend deeper. They continue to develop according to all influencing factors. In many cases, the age of the surface or landform is more important than the age of the parent material itself. There is a very close relationship between soil development, the geological age of the parent material, and the age of the landform (Lovrenčak 1994b). A detailed analysis of this relationship was carried out by Stritar (1965) in determining

Fig. 3.52 Patchwork of agriculture and forestry in eastern Slovenia; an area with Dystric and Eutric Cambisols [(GURS 2011), eTla Web GIS]

Fig. 3.53 Spontaneous afforestation of marginal agricultural land (pastures). Juršče, south-western Slovenia (*Photo* B. Vrščaj)

the chronosequence of the glaciofluvial gravel and sand terraces in the Ljubljana basin.

According to Antić et al. (1980), there is an absolute and relative soil age. Absolute age is expressed in years, from the beginning of soil genesis until the present time. On Slovenian territory and in the Slovenian circumstances, this can be from a few years or a few tens of years for young soils, up to over 1 million years for old soils. The relative age shows the degree or stage of development. Rather than the absolute number of years that have passed, the relative age reveals the intensity of other factors of soil formation. Adapting this to Slovenia, Stritar (1965) distinguishes the following stages:

1. Raw soils (Nudlithic Leptosols);
2. Underdeveloped soils (Fluvisols);
3. Young soils (Eutric and Dystric Cambisols);
4. Fully/well-developed soils (Eutric and Dystric Cambisols);
5. Mature soils (Chromic Cambisols);
6. Degraded soils (Haplic Luvisols and Podzols).

There is very little research on soil age in Slovenia. Soil age information can be extracted from chronosequence research on the soils in the Ljubljana basin, formed on carbonate, glaciofluvial outwash terraces along the Sava River and its tributaries (Jaecks Vidic and Lobnik 1996; Vidic 1998). The duration of soil formation has been determined by 10Be and palaeomagnetic measurements or estimated from the relative topographic height of the geomorphic surface (Vidic et al. 1991). The rates of soil development in the Ljubljana Basin chronosequence determined from the slopes of the chronofunctions decrease with time as logarithmic or power functions, suggesting that the rates of development decrease over time. Soil profile development, using four time/age-dependant properties (rubification, texture, lowering of the pH value, clay films), begins to decrease after 980k years, which can be assumed to be a steady state. This means that the sum of progressive (increasing soil development, i.e. weathering of the parent material) and regressive (decreasing or altering soil development, i.e. erosion) processes equals zero. Soil age was determined on the outwash river terraces from different ice age periods. Their time span varies from 5 K years (Eutric Leptosols) to 1.8 M years (Haplic Luvisols) (Vidic 1998).

Slovenian Soil Classification and WRB

4

Borut Vrščaj and Tomaž Kralj

Abstract

The introductory part of the chapter presents the origin of the Slovenian Soil Classification (SSC), which is linked to the Yugoslav Soil Classification developed in the mid-1970s—a time of intensive soil mapping and field surveying in former Yugoslavia. The SSC taxonomy is based on a hierarchical system in which the soil type represents the central and most important unit. Slovenian soil types are distributed within two divisions: automorphic and hydromorphic soils. The automorphic division has five classes: Undeveloped Soils, Humus-Accumulative Soils, Cambic Soils, Eluvial-Illuvial Soils, and Anthropogenic Soils. The Hydromorphic Soils are divided into the following groups: Fluvisols, Pseudogleys, Gleysols, Peat Soils, and Anthropogenic Hydromorphic Soils. Further, in the chapter, the main soil horizons (A, O, E, G, C, R) are described and the horizon classifiers are briefly discussed and explained. Two tables—automorphic and hydromorphic soil types, list the main Slovenian soil types, the related sub-types, the varieties, and the forms, which are briefly explained. A comparison of the SSC and the WRB is presented in the central part of the chapter. The properties and diagnostic features of 19 SSC soil groups and selected soil types are briefly described and correlated with the WRB classification. The chapter concludes with a table in which the SSC soil types are directly correlated with the WRB soil types.

4.1 The Slovenian Soil Classification

4.1.1 Introduction

The origin of the Slovenian Soil Classification (SSC) is linked to the development of the Yugoslav Soil Classification (Ćirić 1984a; Filipovski 2000; Resulović and Čustović 2002; Škorić 1986b). Until 1991, Slovenia was one of the six constituent Yugoslav republics and is today an independent country. The Yugoslav Soil Classification (YSC) originated in the mid-seventies in a time of intensive soil mapping and field surveying in former Yugoslavia. The initial text on soil classification was published even before that by the Croatian Prof. Michael Gračanin. In his textbook *Pedologija. III dio—Sistematika Tala* [Pedology, Soil Systematics] (Gračanin 1951), he presented a system of soil taxonomy that was based on lithological, chemical, physical, and morphological soil properties and the principles of soil genesis. The development of the soil taxonomy was continued on the federal Yugoslav level within the Yugoslav Soil Science Society (YSSS). The first YSC proposal, which was developed by professors Neugebauer, Ćirić, Filipovski, and Živković (Neugebauer et al. 1963), was based on genetic principles. The first and broadest level of the classification was the soil order. The orders were defined on the basis of the presence of hydromorphic processes and their influences on soil formation. The soil orders were divided on the

B. Vrščaj (✉)
Department of Agricultural Ecology and Natural Resources, Centre for Soil and Environment Research, Agricultural Institute of Slovenia, Ljubljana, Slovenia
e-mail: Borut.Vrscaj@kis.si

T. Kralj
Department of Agricultural Ecology and Natural Resources, Centre for Soil and Environment Research, Agricultural Institute of Slovenia, Gorjuše, Slovenia
e-mail: pedolog@gmail.com

B. Vrščaj et al., *The Soils of Slovenia*, World Soils Book Series,
DOI 10.1007/978-94-017-8585-3_4

second level into soil classes on the basis of the soil profile morphology, mainly the presence and sequence of genetic horizons and soil profile development. The third level accommodates the central and most important units of the YSC—the soil types. This version of the YSC was presented and accepted at the Second Congress of the YSSS in Ohrid in 1963 (Neugebauer et al. 1963). YSC development continued. The classification was improved, corrected, and amended by the YSSS Commission on Soil Genesis, Taxonomy, and Mapping. The new, significantly improved version of the YSC was presented and accepted as the official Yugoslav Soil Taxonomy during the fourth YSSS Congress in 1973 and published (Škorić et al. 1973). Even though an official YSC existed, slightly different soil classifications were applied in the former Yugoslav republics in order to meet some specific needs or, in the author's opinion, due to personal views. The soil classifications of the former Yugoslav republics were further developed. Two important books widely used by the students of agriculture and forestry in former Yugoslavia as source material on soil classification were *Pedologija* [Pedology] (Ćirić 1984a) and *Postanak, razvoj i sistematika tala* [Genesis, Development, and Systematics of Soils] (Škorić 1986b). In Slovenia *Pedologija, kompendij*, a relatively short textbook on soils that includes soil systematics, was used by the students of agriculture as well as students of forestry and geography (Stritar 1991).

Still based on the same YSC, the soil taxonomies of the former Yugoslav republics are still being independently developed and upgraded and thus have become more diverse (Ćirić 1984a, b; Husnjak 2014; Kralj and Grčman 2009a; Resulović and Čustović 2002, 2008; Škorić et al. 1985; Stepančič et al. 1986a; Urbančič et al. 2005). After Slovenia became an independent state, the activities related to the Slovenian Soil Classification merely continued. In a part of the soil science community, the need to upgrade and further improve the SSC is growing. The main reasons are the need for the taxonomy to better reflect productivity, water holding capacity, the relationships with the environmental threats, environmental protection, and comparison to the international classification—the World Reference Base (WRB).

4.1.2 The Slovenian Soil Classification

The contemporary Slovenian Soil Classification (SSC) is based on the soil genesis principle and is still similar to the former YSC from 1973. The taxonomy is based on the hierarchical system in which the soil type represents the central and the most important unit.

The main classification unit—soil type—is upwards grouped into classes and divisions. In order to achieve a more applicable level of soil mapping and soil profile description, soil type is divided into sub-types, varieties, and forms.

SOIL DIVISION ← SOIL CLASS ← **SOILTYPE** → SUBTYPE → VARIETY → FORM

The SSC has 27 major soil types and was adapted to the local needs and specifics and in a number of aspects is similar to the German Soil Classification. The brief presentation of the SSC in the pages below is based on the work of Sušin (1964, 1983) and Stritar (1990b, 1991), Prus (2000) and Zupan et al. (2002a), partially on Urbančič et al. (2005), and especially on the new and revised version of the SSC of Kralj (2008), Kralj and Grčman (2009a, b).

4.1.3 A Brief Description of Soil Horizons

4.1.3.1 The Main Soil Horizons and Sub-horizons Used in the Slovenian Soil Classification

Organic Horizons

An organic horizon marked with **O** is a result of the accumulation of litter on top of mineral horizons in predominantly aerobic conditions. It consists of more than 20% organic C. Three different varieties of O sub-horizons are generally present in woods:

- The **Ol** sub-horizon consists of litter (leaves, needles, moss, dry grass, etc.). Almost intact litter makes up $\geq$80% of the horizon volume. The origin of the plant material is clearly visible, and no visible mycelium is present. The Ol horizon rapidly changes with the time of the year.
- The **Of** sub-horizon is a dark brown, blackish layer of fragmented plant tissue in the initial stage of decomposition. The original plant material is sometimes difficult to identify. In general, mycelia are present and frequently abundant in the event of favourable moisture. At high altitudes, the decomposition processes are especially slow, and consequently, organic horizons can be thick. According to the WRB, it has a fibric character.
- The **Oh** sub-horizon is dark, frequently black to black-brownish. It consists of decomposed and humified organic matter in which the origin is no longer recognizable. According to the WRB, its properties are similar to the sapric properties of Histosols, but not wet.

The thickness and the form of each of the three sub-horizons largely depend on the type of litter (coniferous, broadleaf, grass remains, fern, etc.), aspect, slope, and the micromorphology of the terrain (e.g. microconcave and convex forms), local climatic conditions, and other processes and circumstances that steer the decomposition and humification of the organic matter.

Humus-Accumulative Horizons

Humus-accumulative horizons are marked with an **A**. They are mineral horizons with accumulated and humified organic matter (i.e. the humus-accumulative horizon) that have formed on the soil surface or below an O horizon. The organic matter is well decomposed and humified, showing a colloidal character, and mixed with the mineral particles forming an organo-mineral complex. The morphology of the A horizon is obtained by the formation of soil aggregates of predominantly granular structure, but it does not have the characteristics of the E and B horizons. Organic matter in the A horizon is well decomposed and is well distributed as small particles or formed as a coating on the mineral particles. The A horizon occurs in mollic, umbric, ochric, and histic forms. Regarding the form, it can be of very different colour, mainly different shades of dark to light brown, grey-brown, to complete black; typically, it is darker than the underlying horizon. If the A is closed in parentheses—**(A)**—this represents an initial, weakly developed stage of an A horizon that generally overlies a C or R horizon. The structural aggregates of an (A) horizon are not clearly developed; the colour barely differs from the colour of the C horizon, i.e. fragmented parent material.

Cambic Horizons

The letter **B** marks a cambic horizon that has formed below an A or another B or, rarely, an O horizon. It shows the removal of carbonates, a residual concentration of sesquioxides (coatings), the alteration of forms of silicate clay, and free oxides. If it underlies an E horizon, increased quantities of illuvial clay, iron, carbonates, or humus can be found or detected. The structure is predominantly granular or blocky. It may be of a very different colour: yellowish to brownish, reddish-brown, or intensive red. The B horizon appears in two main forms:

- The **Bv** sub-horizon is a result of weathering of the parent rock and clay formation. Generally, it is yellowish-brown, frequently very porous, and associated with Eutric and Dystric Cambisols. The structure of peds is fine-grained crumby-granular.
- The **Brz** sub-horizon is a clay accumulation horizon—a residue in weathering limestone and/or dolomite. It is clayey, usually with a well-developed and stable sharp-edged blocky structure. In Slovenia, it is associated with Chromic Cambisol.

Eluvial and Illuvial Horizons

The **E** eluvial horizon is as a rule situated below an O, A, or H horizon and overlies a cambic horizon. It contains less clay and/or sesquioxides and is, in general, more acidic. The colour is pale in comparison with the following underlying cambic B horizons:

- The **Bt** horizon, an argiluvic horizon containing increased quantities of illuvial clay. The clay content is at least 20% higher than in the above horizon. The clay coatings are present at the surfaces of structural aggregates.
- The **Bh** horizon, a B horizon in which the accumulation of illuvial humus originated from a superficial A horizon.
- The **Bfe**, a B horizon in which illuvial sesquioxides leached from the E horizon accumulate.

Hydromorphic Horizons

The **G** mineral horizon shows evidence of reduction and secondary oxidation processes due to temporary or permanent anaerobic conditions. According to the WRB, it clearly shows gleyic properties, the matrix is dark grey, and the surface aggregates are bluish and coloured with rust coatings and mottles. Two sub-horizons of the G horizon are used to describe soil profiles:

- The oxymorphic **Go** sub-horizon is temporarily saturated with water and partly oxidized. Rust-coloured mottles here outweigh the grey zones of strong reduction.
- The reductimorphic **Gr** sub-horizon shows strong reductimorphic features and is strongly affected by groundwater or stagnant water. The dark grey and bluish mottles significantly predominate the few oxidized areas, typically and frequently around cracks or root channels.

The **T** horizon, sometimes marked with an **H**, is a peat horizon formed by the accumulation of organic matter under anaerobic conditions. Unless it is artificially drained and partly mineralized, it is saturated with water. It contains more than 18% organic C when the mineral part consists of more than 60% clay, or more than 12% organic C when the mineral fraction contains almost no clay.

Horizons of the Parent Material

The **C** mineral horizon is composed of fragmented and unconsolidated parent material regardless of the degree of weathering. In general, it is barely affected by weathering. It shows no properties of overlying O, A, B, or E horizons.

The letter **R** stands for compact and solid rock regardless of the stage of weathering. It can show no or little surface weathering or can be deeply weathered.

Fig. 4.1 Salt affected soils, Dragonja River, south-western Slovenia (*Photo* T. Kralj)

4.1.3.2 Frequently Used Qualifiers for Horizon Properties

Other frequently used letters added to explain the properties of the main horizons are the following: **b** for a buried horizon (e.g. Ab); **p** for a ploughed or tilled horizon (Ap). The p qualifier should be used only if human activity is clearly visible and present (ploughing, grazing). A **c** marks the accumulation of concretions and an **h** an accumulation of organic matter in the mineral horizon (e.g. Ah, Bh). The qualifiers h and p are mutually exclusive. A **g** is used to indicate gleyic processes in the horizon. It is commonly used for impermeable horizons in Pseudogley (e.g. Bg). A **k** marks the accumulation of calcium carbonate and a **q** the accumulation of quartz (or silicate material).

4.1.4 The Main Soil Types

The basic SSC unit is the soil type. Soil classes are grouped into two soil divisions: **Divison 1—automorphic soils** and **Divison 2—hydromorphic soils**. In comparison with the Yugoslav Soil Classification, the SSC does not recognize salt-affected soils (Kralj and Grčman 2010). They do not exist due to the high precipitation rate in Slovenia (3000 mm in the west, 760 mm on the north-east border, 1500 mm in central Slovenia) in spite of the few hectares of Gleysols in the coastal region (mainly in the Dragonja River valley) which are characterized by the presence of salty groundwater (Fig. 4.1).

As mentioned, soil types are subdivided into three levels: sub-types, varieties, and forms. In comparison with the strict and uniform system of principal and supplementary WRB qualifiers, the SSC qualifiers system is not as strict. Depending on the soil type, the same qualifiers can be used to classify soil type or sub-type, while the same classifiers for a different soil type are used to define the variety or form. This system of qualifiers reflects the importance of the individual properties for general soil properties as well as the productivity of each soil type. A good example can be a dystric–eutric classifier: in the case of Cambisols, the pH,

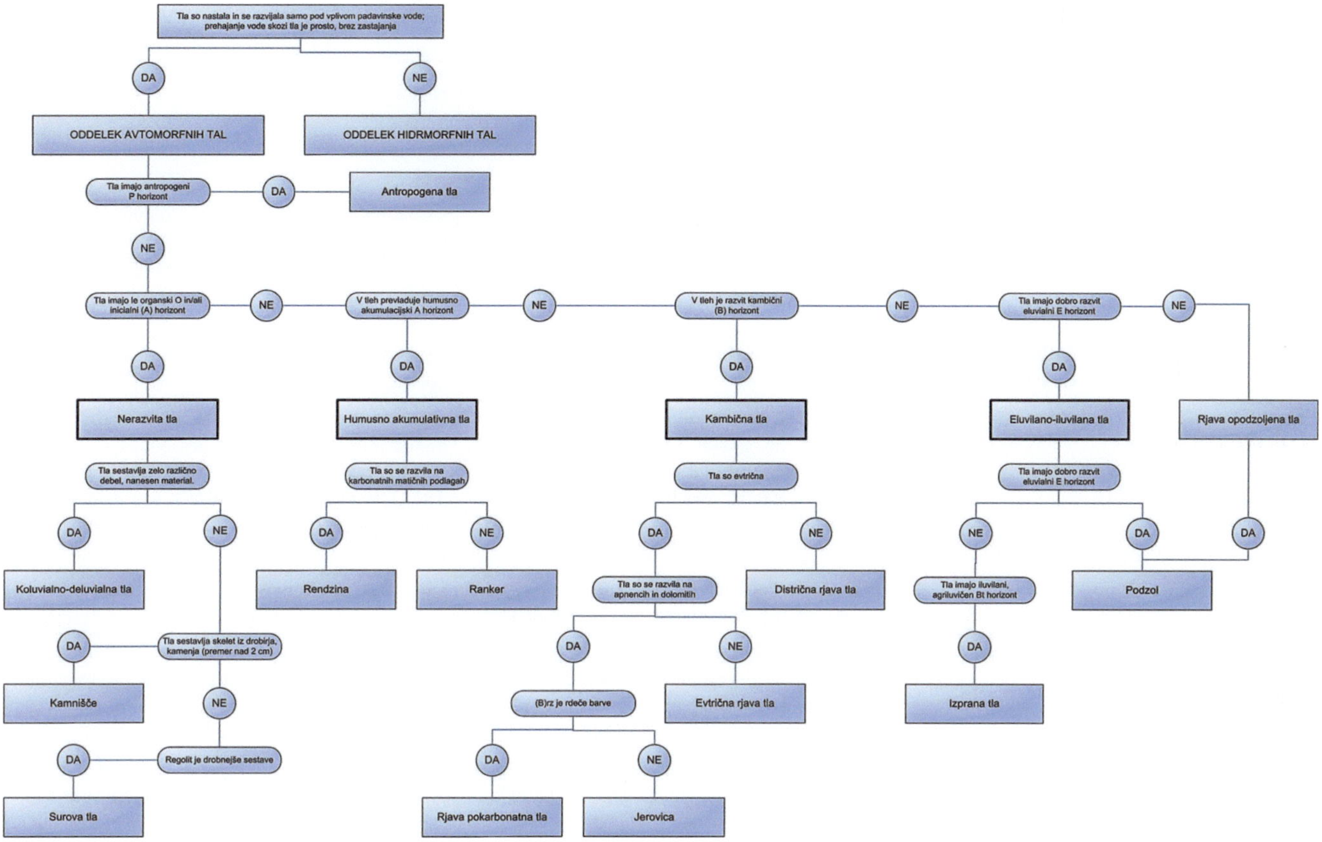

Fig. 4.2 Flow chart for determining soil types in Division 1—automorphic soils (Kralj and Grčman 2009b)

base saturation, and nutrient content make the most important difference in distinguishing between eutric (more productive soils) and dystric (less fertile, acidic) soils. In the case of Gleysols, the pH and base saturation are not as important as far as soil productivity is concerned. In this case, the presence and the depth of gleyic horizons are far more important and therefore used to classify the soil type as Epigley, Hypogley, and Amphigley first of all. In the case of gleyic soils, the pH and base saturation contributions to the agricultural soil quality are of less importance and are therefore used to define the form of the soil type.

4.1.4.1 Division 1: Automorphic Soils

Fourteen main soil types form the group of soils that develop in automorphic conditions—sufficiently aerated with no effect of excessive saturation with water in the soil profile. Automorphic soil types are determined following the flow chart developed by Kralj and Grčman (2009b, 2010) (Fig. 4.2).

Table 4.1 briefly presents the list of soil types and the classifiers/properties used for the classification of sub-types, varieties, and forms of the automorphic soil division.

4.1.4.2 Division 2: Hydromorphic Soils

This division is populated by thirteen main soil types that develop in the permanent or temporary presence of stagnant water or are influenced by high groundwater level. Table 4.2 presents the hydromorphic soil types.

As was the case for Division 1—automorphic soils, the flow chart for the classification of the soil types in Division 2—Hydromorphic soils was also developed by Kralj and Grčman (2009b) (Fig. 4.3).

Unfortunately, the Slovenian Soil Classification has never been officially published in the form of a book and consequently adopted by a broad circle of Slovenian soil scientists. The last official version was developed and to certain extent harmonized as a side product of the more broadly oriented soil research project *The algorithms of digital mapping and pedotransfer functions for the needs of the national soil information system*, in which soil experts from four Slovenian research and educational institutions participated (Vrščaj 2010). Such project consortia to a certain extent guarantee the harmonization and acceptability of the SSC. Yet the SSC remains unpublished and represents a worthy task that has to be completed. An important outcome

Table 4.1 Division 1—automorphic soil types

Class	Soil type	Sub-types	Varieties and forms
Class A1 **Undeveloped Soils** **(A)-[C,R]** profile development	**Lithosol**	Sub-type classification is related to the parent material: primarily to limestone, dolomite, and carbonate moraine; other parent materials are rare	N.A.
	Regosol	Sub-type classification is related to the parent material: carbonate and noncarbonate flysch, noncarbonate soft parent materials such as mudstones and soft shale, sands, and soft sandstone	Classifiers are related to the parent material: flysch and marl; carbonate sands and sandstones; slope gravels; quartz sands, and sandstone
	Colluvial–deluvial soils	Sub-type classification is related to pH and base saturation	Classifiers are related to gleyic properties: nongleyic and gleyic
Class A2 **Humus-Accumulative Soils** **A-[C,R]** profile development	**Rendzina**	Sub-type classification is related strictly to the carbonate parent material: hard and compact limestone and dolomite, carbonate fluvio-glacial sands, and moraine; carbonate slope gravel	Classifiers are related to the type of organic matter, humus forms, and presence of the initial B horizon: terric, mollic, folic, and brunic The following depth qualifiers are frequently used: shallow, medium deep, deep, and colluvial
	Ranker	Classification depends on pH and base saturation of the A horizon: eutric and dystric	Classifiers are lithic, regolithic, eroded
Class A3 **Cambic Soils** **A-[Bv,Brz]-[C,R]** profile development	**Eutric brown soil**	Sub-type classification is based on the parent material: fluvio-glacial sands and moraine; soft carbonate rocks, tertiary clays and loams; basic and neutral rocks	Classifiers are related to the form and diagnostic properties: typical, leached, gleyic, stagnic, and podzolic. The depth qualifiers: shallow; medium deep, deep, and colluvial
	Dystric brown soil	Sub-type classification is based on the parent material: noncarbonate fluvio-glacial sands; other noncarbonate rocks	Variety classification depends on properties and profile development: typical, leached, gleyic, stagnic, podzolic, as well as the depth of the solum: shallow, medium deep, deep, and colluvial
	Kalkocambisol	Depends on profile development and diagnostic properties: typical and leached	The classification depends on soil depth development: shallow, medium deep, deep, and colluvial
	Red-brown soil (Terra Rossa)	Depends on the absence/presence of chert: kremenica (high chert skeleton content in B horizon) and ilovka (loamy-clayey B horizon with little or no chert skeleton)	The classification depends on the presence of luvic properties and depth: colluvial; shallow, medium deep, deep
Class A4 **Eluvial-Illuvial Soils** **A-E-[Bfe, Bh]-[C,R]** profile development	**Luvisol**	Sub-type classification is based on the parent material: developed on limestone and dolomite	Classifiers for variety depend on the parent material, while classification of form is based on acidity and base saturation and stagnic properties: eutric, dystric, acric, Pseudogleyic
	Brown Podzolic Soils	Sub-type classification depends on the type of parent material—different quartz-dominant rocks, sands, and moraine	Classifiers are related to the contact to the parent material and skeleton content
	Podzol	Sub-type classification is based on the presence of the Bh and/or Bfe horizon	Classifiers depend on the thickness of the E horizon, while the form is classified regarding the parent material—different quartz-dominant rocks and sands
Class A5 **Anthropogenic Soils** **P-C** profile development	**Rigosol**	Sub-type classification depends on type—land use: Vitisol; Hortisol; soils of hop fields and orchards	N.A.
	Deposols including various Technosols	N.A.	N.A.

Table 4.2 Hydromorphic soil types

Class	Soil type	Sub-types	Varieties and forms
Class H1 Fluvisols **(A)-I-II**… or **(A)-C** or **(A)-Go** profile development	Undeveloped fluvisols	Sub-type class depends on the pH and base saturation and gleyic properties: calcaric, eutric, dystric, gleyic	The variety depends on the presence and depth of the hydromorphism: nongleyic, hypogleyic, and epigleyic
	Fluvisols	Sub-type class depends on the pH and base saturation: calcaric, eutric, dystric	Classification of variety depends on presence and depth of the hydromorphism (nongleyic, hypogelyic, epigleyic) and the pH/base saturation: eutric and dystric
Class H2 **Pseudogley Soils A-E-Bg-C** profile development	Pseudogley, terraced	Sub-type class depends on the depth of the diagnostic Bg horizon: deep, medium deep, and shallow	The variety depends on the presence and depth of the pH and base saturation and the degree of degradation: eutric, dystric, and retrograded
	Pseudogley, slope	Sub-type class depends on the depth of the diagnostic Bg horizon: deep, medium deep, and shallow	The variety depends on the presence and depth of the pH and base saturation and degree of degradation: eutric and dystric
Class H3 **Gleysols A-G** profile development	Hypogley	Sub-type class depends on the presence of organic matter: mineral and humic	The variety depends on the pH and base saturation: eutric and dystric. The form is classified regarding the depth of the Gr horizon
	Epigley		
	Amphigley		
	Gley on peat	Sub-type class depends on soil depth: mineral and humic	The variety depends on the pH and base saturation: eutric and dystric
Class H4 **Peat Soils T-G** profile development	Peat soils of low moor	Sub-type class depends on the thickness of the T/H horizons	The variety depends on the subsoil type: mineral or gittya
	Peat soils of transitional moor		The variety depends on the presence of mineral subsoil
	Peat soils of high moor		The variety is related to the degree of humification of the T horizons
	Mollic Gley	Sub-type class depends on the subsoil: mineral or gittya	
Anthropogenic Hydromorphic Soil P-G profile development	Hydro-ameliorated soils	Sub-type class depends on the type of hydro-amelioration	N.A.

of the research project was the comparison of the SSC and the WRB 2006 (Kralj and Grčman 2010). An improved and slightly modified version that incorporates the relevant changes with regard to the WRB 2014 is presented in the next section.

4.1.5 A Brief Comparison of the Slovene Soil Classification to the WRB

The following comparison of the Slovene Soil Classification to the World Reference Base 2006 was carried out within a national research project (Kralj and Grčman 2009a, b) and was checked against the new WRB 2014 (FAO and IUSS 2014).

4.1.5.1 Lithosols

The basic definition of Lithosols according to the SSC is that they are bare rock or very thin soils, often discontinued over compact and hard parent material or, in general, sharp-edged rock debris, predominantly limestone and dolomite. Other hard rocks such as grandiorite/tonalite and siliceous breccia or conglomerate are rare. The development of the soil profile remains at a very initial stage; the rock is completely bare, or the humus–accumulative horizon is absent or very thin, fragmented, and discontinued. Comparable soil types in the WRB 2014 are Hyperskeletic Leptosol. In the Alps, Nudilithic forms predominate at altitudes above 1500 m.a.s.l.

4.1.5.2 Regosols

According to the SSC, Regosols are formed on soft rocks, mainly marl or flysch, and rarely on allochthonous transported parent materials. The parent material can be weathered to a various stage. In general, Slovenian regosols are of secondary origin and emerge as hot spots of heavily eroded areas and other denudations, frequently landslides. The humus-accumulative horizon is weakly developed, discontinued, thin, or completely missing. Regosols frequently have a hot spot spatial appearance in western Slovenia on flysch and marl and are a result of erosion due to poor land management. The comparable WRB soil type is Regosol.

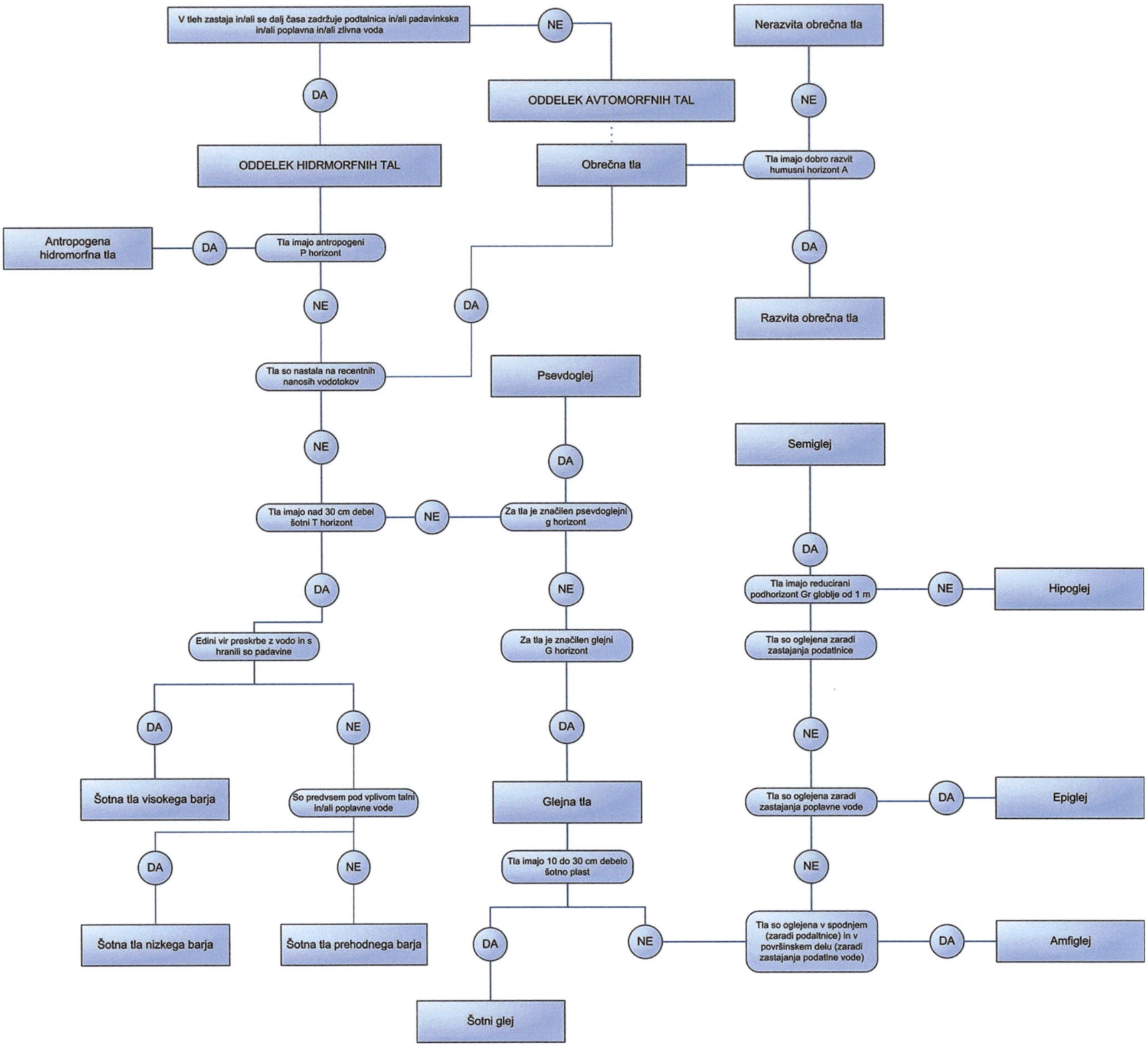

Fig. 4.3 Flow chart for determining the soils in Division 2—Hydromorphic soils (Kralj and Grčman 2009b)

4.1.5.3 Colluvial–Deluvial Soils

The basic criterion according to the SSC is that the soil is formed on lower slopes or on foot slopes out of accumulated, mixed, and heterogeneous medium-to-heavily weathered parent material. The texture is very diverse and can be mixed in the solum. In general, it is unsorted and frequently composed of fine particles and coarse rock debris. The A horizon is weakly developed or present only at an initial stage of development. The main soil-forming process is downslope movement and land sliding—colluvial and/or deluvial processes combined with weathering of the parent material and the influence of vegetation. In the foot slope versions, often fine-texture classes predominate; fewer skeletons are present (30–40%); fragments of alohtonous Ab and B horizons can frequently be detected. Comparable soil types in the WRB are Hyperskeletic Leptosols and Regosols, on which the colluvic qualifier can be used.

4.1.5.4 Rendzinas

The basic criteria for Rendzina are an A-C or A-R type of profile development and carbonate parent material. Rendzina is formed on hard mainly compact limestone and dolomite, on hard carbonate breccia or hard predominantly carbonate conglomerate. The humus-accumulative A horizon or organic A horizon is well developed. The soil structure is well expressed, granular in general. The deeper sub-types,

Fig. 4.4 Shallow Rendzina developed on limestone (*Photo* T. Kralj)

so-called Brown Rendzinas, have an initial cambic horizon, which is thinner than the A above and marked as an A-(B)-C or rarely A-(B)-R profile. Organic forms of Rendzina can have an organic A or an O horizon above a compact R horizon. The total profile depth varies from very shallow (e.g. 5 cm) to deep (A > 50 cm). The direct classification of Rendzina to one of the WRB reference soil groups is difficult (Fig. 4.4).

In general, the A-C and A-R profiles are classified as Eutric Leptosols: hyperskeletic when shallow and formed on a regolith, and nudilithic when shallow on compact and hard limestone or dolomite. According to the WRB, Rendzina with an O-R profile is classified as a Histosol. When the A horizon is thicker than 40 cm or initial development of the B horizon is present, it can be classified as Phaeozem.

4.1.5.5 Ranker

According to the SSC, Rankers have an A-C (regolithic) or A-R (lithic) type of profile development. They are similar to Rendzinas but formed on noncarbonate parent material. Two sub-types, dystric and eutric, are common. The Dystric Ranker is formed on noncarbonate and/or acidic parent material such as quartz breccia, tonalite, and grandiorite. Eutric Ranker overlies calcified or decalcified flysch or a similar soft or friable carbonate rock. Comparable soil types in the WRB are in general classified as Dystric Leptosols. When a deep horizon with umbric properties is present, the Rankers are classified as Umbrisols and Histosols when a thick O horizon or thick and organic A horizon is present. According to the WRB, Eutric Rankers are classified as Eutric Leptosols.

4.1.5.6 Eutric Brown Soils

The basic criteria according to the SSC are an A-Bv-C or A-Bv-R profile development and high base saturation. The diagnostic cambic Bv horizon is formed on carbonate (flysch) or basic rocks (marl, diabase) and soft limestone with a high degree of impurity (clay minerals). The pH of the Bv horizon must be >5.5 measured in $CaCl_2$, or the base saturation is >50%. Comparable soil types in the WRB are Eutric Cambisols when base saturation is >50%. Depending on the texture, a clayic or siltic qualifier can be used.

4.1.5.7 Dystric Brown Soils

The basic criteria that define Dystric Cambisols are, similar to Eutric Cambisols, an A-Bv-C or A-Bv-R profile development and, differently than Eutric Cambisols, a mandatory low pH and low base saturation. Such soil is formed on noncarbonate and siliceous parent material. The pH of the Bv horizon must be ≤ 5.5 measured in $CaCl_2$ or KCl, while

Fig. 4.5 Shallow Terra Rossa (*Photo* T. Kralj)

the base saturation is $\leq 50\%$. Comparable soil types in the WRB are Dystric Cambisols; clayic or siltic qualifiers can be used. In the event the Bv shows umbric properties, it can be classified as Umbrisol.

4.1.5.8 Kalkocambisol

Basic criteria according to the SSC: soil with an A-Brz-C or A-Brz-R profile development. The diagnostic heavy-textured Brz has a well-expressed blocky sharp-edged structure and overlies hard limestone or dolomite. It is brown or yellowish-brown; in any case, the colour is less red than 2.5YR, with a value and chroma >3. The pH of the Brz horizon is >5.5 measured in $CaCl_2$ and base saturation is >50%. Soil depth varies from very shallow (e.g. 25 cm) to very deep (a few metres) over very short distances. Comparable soil types in the WRB are Cambisols; clayic or siltic; chromic or rhodic; leptic, endoleptic, or epileptic qualifiers can be used.

4.1.5.9 Jerina/Terra Rossa

According to the SSC, a Jerina soil has an A-Brz-C or A-Brz-R profile development and has formed on hard limestone or dolomite in Mediterranean or sub-Mediterranean climatic conditions. The diagnostic Brz frequently has a very well-expressed blocky structure; it is reddish or intensively red—the colour is ≥ 2.5YR or redder, with value and chroma <3.5. The pH of Brz is generally >5.5 ($CaCl_2$), and base saturation is >50%. The soil depth varies over short distances. Two sub-types of Jerina were identified by Kodrič (1958) and Sušin (1964): "Kremenica" with a substantial part of chert skeleton—typically apx. 25% (Vrščaj 2014a), which originate from limestone and accumulate within the profile, and "Ilovka" with little or no chert material. The colour as the Terra Rossa diagnostic property is not compatible with the SSC criterion. Comparable soil types in WRB are Cambisols, mainly clayic and rarely siltic. Chromic or rhodic, eutric or dystric, leptic, endoleptic, and epileptic qualifiers can be used (Fig. 4.5).

4.1.5.10 Luvisols

The SSC defines Luvisols as soils with a diagnostic eluvial E horizon overlying an illuvial horizon that is enriched by clay, iron, nutrients, and humic substances. The illuvial Bt, Bfe, or Bh horizons have higher base saturation and higher pH. Luvisols frequently develop on various carbonate (limestone and dolomite) and noncarbonate rocks. The soil depth can vary over short distances. According to the WRB, the SSC's Luvisols correspond to the WRB's Luvisol and Acrisol, and rarely to Alisol. When translating, the argic qualifier can frequently be used (Figs. 4.6, 4.7, and 4.8).

Fig. 4.6 Luvisol, Leskova valley, central Slovenia (*Photo* T. Kralj)

Fig. 4.7 Luvisol, Savinja valley, central Slovenia (*Photo* B. Vrščaj)

4.1.5.11 Podzols

The basic criteria according the SSC: soil with a diagnostic eluvial E horizon composed of chert or some other siliceous sand or skeleton and illuvial horizon(s) with accumulated Fe oxides and/or organic matter. The soil is formed on siliceous, fractured, and well-drained parent material in cold and humid conditions, present on high plateaus >1200 m.a.s.l. The soil type corresponds to the WRB Podzol when albic and spodic horizons are well developed. Less developed forms of podzols can be classified as Albeluvisols (FAO 2007) or Retisols (FAO and IUSS 2014).

4.1.5.12 Brown Podzolized Soils

The basic criteria according the SSC: a weakly developed E horizon shows medium-to-poorly expressed eluvial properties. The E horizon overlies at least one spodic horizon, Bfe or Bhe. Depending on the degree of leaching and the development of the E horizon, the soil can be classified as Albeluvisol or Podzol.

4.1.5.13 Rigosols

The SSC defines Rigosols as deeply tilled (ploughed) agricultural soils. The impact of tillage must be strong and consists of 50–70 cm deep mixing of topsoil. Rigosols are developed on soft, frequently carbonate parent material, typically marl and flysch. The mixed P horizon has a significant amount of medium- to well-weathered skeleton. In rare cases, such soils have been ploughed to a depth of 80 cm. The SSC's Rigosol can be classified according to the WRB as Anthrosol, Regosol, or, in some cases, as Cambisol with terric, hortic, or plaggic qualifiers.

4.1.5.14 Deposols

The basic criteria according to the SSC: the soil is developed on landfills of deposited waste materials of various origins. Frequent materials include mine tailings, quarry deposits, construction waste, thermal power plant ashes, and municipal waste. The appropriate WRB classification is Technosol, and depending on the deposited material, it has urbic, garbic, hortic, plaggic, and pretic qualifiers.

4.1.5.15 Undeveloped Fluvisols

The basic criteria according to the SSC: young weakly developed soil on gravely to sandy alluvial material with weak or very weak profile development. Mainly present on

Fig. 4.8 Acidophilus vegetation of the Luvisol: *Pteridium aquilinum*, *Luzula luzuloides*, *Vaccinium myrtillus*, *Castanea sativa*, and *Pinus sylvestris* (*Photo* B. Vrščaj)

first river terraces, the soil is influenced by groundwater and frequent floods. Following the WRB, it should be classified as Fluvisol. Due to high or even predominant skeleton content, the hyperskeletic qualifier is frequently used.

4.1.5.16 Fluvisols

The basic criteria according to the SSC: genetically young soils that develop on fluvial deposits. The soil profile is deep with high humus content in the topsoil. It features strata of silty to silty sand and sometimes clayey alluvial material that can be intersected by layers of carbonate, siliceous, or mixed sand, with the influence of groundwater or floods. The profile generally shows visible stratification of fluvic deposits of different textures. Frequently, the fine-textured and coarse-textured material exchange has an irregular pattern. If classified according to the WRB, the frequently used qualifiers in Slovenian would be mollic, eutric, dystric, humic, gleyic, skeletic, siltic, and folic.

4.1.5.17 Pseudogleys

The basic criteria according to the SSC: stagnation of water in the upper part of the profile that overlies a predominantly silty, loamy, or clayey, dense, and compact, very low to practically impermeable Bg horizon. In the topsoil, the aerobic and anaerobic processes alternate following the wet or dry season. In summer, they can be dry and hard with deep cracks, while in the wet period of the year, it appears similar to Gley, being wet and frequently saturated with the water. The SSC distinguishes between Pseudogley formed on gentle slopes and planar Pseudogley formed on Pleistocene and Pliocene silty-clayey alluvial terraces. If clay particles dominate in the profile, the Pseudogley corresponds to Planosol or Stagnosol in the case of a dominant silty component (Fig. 4.9).

4.1.5.18 Gleys

The basic criteria according to the SSC: soil saturated with water long enough to develop reducing conditions. Depending on the type of the water, the Gley is classified into three sub-types. Hypogley is saturated by high-level groundwater in the bottom of the profile; Epigley has predominantly moist or wet topsoil due to slow run-off of surface waters due to floods or accumulated rainwater. Amphigley is saturated by both surface and groundwater at

Fig. 4.9 Mottled layer with oxides predominantly inside of soil aggregates (*Photo* B. Vrščaj)

the same time; thus, two Gr horizons develop. In general, all three types of Gley in Slovenia are heavy-textured, low permeable, dense, and frequently plastic and sticky when wet or moist and hard and compact when dry. The WRB does not distinguish Gleysols on the basis of the origin of stagnant water. All three forms of Gley are qualified as Gleysol. Additional qualifiers to define other important properties such as eutric, dystric, mollic, folic, and in rare cases histic should be used.

4.1.5.19 Gleys Developed on Peat

The basic criteria according to the SSC: a combination of mineral Gley and peat soil material. These are anthropogenized soils that have developed in a large area of drained and harvested peat of the Ljubljana Moor. The area is still frequently flooded by rivers bringing and depositing silty and clayey material. An appropriate translation to the WRB classification depends on the dominant component peat or mineral component. If it is peat, the soil is classified as Histosol, and when the mineral component prevails, the WRB Gleysol group suits this SSC group the best.

4.2 Comparison of SSC Soil Types to the WRB Reference Soil Groups

The WRB separates the 32 reference soil groups in the hierarchy into their own classifications comparable to the SSC soil types. The SSC distinguished 27 soil types. Table 4.3 presents the rough classification of SSC soil types in the reference basis of the WRB classification. For a more precise classification, it is necessary to check the properties of the specific profiles.

4.2.1 Summary

The SSC classification is predominantly based on the stage of soil profile development. The soil properties are described by qualifiers on the level of the soil sub-type, variety, and form. The SSC distinguishes 27 soil types in six classification levels; the central unit and the central level of the classification is the soil type. The WRB distinguishes 32 basic units—reference soil groups. The WRB classifies

Table 4.3 Comparison of the SSC soil types to the WRB 2104 Reference Soil Groups

SSC soil types	Corresponding WRB soil types
Litosol	Nudilithic Leptosols
	Hyperskeletic Leptosols
Regosol	Regosols
Koluvialno-deluvialna tla [Colluvial–Deluvial Soil]	Hyperskeletic Leptosols
	Regosols
Rendzina	Leptosols (Eutric)
Rendzina	Hyperskeletic Leptosols
Sprsteninasta rendzina	Mollic Leptosols
Rendzina	Rendzic Leptosols
Prhninasta rendzina	Folic Leptosols (Eutric)
Rendzina	Cambic Leptosols (Eutric)
Rjava rendzina [Brown Rendzina]	Folic Histosols
Rjava rendzina [Brown Rendzina]	Phaeozems
Rjava rendzina [Brown Rendzina]	Cambisols (Eutric)
Ranker	Dystric Leptosols
Ranker	Cambic Dystric Leptosols
Ranker	Eutric Leptosols
Ranker	Cambic Eutric Leptosols
Ranker	Umbric Leptosols
Ranker	Folic Histosols
Ranker	Umbrisols
Evtrična rjava tla [Eutric Brown Soil]	Eutric Cambisols
Distrična rjava tla [Dystric Brown Soil]	Dystric Cambisols
	Umbrisols
Rjava pokarbonatna tla [Kalkocambisol]	Eutric Cambisols
	Chromic Cambisols
Jerina—kremenica [Terra Rossa]	Chromic Skeletic Cambisols
Jerina—ilovka [Terra Rossa]	Chromic Cambisols
Izprana tla [Leached Soils]	Luvisols
Izprana tla, akrična [Leached Soils, acric]	Acrisols
Podzol	Retisols
	Podzols
Rjava opodzoljena tla [Brown Podzolic Soils]	Podzols
	Retisols
Rigolana tla [Rigosols]	Regosols
	Anthrosols
Tla deponij [Landfill Soils]	Technosols
Nerazvita obrečna tla [Undeveloped Fluvisols]	Fluvisols
	Fluvisols (Hyperskeletic)
	Fluvisols (Siltic)
	Fluvisols (Arenic)
Obrečna tla [Fluvisols]	Fluvisols
	Gleysols

(continued)

Table 4.3 (continued)

SSC soil types	Corresponding WRB soil types
Psevdoglej, pobočni [S developed on gentle slopes] Psevdoglej, ravninski [developed on plains or terraces]	Stagnosols
	Planosols
Hipoglej	Gleysols (Dystric/Eutric)
Amfiglej	Gleysols (Dystric/Eutric)
Epiglej	Gleysols (Dystric/Eutric)
Glej na šoti [Gley formed on peat]	Histosols
	Gleysols
Šotna tla nizkega barja [Peat soils of low moor]	Rheic Histosols
Šotna tla prehodnega barja [Peat soils of transitional moor]	Histosols
Šotna tla visokega barja [Peat soils of high moor]	Ombric Histosols
Mineralno organska tla [Mineral–organic soil]	Mollic Gleysols
Hidromeliorirana tla [Hydro-ameliorated soil]	Histosols (Drainic) or Gleysols (Drainic)

Adapted from Kralj and Grčman (2009a)

soils on the basis of diagnostic horizons, diagnostic properties, and diagnostic materials—soil features. In comparison with the SSC and due to the well-defined diagnostic properties and diagnostic materials, the WRB soil name is more explanatory in comparison with that of the SSC. On the other hand, the WRB has no classifiers to distinguish different soils formed on hard limestone and dolomite. The latter parent material forms the largest part of the Slovenian lithology. In combination with the topography and climatic soil-forming factors, a variety of different soil types develop. That is why the SSC distinguishes between different forms of Cambisols and separates them into so-called post-carbonate soils, Eutric Brown soils, and Terra Rossa. The SSC soil types can be translated into the WRB reference soil groups. A direct one-to-one translation from the SSC soil types to the WRB soil types is not feasible. Individual soil properties and the presence of individual diagnostic horizons and their features have to be considered for a suitable translation of the SSC soil types into the WRB taxonomy. In some cases, an SSC soil type can be classified into a WRB soil type with several principal and supplementary qualifiers. More frequently, only the reference WRB soil group can be defined due to a lack of information on the morphological properties, profile development, adequate definitions of the SSC soil type, and/or analytical data.

5 An Overview of Soils of Slovenia

Borut Vrščaj

Abstract

The chapter describes the most frequent/dominant soil types in Slovenia. Lithosol and Regosol represent the class of Undeveloped Soils that feature an (A)-C or (A)-R sequence of horizons. Rendzina and Ranker are the most dominant soil types belonging to the class of humus-accumulative soils (with an A-C or A-R horizon structure). The main Cambisols (with either an A-B-C or A-B-R sequence of horizons) are Dystric Brown Soil, Eutric Brown Soil, and Kalkocambisol. The order of Hydromorphic Soils is presented by a description of Pseudogley, Hipogley, Amphigley, and Epigley. For each soil, the genesis and soil profile development is explained, and the main physical and chemical properties are described. The statistics on selected soil properties are derived from the measured soil profile datasets of the presented soil types. The representative examples of measured soil profile datasets are included in the soil-type descriptions. The spatial distribution of soil types is provided by maps. This chapter is illustrated with photographs of land, soil profiles, and soil features.

5.1 Representative Slovenian Soil Types

This chapter presents the main soil types of Slovenia. For each presented soil, a brief description is given as well as the measured characteristic soil profile analytical dataset. The median values for individual soil properties per diagnostic horizon are derived from the soil profile database ("Talni profili, digitalna baza podatkov (TP) [Soil profile; digital database (SP)]", 2001). Hereinafter, this source of data is further abbreviated as "SP database". The order of soils follows the soil classification. First of all, the soils from the order automorphic are presented, starting with Lithosol. Secondly, Pseudogley, Hypogley, Amphigley, and Epigley are described to present the most important soils in terms of extent that belong to the order of Hydromorphic soils.

5.1.1 Order A. Automorphic Soils

5.1.1.1 Class A1. Undeveloped Soils [(A)-C (A)-R Sequence of Horizons]

Lithosol

Genesis and Soil Profile Development Lithosol is a soil in a very initial stage of formation. According to the Slovenian Soil Classification (SSC), Lithosol is practically bare rock, a soil similar to Nudilithic Leptosol in the WRB. In Slovenia, Lithosol forms almost entirely on compact Triassic limestone and dolomite. A bare rock is marked as an R when it is compact or with a C horizon when the parent material is fragmented (i.e. a gravel of talus). Lithosol is also

B. Vrščaj (✉)
Department of Agricultural Ecology and Natural Resources, Centre for Soil and Environment Research, Agricultural Institute of Slovenia, Ljubljana, Slovenia
e-mail: Borut.Vrscaj@kis.si

B. Vrščaj et al., *The Soils of Slovenia*, World Soils Book Series,
DOI 10.1007/978-94-017-8585-3_5

a very shallow form of soil with an initial (A) or Oh horizon rich in the organic matter (>25%), or even organic (>35% of organic matter). The soil profile has no clearly expressed morphology and the solum is incomplete. Lithosol with an (A) or undeveloped Oh would be classified as Lithic Leptosols according to the WRB. The structure of the soil profile can vary over short distances and the (A) or Oh horizon is frequently discontinued. If an organic or mineral layer is present on top of a C or R, the sequence of horizons is Oh-R, (A)-C, or (A)-R, and (A)-C-R.

The development of the soil is terminated due to harsh climatic conditions (long winters, snow cover) and especially due to the strong erosion. The high annual precipitation rate (2500–3500 mm in the Slovenian Alps), steep slopes, and very slow decomposition (dissolution) of the limestone/dolomite practically halt profile development. If a thin Oh or (A) horizon exists, it is predominantly discontinued with patches of bare rock and frequently penetrated with rock fragments and outcrops. The discontinuity of an (A) horizon increases with altitude to areas of completely bare rock surface. At lower altitudes, it is related to relief forms. On shoulders and ridges, the discontinuity is total, while in concave forms of relief, a thick, colluvial, and frequently hyperskeletic and/or sandy (A) horizon that is larger in terms of extent can accumulate. The main processes that steer the soil formation are the slow dissolution/weathering of the parent material and strong erosion. In a stable topographic position and favourable climate, Lithosol gradually evolves into an Organogenic Rendzina and Kalkocambisol (Stepančič et al. 1987a, 1986b) (Fig. 5.1).

Main Properties Lithosol is to a large extent a bare rock in some places with a thin Oh horizon or mineral sandy and gravelly (A) horizon. The typical chemistry of Oh horizons in Alpine areas is the following: the pH is 6.5–6.9 (KCl), P_2O_4 apx. 3.6 mg/100 g, and the K_2O content is apx. 20 mg/100 g of dry soil. The SOM content frequently exceeds 30% and reaches 50–60% at high altitudes. The N content is apx. 2%. The base saturation is 60–75% and the CEC is typically between 80 and 85 $mmol_c$/100 g of dry soil (Fig. 5.2).

Distribution Lithosol is located primarily in the Julian and Savinja-Kamnik Alps in the north-west and north of Slovenia. Below 2000 m.a.s.l., it is present to a limited extent and primarily in a patchy pattern in an overall area of limestone and dolomite parent material on locations where the relief is rough. It appears on ridges, peaks, and edges of tectonic faults. At low altitudes, large areas of Lithosol can be found on the south ridges of Trnovski forest and the Trst-Komen plateau as well as the steep slopes north of the Kolpa River. According to the DSM25, the total extent of Lithosol is 2.07% of the country. As a dominant soil type, it extends over 1.3% of the Slovenian territory.

Fig. 5.1 Lithososl is the dominant soil type in the mountainous areas of Slovenia (*Photo* T. Kralj)

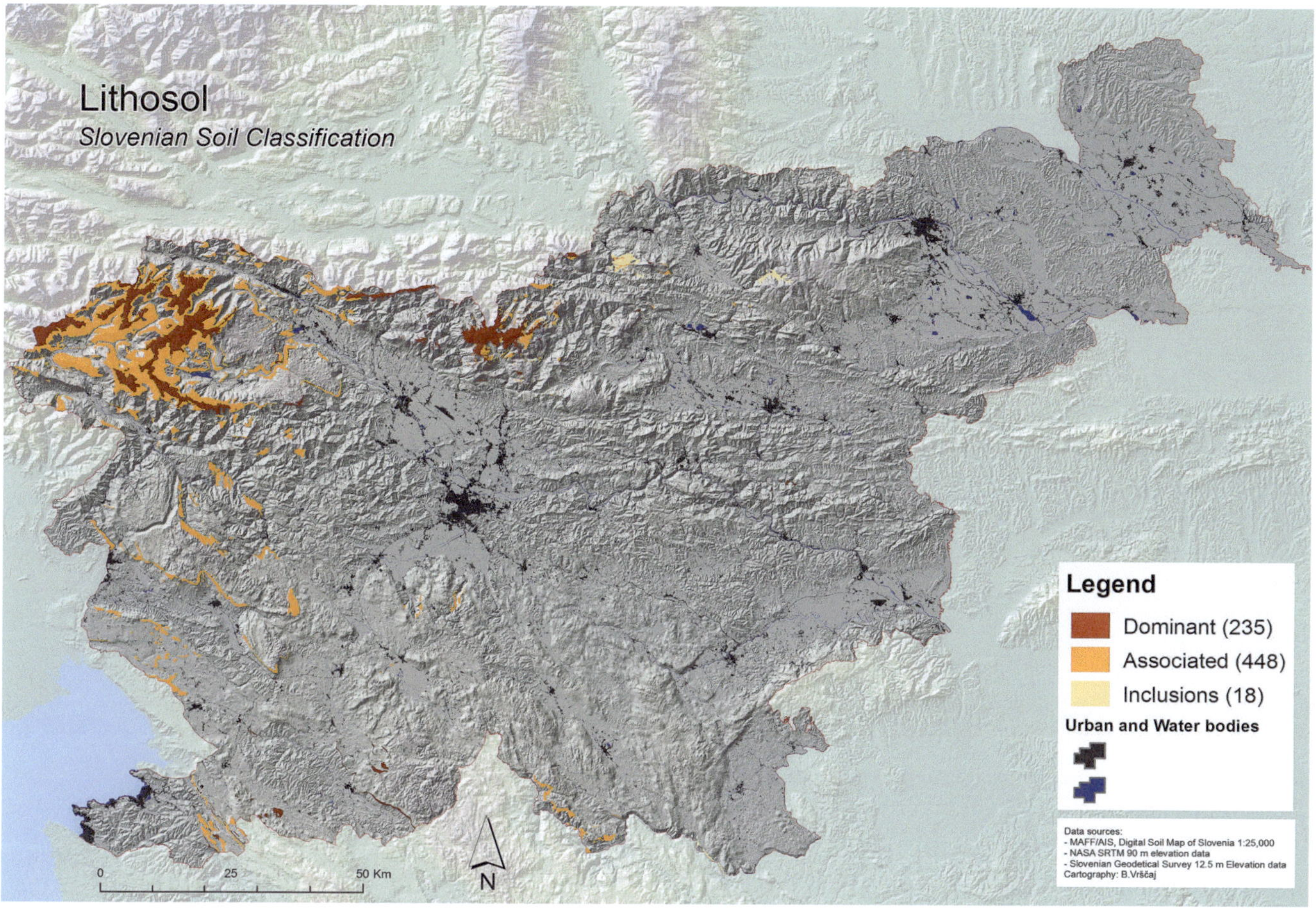

Fig. 5.2 The distribution of the Lithosol

Land Use/Land Cover Largely a bare rock, the land cover can be a mountain grassland with some mountain pine *(Pinus mugo)* or, rarely and at lower altitudes, a sparse forest (*Larix decidua*).

Regosol

Genesis and Soil Profile Development According to the SSC, Regosol belongs to the class of undeveloped automorphic soils and has an (A)-C profile development. It represents an initial stage of soil development on loose, soft parent material that is prone to mechanical and chemical weathering. The main Regosol parent substrates are marl, carbonate mudstone, slate, and, especially the flysch formations composed of varying strata of marl or siltstone and quartz sandstone which in some areas of Slovenia is cemented with $CaCO_3$. Regosol largely inherits the properties of the parent material. The weathered material of the Regosol that develops on flysch and marl is abundant in terms of volume and rich in carbonates and some plant nutrients. The potentially rapid development of the soil profile is mainly stopped by erosion, and the soil remains in an initial stage. In the event erosion is stopped, the development of the soil profile continues in the direction of the formation of an A-C profile or an A-R profile to form a Eutric Ranker (Lobnik et al. 1996b, 1995a; Rupreht et al. 1985).

The parent material, which is soft and weathers easily, enables the development of a fertile soil, rich in Ca and other nutrients (Fig. 5.3).

Main Properties As an example, we present the chemistry of soil profile No. 860 excavated near Dolenja Vas, south-western Slovenia. The analysed profile has the (A)C1–C2 sequence of horizons. The pH is 8.4 (H_2O) and 7.6 (KCl), the texture class is loam, the P_2O_4 is <2 mg/100 g, and the K_2O content is 16.3 mg/100 g of dry soil. The SOM content is 0.5%; C_{org} is 0.5%, the N content is 2%, the base saturation is 98%, and the CEC is 44 $mmol_c$/100 g of soil. The median values for saturation with Ca, Mg, K, N, and H are 96.4, 1, 0.5, 0.9, and 1.1%, respectively.

Distribution Almost as a rule, Regosol in Slovenia appears to a very small extent on locations where erosion is severe. It is most commonly triggered by humans often due to poor land

Fig. 5.3 Regosol developed on flysch, Koper hinterland, south-western Slovenia (*Photo* B. Vrščaj)

management. The total extent of Regosol as a dominant soil type is very small. In general, it is associated with calcaric Eutric Brown soil and cannot be separately mapped on small-scale maps as an individual soil type. This is not the case in Istria (south-eastern Slovenia), where larger areas of Regosol are situated on steep slopes and associated with Rendzina. Thirty % of Regosols predominate within the DSM25 soil mapping units (SMU). According to the DSM25, the total area of Regosol is only 870 ha or 0.04% of the country. As a dominant SMU soil type, it is present on 67 ha. In general, it is an associated soil type (apx. 803 ha) (Fig. 5.4).

Land Use/Land Cover The common land cover in Regosol areas is sparse vegetation, bush, and forest, often in a transitional stage of spontaneous afforestation.

5.1.1.2 Class A2. Humus-Accumulative Soils [A-C, A-R Sequence of Horizons]

Specifics of Limestone and Dolomite Relevant for Pedogenesis

Limestone and dolomite are similar rocks. The main limestone mineral is calcite ($CaCO_3$), and the main mineral of dolomite rock is the same-named mineral dolomite ($CaMg(CO_3)_2$). Limestone and dolomite rocks share similar colours. It is mainly white, white to pale grey, white to dark grey. Pink, reddish, and very rarely a pale green limestone and dolomite can be found in Slovenia. The rock hardness according to Mohs is similar, frequently 3–3.5. Both of them are soluble in hydrochloric acid diluted to a ratio of 1:3. They are both known for their ability to neutralize acids. Limestone is (inter alia) frequently crushed, grained, and used as a soil liming material.

One difference between limestone and dolomite rock that is important for soil genesis and the soil profile sequence is the way they weather. The limestone is primarily prone to chemical weathering. Limestone rock is in general compact and hard. The main weathering takes place on the surface, while the inner part of the rock can be unaltered. The surface of the rock is relatively smooth, rounded, and reshaped in a number of forms specific to the Karst limestone. The contact of the soil horizons with limestone is sharp, lithic. For areas of relatively pure limestone, a high degree of surface rockiness and rock outcropping is common or even characteristic. In contrast to limestone, dolomite is prone to physical weathering in the first place. The surface of dolomite rock is often, not to say predominantly, rough, friable, gravelly, and deeply fragmented. The physical weathering can be high and accompanied by a chemical weathering deep in the cracks. In general, a rock outcropping is far less frequent in a dolomite area. The contact of soil horizons to dolomite rock is primarily gradual, regolithic (Fig. 5.5).

The presence of the available magnesium facilitates the conversion of calcite into dolomite. This chemical process

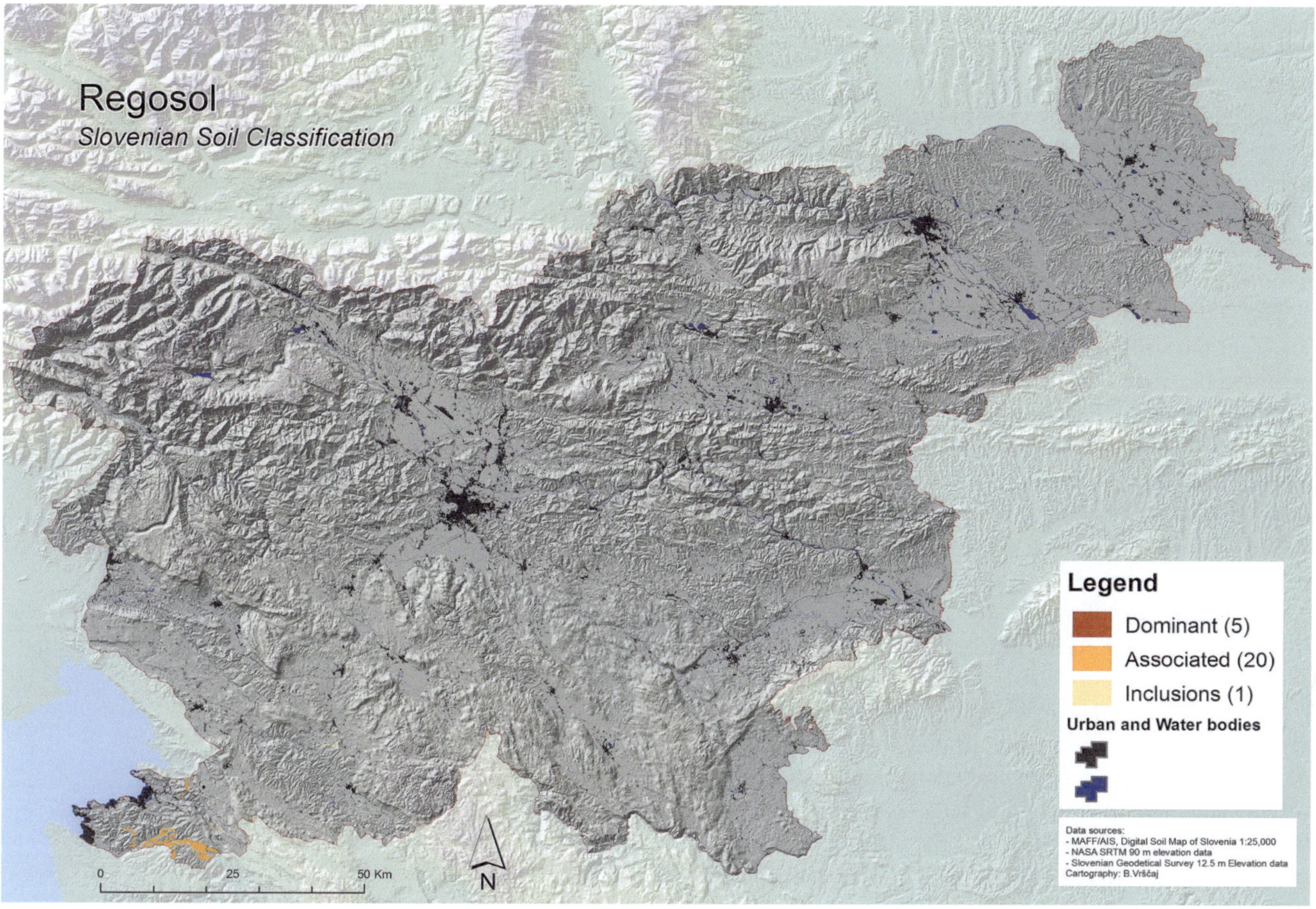

Fig. 5.4 The distribution of Regosol in Slovenia

Fig. 5.5 Rendzina, the contact to dolomite rock (*Photo* T. Kralj)

(known as dolomitization) can significantly alter a limestone and gradually change it into a dolomite to form a parent material known as *dolomitic limestone*. The ratio of $CaCO_3$ and the $CaMg(CO_3)_2$ may vary in such a rock. In Slovenia, the transitional forms of dolomitic limestone are frequent; therefore, the weathering, rock surface properties, sequence of horizons, stoniness of the horizons, and contact to the parent material vary and might not be as characteristic as is described for pure limestone or dolomite.

The soil genesis of soils in these substrates strongly differs from the soil genesis on other sedimentary, igneous, or metamorphic rocks except marble. In both cases, the predominant weathering is dissolution of the $CaCO_3$ and the $CaMg(CO_3)_2$. In other words, the total volume of rock (>95%) is dissolved and washed away by waters. By this, the total volume of the rock is almost completely reduced and large empty spaces are created (dolinas, poljes, and underground caves). If not washed away by erosion, the residual clay and fine silt (i.e. the impurities in the limestone/dolomite) forms the bulk volume of the mineral component of soil horizons.

The intensity of the weathering to a large extent depends on the impurity of the rock (the quantity of noncarbonate minerals), the quantity of precipitation, the presence of acids in rainwater, soil moisture, and temperatures. The development of the soil is very slow. The time needed for the accumulation of a 1-cm-thick residue of clay is estimated to be apx. 8–12 ka.

The soil formation on dolomite and limestone is specific in terms of the very low volume of weathered material, the diverse composition of residual clays, and the very long time needed for soil to form.

Rendzina

Genesis and Soil Profile Development Rendzina is a young soil that belongs to the group of humus-accumulative soils. According to the SSC, the Rendzina develops on hard limestone and dolomite or any transitional form of rock between dolomite and limestone. The Triassic, Jurassic, and Cretaceous limestones/dolomites in Slovenia are very pure and compact. The content of the $CaCO_3$ is high, 95–99%. Residual clay and fine silt constitute the main volume of the mineral component of the A horizon of Rendzina. A Rendzina that develops on compact limestone has an A-R sequence of horizons; therefore, contact to an R horizon is lithic. In the A horizon, an individual rounded limestone skeleton (e.g. 5–10 cm) can be present. The sequence of horizons of Rendzina that develops on a dolomite is primarily A-C, A-C-R, or A-AC-C-R. In an A horizon, the skeleton might pose a significant volume in the form of small stones (0.5–1 cm). In forests and unmanaged grasslands where the grass is not harvested, the A horizon is overlaid by Ol, Of, and Oh horizons of various thickness (Prus et al. 1985b; Stepančič et al. 1987a, b, 1986b) (Fig. 5.6).

Main Properties The SSC recognises three sub-types of Rendzina: Organogenic Rendzina, Organo-mineral Rendzina, and Brown Rendzina. **Organogenic Rendzina** is a soil characterized by the accumulation of organic material due to low mineralization and the formation of the Oh or Ah horizon. Both horizons are rich in organic matter and low in the mineral component that is residual clay and fine silt, frequently mixed with coarse fragments of parent rock.

Fig. 5.6 Ol and Of horizons overlying an A horizon in unmanaged grassland (*Photo* B. Vrščaj)

Fig. 5.7 Organogenic Rendzina formed on limestone (*Photo* T. Kralj)

According to the WRB, Organogenic Rendzina can be classified as a Histosol or Folic Leptosol. The soil is predominantly shallow, well aerated, relatively rich in nutrients, and has a low available water capacity (AWC). The soil depth varies in relation to the micromorphology of the relief, the structure of the parent material, the steepness of the slope, and the intensity of the erosion processes. Changes in the soil depth and the thickness of the individual horizons typically occur within a few meters or less. In the concave forms of the relief, the profile can be deep and colluvic (i.e. mixed with coarse skeleton or rocks). The properties and the profile development to a large extent depend on the microclimatic conditions (Fig. 5.7).

Representative Measured Profile The measured profile dataset presents the typical chemical data of the Organogenic Rendzina developed on limestone. A relatively thick profile with an Oh-OhC sequence of horizons was sampled on a northern slope at an altitude of 1300 m.a.s.l. (SP database) (Table 5.1).

The Organo-Mineral Rendzina represents a parallel stage of Rendzina development. The sequence of horizons can be A-C, A-C-R, and A-R. They develop at lower altitudes in more favourable microclimatic conditions (i.e. higher temperatures and longer periods without frost). The A horizon is dark, rich in organic matter and has a well-developed granular, sub-granular, or sub-block structure. The microbial activity is increased and the mineralization of the organic matter is more intensive. The share of organic matter in the A horizon is decreased while the share of the mineral component is increased. The C/N ratio is closer to the optimal (10–12). The base saturation is medium-high to high. In grasslands, two or three A horizons that mainly differ as to morphological properties can be identified. The A_1 on the top is well developed, loose, friable, crumbly, abundant on fine grass roots, and richer in organic matter than horizons below. A typical depth of the A_1 is 8–10 cm. The A_2 and a potential A_3 are denser, more compact but still friable, more mineral and fewer roots are present. By deepening the soil profile, the AWC increases. The colluvial forms of Organo-mineral Rendzina that are deeper than 50 cm are especially valuable due to the high fertility. The AWC is sufficient in years with normal rainfall. In accordance with the WRB, Organo-mineral Rendzina is classified as Rendzic Leptosol, Mollic Leptosol, or Phaeozem in the case of a deep mollic A horizon (Stepančič et al. 1987a, 1986b).

Representative Measured Profile Organo-mineral Rendzina developed on a dolomite—a representative soil profile dataset (Source: SP database) (Table 5.2).

Brown Rendzina
represents the next stage of Rendzina development. The profile is deepened and the cambic Brz horizon is developing or is developed. The Brz can be thin and in an initial stage of development. The weathering limestone/dolomite appears sandy and silty and still contains $CaCO_3$. The sequence of

Table 5.1 The typical chemical data of the Organogenic Rendzina developed on limestone. A relatively thick profile with an Oh-OhC sequence of horizons was sampled on a northern slope at an altitude of 1300 m.a.s.l. (SP database)

	Depth		pH		Particle size distribution						P2O5	K_2O	N	C	*OS*	*C/N*
	From	To	KCl	$CaCl_2$	Fine sand (0.2–0.05 mm)	Coarse silt (0.05–0.02 mm)	Fine silt (0.02–0.002 mm)	Total silt (0.02–0.002 mm)	*Clay* (<0.002 mm)	Text. Class	Al extr. (plant-available)		Kyehl-dahl	Walkely Black		
Horizon	cm	cm	-	-	%	%	%	%	%	Class	mg/100g		%	%	%	*Ratio*
Oh	0	25	6.6	–	–	–	–	–	–	–	10.6	12.6	62.1	*1.7*	36.0	*20.9*
OhC	25	50	–	–	–	–	–	–	–	–	–	–	–	–	–	–

	Depth		Basic cations				Acidity	*Bases*	*CEC*	*Base saturation (%)*					
	From	To	Ca	Mg	K	Na	H	*sum.*		*Total*	*Ca*	*Mg*	*K*	*Na*	*H*
Horizon	cm	cm	meq/100 g soil (1M NH_4OAc, pH7)							%					
Oh	0	25	77.1	1.6	0.2	0.3	16.2	*79.3*	*95.5*	*83.0*	*80.7*	*1.7*	*0.2*	*0.4*	*17.0*
OhC	25	50	–	–	–	–	–	–	–	–	–	–	–	–	–

Table 5.2 Organo-mineral Rendzina developed on a dolomite—a representative soil profile dataset (*Source* SP database)

	Depth		pH		Particle size distribution						P2O5	K_2O	N	C	*OS*	*C/N*
	From	To	KCl	$CaCl_2$	Fine sand (0.2–0.05 mm)	Coarse silt (0.05–0.02 mm)	Fine silt (0.02–0.002 mm)	Total silt (0.02–0.002 mm)	*Clay* (<0.002 mm)	Text. Class	Al extr. (plant-available)		Kyehl-dahl	Walkely Black		
Horizon	cm	cm	–	–	%	%	%	%	%	class	mg/100 g		%	%	%	*Ratio*
A	0	9	6.8	–	29.7	14.6	29.5	*44.1*	26.2	Lo-ClLo	14.5	22.5	18.9	*0.7*	10.9	*15.4*
AC	9	34	7.1	–	30.8	15.0	31.7	*46.7*	22.5	Lo	8.0	15.1	10.8	*0.5*	6.2	*13.0*
C	34	–	–	–	–	–	–	–	–	–	–	–	–	–	–	–

	Depth		Basic cations				Acidity	*Bases*	*CEC*	*Base saturation (%)*					
	From	To	Ca	Mg	K	Na	H	*Sum.*		*Total*	*Ca*	*Mg*	*K*	*Na*	*H*
Horizon	cm	cm	meq/100 g soil (1 M NH_4OAc, pH7)							%					
A	0	9	31.3	14.0	0.5	0.1	7.0	*45.9*	*52.9*	*86.8*	*59.2*	*26.4*	*0.9*	*0.2*	*13.2*
AC	9	34	24.9	10.8	0.3	0.1	4.1	*36.0*	*40.1*	*89.8*	*62.0*	*26.9*	*0.7*	*0.2*	*10.2*
C	34	–	–	–	–	–	–	–	–	–	–	–	–	–	–

Fig. 5.8 Shallow, lithic, or regolithic Rendzina is the dominant soil type in the mountainous areas of Slovenia. The photograph shows the Zajamniki mountain pastures, 1300 m.a.s.l., Bohinj valley (*Photo* B. Vrščaj)

horizons is A-C-(Brz)-R or A-(Brz)-R. The Brz may be well developed with a well-formed soil structure of peds but thin. The thickness of the Brz is less than 50% of the A horizon. The properties of the Brown Rendzina Brz horizon are quite similar to the properties of the Brz horizon in Kalkocambisol, the next stage of soil profile development. In accordance with the WRB, Brown Rendzina with a deep mollic horizon can be classified as Mollic Leptosol or Phaeozem.

Distribution Rendzina is present in a variety of forms and climates, at all altitudes and under many different land uses in central, western, and southern Slovenia (Fig. 5.10). The distribution of Rendzina clearly indicates the extent of hard limestone and dolomite in Slovenia. Organogenic Rendzina is primarily situated in the Alps, north-western Slovenia, and on Snežnik Mountain in the south, where it is associated with Lithosol. Organogenic Rendzina primarily appears at high altitudes, on steep slopes, on northern slopes at lower altitudes, cold microclimatic locations, and carbonate moraine deposits in the cold Alpine valleys of Krma, Kot, Logarska dolina, and others. On the high plateaus of Pokljuka and Jelovica, it is present mainly above 1000 m.a.s.l (Figs. 5.8 and 5.9).

Organo-mineral Rendzina and Brown Rendzina predominate on the elevated plateaus of Trnovski forest and Banjščice (apx. 800–1200 m.a.s.l). It is associated with the Organogenic Rendzina that can be found on steep slopes, cold and humid microlocations, and with Brown Rendzina at lower altitudes. At low altitudes and in warmer climates, Organo-mineral Rendzina primarily alternates with Kalkocambisol in flat areas and on moderate slopes, and with Lithosol on steep slopes and rough terrain forms. The main areas where Organo-mineral Rendzina is present as a

Fig. 5.9 Brown Rendzina formed on carbonate moraine, Gorjuše, 950 m.a.s.l., the Pokljuka plateau (*Photo* B. Vrščaj)

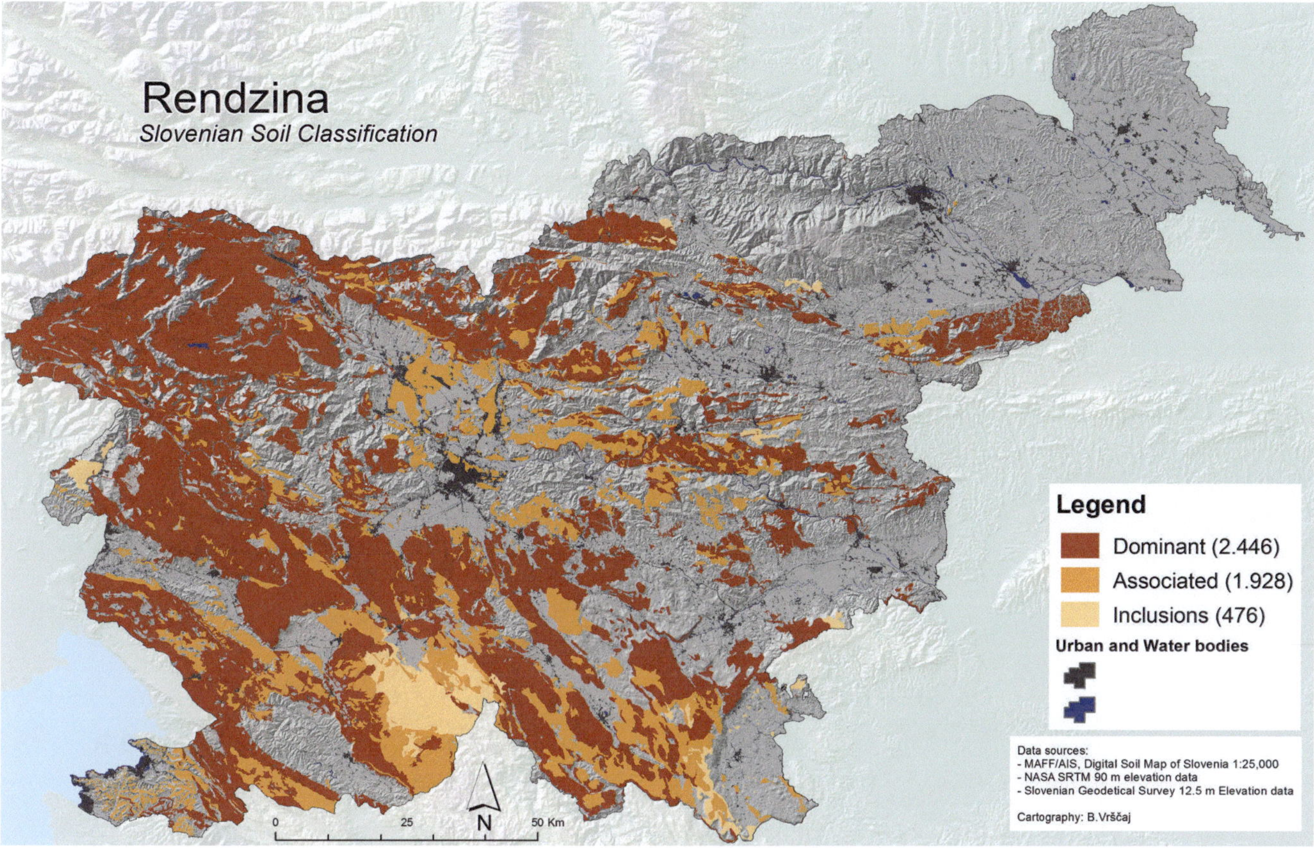

Fig. 5.10 The distribution of Rendzina in Slovenia

dominant soil type are the Trst-Komen plateau, the Dolenjska region, and the woods in the hilly hinterlands of the city of Kočevje, south-eastern Slovenia, and south-west of the city of Novo Mesto. Detached areas of Rendzina are present in east-central Slovenia in the hilly region south of the city of Celje, and north of Donačka Mountain close to the Croatian border. In the large basins of Ljubljana and the Savinja River valley, Rendzina is formed on young Quaternary deposits of limestone/dolomite gravel and coarse sands. In these areas, Organo-mineral Rendzina and Brown Rendzina are associated with Eutric Brown soils. According to the DSM25, the total extent of the Rendzina is 26.7% of the country. As a dominant soil type, it extends over 18.6%, and as associated over 8.1% of the territory (MKGP and CPVO 2001) (Fig. 5.10).

Land Use/Land Cover The main land use on Rendzinas is forest, grassland in lowlands, and Alpine meadows and pastures in mountainous and hilly areas (Fig. 5.11). On the flats of Ljubljana basin and the Savinja River valley, Organo-mineral Rendzina is ploughed.

Ranker

Genesis and Soil Profile Development Ranker is similar to Rendzina: it is a young humus-accumulative soil. The Ranker sequence of horizons is predominantly A-C, A-C-R, or A-AC-C-R. A Ranker with a sequence of organic horizons on top of compact rock (Ol-Oh-R) is rare. The contact with the parent rock is mainly regolithic and less frequently lithic. In the forests, the A horizon is overlaid by an Ol, Of, or Oh horizon of various thicknesses.

The SSC distinguishes two sub-types of Rankers: Dystric and Eutric Ranker. Dystric Ranker develops on noncarbonate parent material of very different igneous, metamorphic, and sedimentary rocks such as tonalite, quartz conglomerate, slate, gneiss, as well as silicate gravel and sands. Eutric Ranker is very similar to Dystric Ranker. It develops on carbonate parent material, in Slovenia mainly on flysch, marl, carbonate mudstone, and other soft carbonate rocks. The pH_{water} is higher than 5.5 and the base saturation is higher than 50% (Prus 2000).

According to the WRB, deep Dystric Ranker is classified as Umbric Leptosol and Eutric Rankers as Eutric Leptosol.

Fig. 5.11 Rendzina on carbonate gravel deposits, central Slovenia (*Photo* B. Vrščaj)

Main Properties The mineral component of the A horizon is composed of well-weathered and changed parent material of various textures. The properties of the A horizon largely correspond to the *umbric* qualifier. It is a dark, blackish-greyish colour, thick, acidic, and low in base saturation. The organic matter content is medium to high. The median values of the A horizon for pH (KCl), organic matter content, and base saturation are 4.0, 6.1, and 18.8%, respectively. The texture depends on the parent material, but in general coarse-textured Rankers prevail. The AWG is moderate to low, and in regolithic forms, the skeleton content can be low to high.

Representative Measured Profile A Ranker developed on fluvial quartz gravel and coarse sands. Dravsko-Ptujsko plain, north-eastern Slovenia (SP database) (Table 5.3).

Distribution According to the DSM25, Rankers are largely associated with Dystric Brown soils in Central Slovenia and in the area of Čičarija to the south-west. Rankers are dominant in the north of the Pohorje massif and north of Drava River and in isolated locations in the area of Donačka Gora, in eastern Slovenia. In these areas, they are associated with Dystric Brown soil. Ranker develops on Quaternary quartz gravel and sand deposits to the east and west of the city of Ptuj. The total extent of Ranker amounts to 4.5% of the country. As a dominant soil type, it extends over 1.5% of Slovenia and over 3.0% of Slovenia when associated with Dystric Brown soil. The extent of Eutric Ranker is associated with Eutric Cambisol (Stepančič et al. 1986a, 1984a) (Fig. 5.12).

Land Use/Land Cover Slovenian Dystric Ranker is predominantly covered with woods and to a lesser extent by grassland, frequently pastures and meadows of lower quality. On Dravsko-Ptujsko plain and Murska Sobota plain Dystric Rankers are under the arable land. The land use of Eutric Ranker is mainly agricultural: orchards, vineyards, and meadows.

5.1.1.3 Class A3. Cambic Soils [A-B-C, A-B-R Sequence of Horizons]

Dystric Brown Soil

Genesis and Soil Profile Development According to the SSC, Dystric Brown soil belongs to the class of Cambic soils, a third group in the evolution of automorphic soils. A usual sequence of horizons is A-Bv-R or A-Bv-C-R. The transitional horizons such as ABv, BvC, or horizons of (Bv) in an initial stage mixed with a large portion of regolith (C) are frequent. In forests, the Ol, Of, and Oh horizons of various depth form on top of the A horizon. In accordance with the SSC, Dystric Brown soil is first of all classified as an acidic soil with low base saturation. The base saturation should be lower than 50% and the pH_{water} lower than 5.5 (Ćirić 1984a; Prus 1983; Urbančič et al. 2005). The genesis of Dystric Brown soil is highly correlated with the parent material. It develops on a variety of acidic, noncarbonate rocks of igneous, metamorphic, and sedimentary origin. Frequent parent materials in Slovenia are tonalite, granodiorite, quartz sandstone, quartz breccia and conglomerate, decalcified flysch, slate, gneiss, as well as Quaternary alluvial deposits of quartz sand and gravel (Prus et al. 1985b; Stepančič et al. 1985). The contact to the parent rock is mainly regolithic and rarely lithic. The primary pedogenetic process is the decomposition of the primary and the formation of the secondary minerals (argilogenesis). The corresponding WRB soil type is Dystric Cambisol (Fig. 5.13).

Table 5.3 A Ranker developed on fluvial quartz gravel and coarse sands. Dravsko-Ptujsko plain, north-eastern Slovenia (SP database)

	Depth		pH		Particle size distribution						P_2O_5	K_2O	N	C	*OS*	*C/N*
	From	To	KCl	$CaCl_2$	Fine sand (0.2–0.05 mm)	Coarse silt (0.05–0.02 mm)	Fine silt (0.02–0.002 mm)	Total silt (0.02–0.002 mm)	*Clay* (<0.002 mm)	``Text. Class	Al extr. (plant-available)		Kyehl-dahl	Walkely Black		
Horizon	cm		cm	-	-	%	%	%	%	%	class	mg/100g	%	%	%	*Ratio*
Ap	0	18	4.4	–	51.8	12.8	21.1	*33.9*	14.3	SaLo-Lo	30.5	12.6	6.3	*0.2*	3.6	*15.0*
AC	18	28	4.5	–	54.5	11.4	20.2	*31.6*	13.9	SaLo-Lo	28.5	11.6	6.3	*0.2*	3.6	*15.0*

	Depth		Basic cations				Acidity	*Bases*	*CEC*	*Base saturation (%)*					
	From	To	Ca	Mg	K	Na	H	*Sum.*		*Total*	*Ca*	*Mg*	*K*	*Na*	*H*
Horizon	cm	cm	meq/100 g soil (1 M NH_4OAc, pH7)							%					
Ap	0	18	2.2	0.1	0.3	0.0	17.1	*2.6*	*19.8*	*13.3*	*11.1*	*0.6*	*1.4*	*0.2*	*86.7*
AC	18	28	2.5	0.1	0.3	0.0	16.9	*2.9*	*19.8*	*14.8*	*12.8*	*0.6*	*1.3*	*0.2*	*85.2*

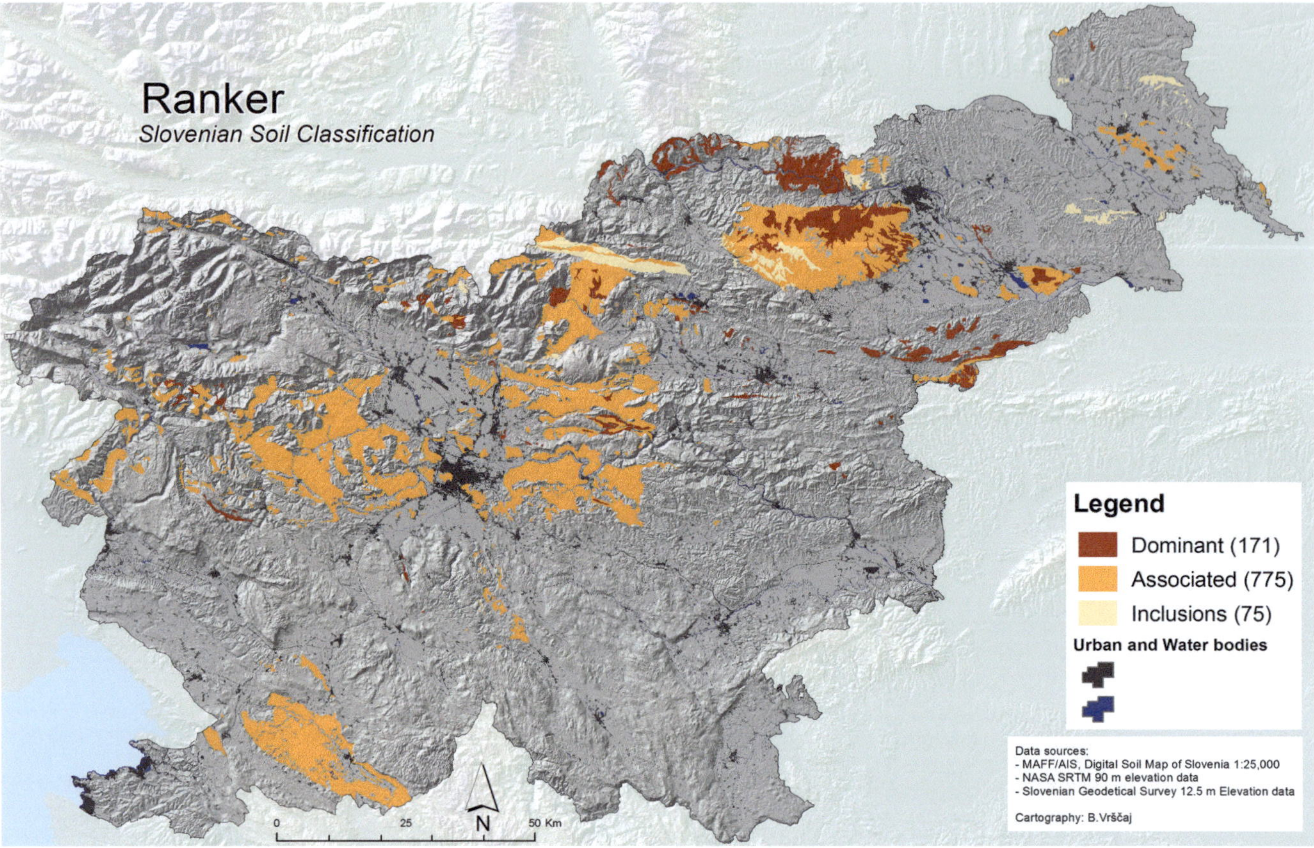

Fig. 5.12 The distribution of Ranker

Dystric Brown soil is a central stage of the soil evolution on noncarbonate parent material: *Lithosol → Ranker → Dystric Brown soil → leached Dystric Brown soil → Brown Podzolized Soil → Podzol.*

Main Properties The high mean annual precipitation in Slovenia (typically 3600–760 mm), less favourable climatic conditions at higher altitudes, and lower mineralization lead to the accumulation of organic matter in A horizons that may become thick. In general, the A horizon of Dystric Brown soil is relatively thin (the median is 11 cm). In forests, the thickness of an A horizon may be thinner, typically 2–4 cm. The thickness of the Bv horizon of shallow Dystric Cambisol is 25 cm. In deep forms, the Bv horizons can be separated into Bv1, Bv2, …, Bv-n mainly due to the morphological properties. In this case, the thickness of Bv1, Bv2, and Bv3 where they exist, is 26, 28, and 22 cm, respectively. The texture of the A and Bv horizons varies significantly in relation to the parent material. The structure of the Bv horizon is in general granular-crumbly and the porosity is high. In the case of long and abundant precipitation, landslides on steep slopes may occur due to the heavy mass of topsoil saturated with water and soggy subsoil. According to the Source: SP database, the median values of the A horizon for pH (KCl), base saturation, and organic matter content are 4.0, 11.5, and 7.6%, respectively. For the Bv horizon, the median values for pH (KCl), base saturation, and organic matter content are 4.1, 12.7, and 1.1%, respectively.

The high mean annual precipitation in the Alps (3600–2000 mm) in combination with cold climatic/microclimatic conditions, the porous and permeable structure, and low base saturation induce the podzolization process. By this, Dystric Brown soil gradually changes to a Dystric Brown soil—Podzolized, Brown Podzolized soil, and, in rare locations, to Podzol. In a large part of the Dravsko-Ptujsko plain and the Murska Sobota plain, Dystric Cambisol forms on Quaternary fluvial, predominantly quartz gravel and sand (Stepančič et al. 1986a, 1984a) (Fig. 5.14). The soil in these areas is shallow, sandy/gravelly, highly permeable, low in organic matter, and acidic; the typical pH (KCl) is 4.4–4.7 in the A horizon, and 4.8–5.1 in the Bv horizon.

Fig. 5.13 Dystric brown soil formed on tonalite, Pohorje, north-eastern Slovenia (*Photo* B. Vrščaj)

Representative Measured Profile Data on a Dystric Brown soil developed on quartz conglomerate, south of the city of Slovenj Gradec (SP database) (Table 5.4).

Distribution According to the DSM25, Dystric Brown soil is largely present as a dominant soil in association with a Dystric Ranker. It mainly appears in central Slovenia. West of the capital, between the cities of Žiri and Kranj, it develops on shale, quartz conglomerate, porphyry, and keratophyre. In a large area east of Ljubljana, it develops on Permian-Carbone slate (Prus et al. 1985b; Stepančič et al. 1985). The main Dystric Brown soil area is north-eastern Slovenia. It is dominant on the Pohorje massif, the Dravsko-Ptujsko plain, and the Murska Sobota plain. As an associated soil type, it is present in the area of Goričko, north-eastern Slovenia and in the hilly area north of Ptuj. In south-western Slovenia in the area of Čičarija, it is the dominant soil in associations with Ranker (Fig. 5.15).

In total, Dystric Brown soil covers 18.6% of Slovenia. As a dominant soil type, it extends over 14.1% of the territory and on 4.5% when it associates with other soils (Fig. 5.16).

Land Use/Land Cover The major land use that covers Dystric Brown soil is forest. Among agricultural land uses, meadows and pastures prevail in hilly and mountainous areas where the farms are oriented to diary production. In two large parts of the Dravsko-Ptujsko plain and the Murska Sobota plain, thin and permeable Dystric Brown soil is ploughed for cereals and maize production (Fig. 5.17).

Fig. 5.14 Fields in the Dystric Cambisol area, Savinja River valley (*Photo* B. Vrščaj)

Table 5.4 Data on a Dystric Brown soil developed on quartz conglomerate, south of the city of Slovenj Gradec

	Depth		pH		Particle size distribution						P2O5	K_2O	N	C	*OS*	*C/N*
	From	To	KCl	$CaCl_2$	Fine sand (0.2–0.05 mm)	Coarse silt (0.05–0.02 mm)	Fine silt (0.02–0.002 mm)	Total silt (0.02–0.002 mm)	*Clay* (<0.002 mm)	Text. Class	Al extr. (plant-available)		Kyehl-dahl	Walkely Black		
Horizon	cm	cm	–	–	%	%	%	%	%	class	mg/100 g		%	%	%	*Ratio*
A	0	8	4.4	–	31.1	17.2	36.8	*54.0*	14.9	SiLo	1.9	6.9	8.9	*0.3*	5.2	*16.1*
ABv	8	24	3.8	–	33.8	14.6	34.5	*49.1*	17.1	SiLo	–	2.9	4.2	*0.2*	2.4	*12.2*
Bv1	24	50	3.8	–	33.6	12.0	32.7	*44.7*	21.7	Lo	–	2.4	2.4	*0.1*	1.4	*12.6*
Bv2	50	70	3.9	–	33.7	15.1	28.5	*43.6*	22.7	Lo	–	–	–	–	–	–
BvC	70	90	–	–	–	–	–	–	–	–	–	–	–	–	–	–

	Depth		Basic cations				Acidity	*Bases*	*CEC*	*Base saturation (%)*					
	From	To	Ca	Mg	K	Na	H	*Sum.*		*Total*	*Ca*	*Mg*	*K*	*Na*	*H*
Horizon	cm	cm	meq/100 g soil (1 M NH_4OAc, pH7)							%					
A	0	8	7.1	2.3	0.2	0.3	17.9	*9.9*	*27.8*	*35.7*	*25.6*	*8.4*	*0.6*	*1.1*	*64.3*
ABv	8	24	3.4	1.1	0.1	0.3	19.2	*4.8*	*24.0*	*20.1*	*14.0*	*4.7*	*0.3*	*1.0*	*79.9*
Bv1	24	50	2.0	0.8	0.1	0.1	17.0	*3.0*	*20.0*	*15.1*	*10.1*	*4.2*	*0.3*	*0.5*	*84.9*
Bv2	50	70	1.5	0.8	0.1	0.1	14.1	*2.5*	*16.6*	*15.3*	*9.3*	*5.0*	*0.4*	*0.6*	*84.7*
BvC	70	90	–	–	–	–	–	–	–	–	–	–	–	–	–

Fig. 5.15 Dystric Brown soil (*Photo* T. Kralj)

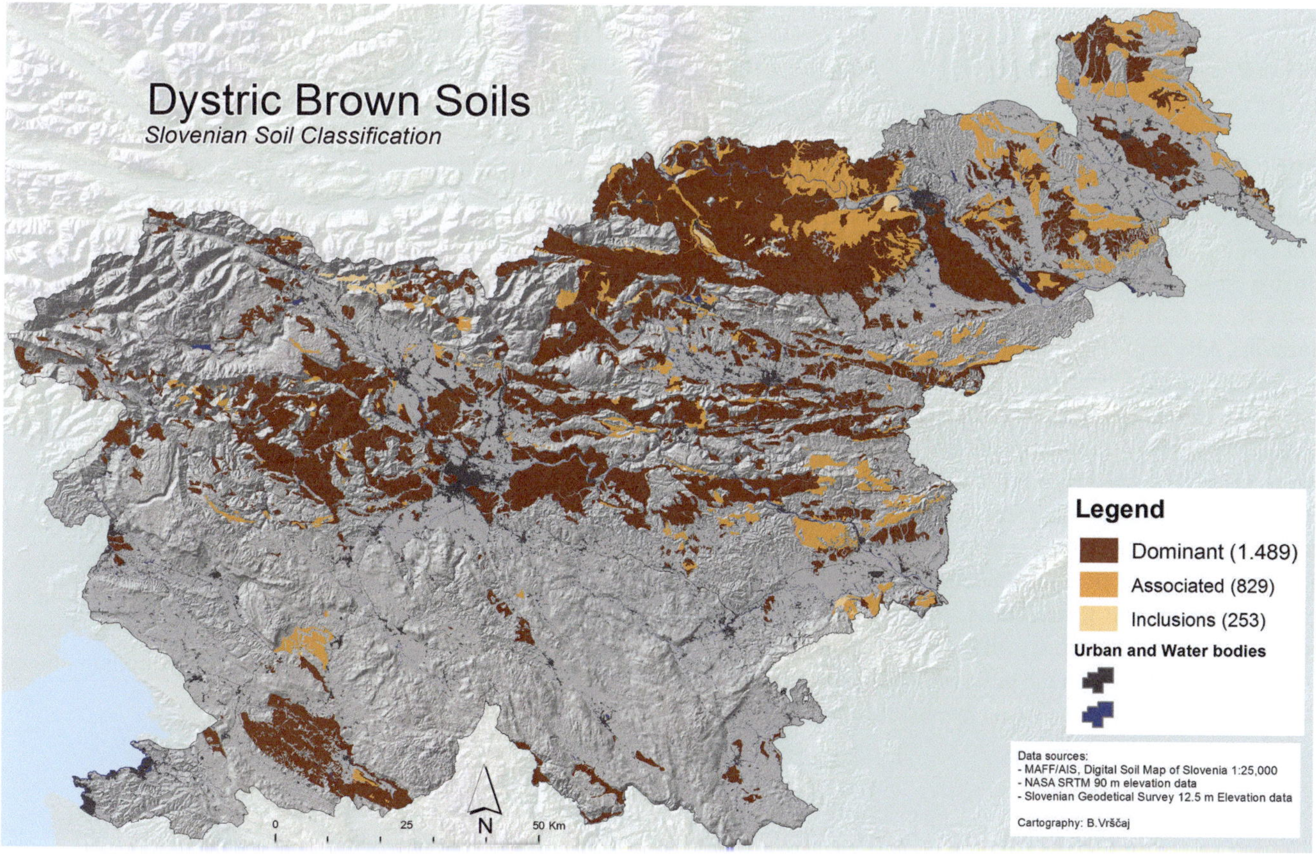

Fig. 5.16 The distribution of the Dystric Brown soil

Fig. 5.17 Dystric Brown soil on Pleistocene loamy-gravely deposits, Savinja River valley (*Photo* B. Vrščaj)

Eutric Brown Soil

Genesis and Soil Profile Developments Eutric Brown soil belongs to the class of cambic soils. Similarly as Dystric Brown soil, it has the following sequence of horizons: A-Bv-R, or A-Bv-C-R, or, rarely, A-Bv-R. The transitional horizons ABv and BvC are frequent. Eutric Brown soil is a neutral to slightly basic soil rich in basic cations. The base saturation should be higher than or equal to 50%, and the pH_{water} is higher than or equal to 5.5. In the forests, a sequence of O horizons forms on the soil surface. As is the case with Dystric Brown soil, Eutric Brown soil highly correlates with the properties of the parent material. It develops on different carbonate or basic rocks. The main parent materials are flysch, marl, carbonate mudstone, and carbonate shale, Quaternary alluvial deposits where limestone and dolomite form the bulk volume of the gravelly and sandy material. It is present on base-rich metamorphic and igneous rocks (diabase). The contact with the parent material is in general regolithic. The primary pedogenetic process is decomposition of the primary and formation of the secondary minerals and leaching of the $CaCO_3$, which is accelerated in the areas with high mean annual precipitation. According to the WRB, the corresponding soil type is Eutric Cambisol.

Eutric Brown soil is a central soil type in the soil evolution on carbonate parent material: *Lithosol* → *Eutric Ranker* → *Eutric Brown soil* → *leached Eutric Brown soil,* → *Luvisol* → *Pseudogley*

Main Properties Eutric Brown soil has a medium thick, mollic A horizon (the median is 16 cm). The median thickness of the first Bv horizon is 16 cm. In the case of deep soil, Bv1, Bv2, … Bv3 horizons that slightly differ as to individual morphological properties are 25, 24, and 29 cm deep, respectively. The texture of the A and Bv horizons largely depends on the parent rock, colluvial processes, and microrelief. A granular and crumbly structure prevails in the Bv horizon. These horizons are loose, and easily friable. Individual peds are distinctively 0.5 cm big and can easily separate—they do not stick together. The porosity of the Bv horizon is high. Abundant precipitation fills the pores and when the subsoil is wet enough it slides down slope. Landslides are frequent in north-eastern Slovenia where Eutric Brown soil forms on clayey substrates such as marl and mudstones. According to the Source: SP database, the median values of the A horizon for pH (KCl), base saturation, and organic matter content are 5.8, 72, and 4.6%, respectively, while for the Bv horizon, they are 6.5, 88, and 1.6%, respectively. In the Ljubljana basin and the Upper and Lower Savinja River valley, Eutric Cambisol forms on Quaternary fluvial predominant quartz gravel and sand. The soil in these areas is shallow, sandy and gravelly, highly permeable, and thus very sensitive to drought (Prus et al. 1985b; Radišek et al. 2014a, b; Rupreht et al. 1985; Stepančič et al. 1986a, 1984a).

Representative Measured Profile The data on a Eutric Brown soil, regolithic, shallow to medium deep, developed on flysch, Goriška Brda, western Slovenia (SP database) (Table 5.5).

Distribution Eutric Brown soil is present within small areas across the country and predominates in the Primorska Region (Vipavska valley and the Brda vineyard area), in the hilly region between the cities of Maribor, Murska Sobota, and Ptuj, north-eastern Slovenia, and in the Ljubljana basin, and eastern Slovenia (Figs. 5.18 and 5.19).

Table 5.5 The data on a Eutric Brown soil, regolithic, shallow to medium deep, developed on flysch, Goriška Brda, western Slovenia

	Depth		pH		Particle size distribution						P2O5	K_2O	N	C	*OS*	*C/N*
	From	To	KCl	$CaCl_2$	Fine sand (0.2 - 0.05 mm)	Coarse silt (0.05–0.02 mm)	Fine silt (0.02–0.002 mm)	Total silt (0.02–0.002 mm)	*Clay* (<0.002 mm)	Text. Class	Al extr. (plant-available)		Kyehl-dahl	Walkely Black		
Horizon	cm	cm	–	–	%	%	%	%	%	Class	mg/100 g		%	%	%	*Ratio*
A	0	13	7.1	–	19.3	36.4	13.7	*50.1*	30.6	–	5.4	29.1	6.2	*0.3*	3.6	*10.6*
BvC	13	40	7.3	–	14.8	41.9	10.1	*52.0*	33.2	–	–	–	1.6	*0.1*	0.9	*7.3*
CBv	40	85	7.5	–	20.6	42.6	8.8	*51.4*	28.0	–	–	–	0.6	*0.1*	0.4	*5.4*

	Depth		Basic cations				Acidity	*Bases*	*CEC*	*Base saturation (%)*					
	From	To	Ca	Mg	K	Na	H	*Sum.*		*Total*	*Ca*	*Mg*	*K*	*Na*	*H*
Horizon	cm	cm	meq/100 g soil (1 M NH_4OAc, pH7)							%					
A	0	13	35.8	1.1	0.5	0.3	2.8	*37.6*	*40.4*	*93.1*	*88.6*	*2.6*	*1.2*	*0.6*	*6.9*
BvC	13	40	34.7	0.7	0.3	0.3	1.0	*36.0*	*37.0*	*97.2*	*93.8*	*1.8*	*0.8*	*0.7*	*2.8*
CBv	40	85	33.3	0.6	0.3	0.2	1.1	*34.3*	*35.4*	*96.9*	*93.9*	*1.6*	*0.8*	*0.6*	*3.1*

Fig. 5.18 The vineyard area of Goriška Brda (*Photo* T. Kralj)

As a dominant soil type, Eutric Brown occupies 7.9% of the national territory, and 5.8% as an associated soil type. In total, it represents 13.7% of the territory and ranks as the fourth most common soil type in Slovenia (Figs. 5.21 and 5.22).

Land Use/Land Cover The prevailing land use on Eutric Brown soil is agriculture. Being rich in calcium, and relatively rich in nutrients, it is mainly ploughed or tilled in another way (Figs. 5.20 and 5.23). In favourable climatic

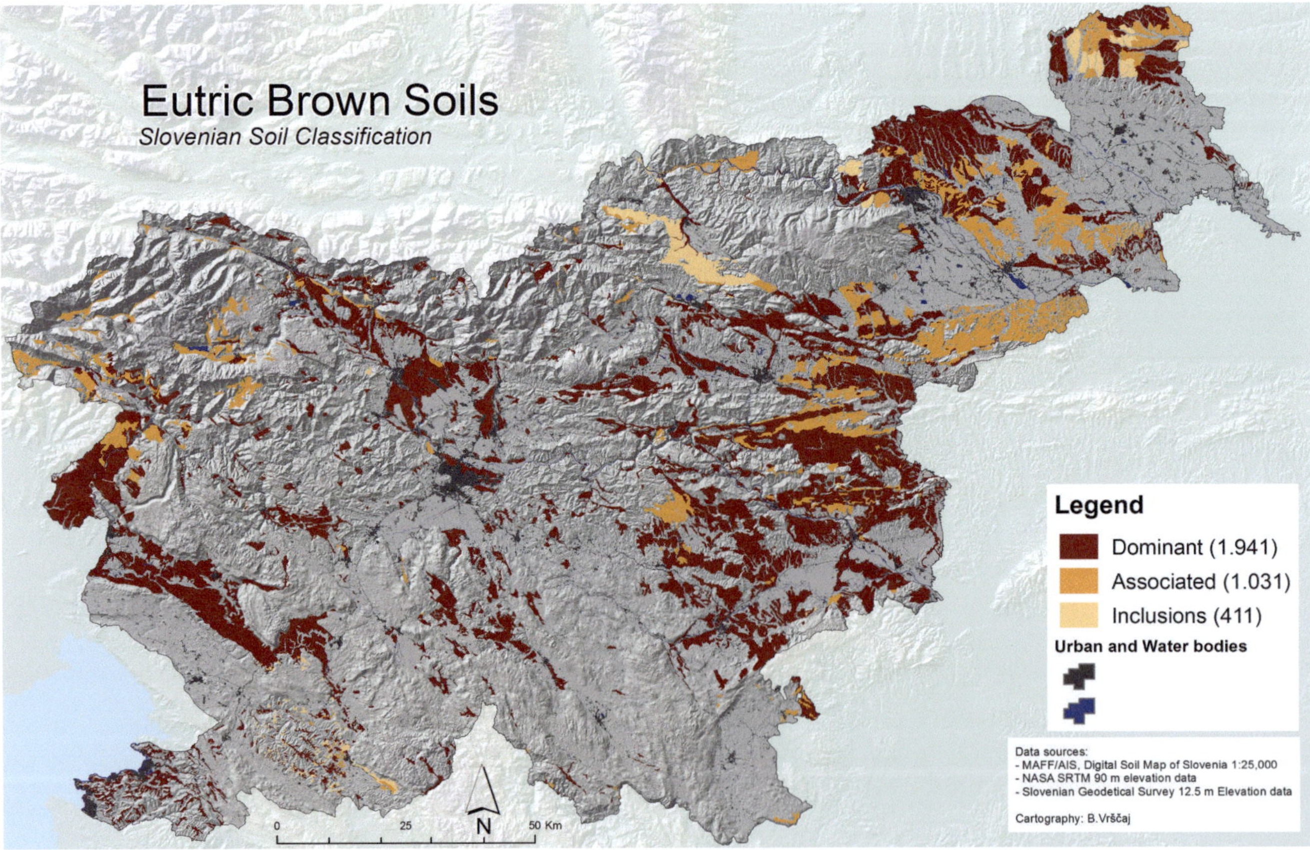

Fig. 5.19 The distribution of the Eutric Brown soil

Fig. 5.20 Shallow Eutric Brown soil on quaternary carbonate gravel (*Photo* B. Vrščaj)

Fig. 5.21 The terrain diversity is high in the hilly Pre-Alpine areas. Very different soil types vary over short distances. The background of the photograph shows a ridge with rock outcrops and deep, colluvial Eutric Brown soil in front (Fig. 5.22) (*Photo* B. Vrščaj)

Fig. 5.22 Eutric Brown soil, deep, colluvial (*Photo* B. Vrščaj)

Fig. 5.23 Eutric Brown soil, Vipava River valley (*Photo* B. Vrščaj)

Fig. 5.24 Secondary carbonates in Calcaric Cambisol, hinterland of the city of Koper (*Photo* B. Vrščaj)

areas and depending on the topography, it is primarily used for vineyards and orchards. Eutric Cambisol is by extent the main vineyard soil in Slovenia. Frequently it is highly anthropogenized and turned into Regosol (Rupreht et al. 1985; Stepančič et al. 1984a, 1986a; Vrščaj 2014b) (Figs. 5.24, 5.25, 5.26 and 5.27).

Kalkocambisol

Genesis and Soil Profile Development Kalkocambisol is the fourth most important Slovenian soil type that belongs to the Cambic class of soils. According to the SSC, Kalkocambisol is a noncalcareous soil that develops on hard limestone and dolomite or any transitional stage of dolomitic limestone. It is neutral to slightly basic soil and rich in base cations. The base saturation is higher than or equal to 50% and the pH_{water} is higher than or equal to 5.5. As already mentioned, the Triassic, Jurassic, and Cretaceous limestone/dolomite in Slovenia are quite pure. The content of insoluble particles is very low, frequently less than 2%. The diagnostic horizon of the Kalkocambisol is a Brz, where the *rz* comes from the Latin *residuum*—i.e. an insoluble and residual mixture of clay and fine silt. After the dissolution of the $CaCO_3$, the residual material forms the main volume of the mineral component of the Brz and A horizons. When it develops on compact and pure limestone, the contact to the parent rock is primarily lithic. The general sequence of horizons in this case is A-Brz-R. Throughout the profile individual rounded and slowly weathering, limestone skeleton is often present. The sequence of horizons of Kalkocambisol that develops on dolomite in general slightly differs. The total skeleton content can be high, but the size of the skeleton is in general smaller. The common sequence of horizons of the Kalkocambisol formed on dolomite is A-Brz-C-R. In the mollic A horizon, the skeleton might be of significant volume (Vrščaj 2014a). In the forests and unmanaged grasslands where the grass is not harvested, the A horizon is overlaid by Ol, Of, and Oh horizons of various thickness (Lobnik et al. 1996b, 1987; Prus et al. 1985b; Radišek et al. 2014a; Stepančič et al. 1987a, b, 1984b; Stritar 1990b).

Fig. 5.25 Cornfield on Eutric Brown soil, Gorenjska region (*Photo* B. Vrščaj)

Fig. 5.26 Earthworm excrement on Ca-rich Eutric Brown soil (*Photo* B. Vrščaj)

The SSC recognizes two sub-types: typical and leached Kalkocambisol. In the typical Kalkocambisol, the upper Brz horizon shows no or very limited signs of eluviation. In the leached Kalkocambisol, the top Brz horizon can have less clay and/or is paler than the underlying horizons and/or has a lower pH and a lower base saturation. The underlying illuvial versions of Brz or Bt horizons accumulate the leached material and substances (Fig. 5.34).

The primary pedogenetic process is the dissolving of the limestone/dolomite, accumulation, and alteration of clay

Fig. 5.27 Allolobophora chlorotica on Eutric Brown soil (*Photo* T. Kralj)

minerals, leaching of clay, organic matter, and to a certain extent Mn and Fe. In the late stage of development, black Fe and Mn accumulations on the surface of peds occur. The next stage of the profile development is Luvisol.

Kalkocambisol is the central soil type of the evolutionary series of soils that develop on hard limestone and dolomite: *Lithosol* → *Rendzina* → *Kalkocambisol* → *Leached Kalkocambisol* → *Luvisol* (*Chromic Luvisol*) (Fig. 5.28).

Fig. 5.28 Leached Kalkocambisol—typical rounded forms of limestone weathering. Draganja sela, Gorjanci, south-eastern Slovenia (*Photo* T. Kralj)

Main Properties The Brz horizon features an important property that crucially characterizes this soil type and steers the eluviation-illuviation process, the air-water regime, AWC, and other properties including fertility, productivity, and cultivation practices of the Kalkocambisol. The clay content of the Brz is high. According to the Source: SP database, the median values for sand, silt, and clay in the Brz horizon are 8.8, 47, and 41%, respectively. In the silt fraction, fine silt prevails. The Brz has a very characteristic polyhedral blocky structure (Fig. 5.29). The peds are angular and the edges are sharp. They have been created by the wetting and drying of the Brz horizon. The structure of peds is stable mainly due to a high clay and Ca^{2+} content. The structure of peds resists precipitation throughout the profile. A stable and permanent system of macropores is established that drains the entire soil profile in a short period of time. Even though mineral and clayey, the subsoil is permeable and well aerated. In total, a highly clay content soil shows no traces of hydromorphism. Very good percolation and rapid removal of water enable tillage a few hours after

Fig. 5.29 The characteristic polyhedral blocky structure of Kalkocambisol (*Photo* B. Vrščaj)

precipitation stops. The high clay content suggests a good AWC. In reality, in dry periods the clayey Brz horizons become cracked with numerous fine cracks that go deep into the soil profile.

One of the consequences of the specifics of the Brz horizon and Kalkocambisol is reflected in the specific cultural landscape. Namely, the slopes are terraced. The terraces that have been established over centuries by ploughing the Kalkocambisol need no support; no walls are built. The terraces are stable and practically no maintenance is needed. The erosion is minimal to modest especially when compared to the erosion in areas of Eutric or Dystric Cambisols. The Dolenjska region is known for such terraces (Figs. 5.30 and 5.31) and dolines (Fig. 5.32).

The Brz is low in phosphorus and medium-low in potassium. The median values for plant-available phosphorus and potassium in the Brz horizon are 1.1 and 7.9 mg/100 g dry soil, respectively. The median values for the pH are 6.3, organic matter content 1.1%, and the base saturation 75%.

The soil depth is another very characteristic feature. Namely, within a very short distance (i.e. 10–20 m) the soil depth can vary from a rock outcrop to several-metre-deep pockets filled with residual fine silt and clay (Fig. 5.35). These pockets act as storage for water and nutrients. On the surface, the area shows marginal production potentials, but the vegetation with a deep rooting system successfully utilizes the supply of water and nutrients. In such areas, the forests can be very productive (Table 5.6).

Distribution Kalkocambisol is the main soil type in southern Slovenia. As a dominant soil type, it is present in the Notranjska and Dolenjska regions. On the elevated plateau of Trnovski forest and the Trst-Komen plateau, it is associated with Rendzina. In east-central Slovenia, Kalkocambisol is the dominant soil type and varies with the associated Rendzina within patches that are small in extent (Fig. 5.33).

As a dominant soil type, Kalkocambisol covers 7.9% of the territory and 5.8% when it is associated. In total, Kalkocambisol represents 13.7% of the territory, making it the fourth most common soil type in Slovenia.

Land Use/Land Cover The prevailing land use depends mainly on the slope, altitude, soil depth, and surface rockiness. Deep Kalkocambisols are frequently terraced and used

Fig. 5.30 Terraces in the Kalkocambisol area (*Photo* T. Kralj)

Fig. 5.31 Terraces in the Kalkocambisol area (*Photo* T. Kralj)

as arable land. Skeletic, shallow, and located on unfavourable terrain, Kalkocambisol is used primarily for meadows (Stritar 1990b). The areas that feature a high degree of rock outcropping are under forests. Kalkocambisol as well as Rendzina are the soil types regarding which land abandonment and spontaneous afforestation are intensive.

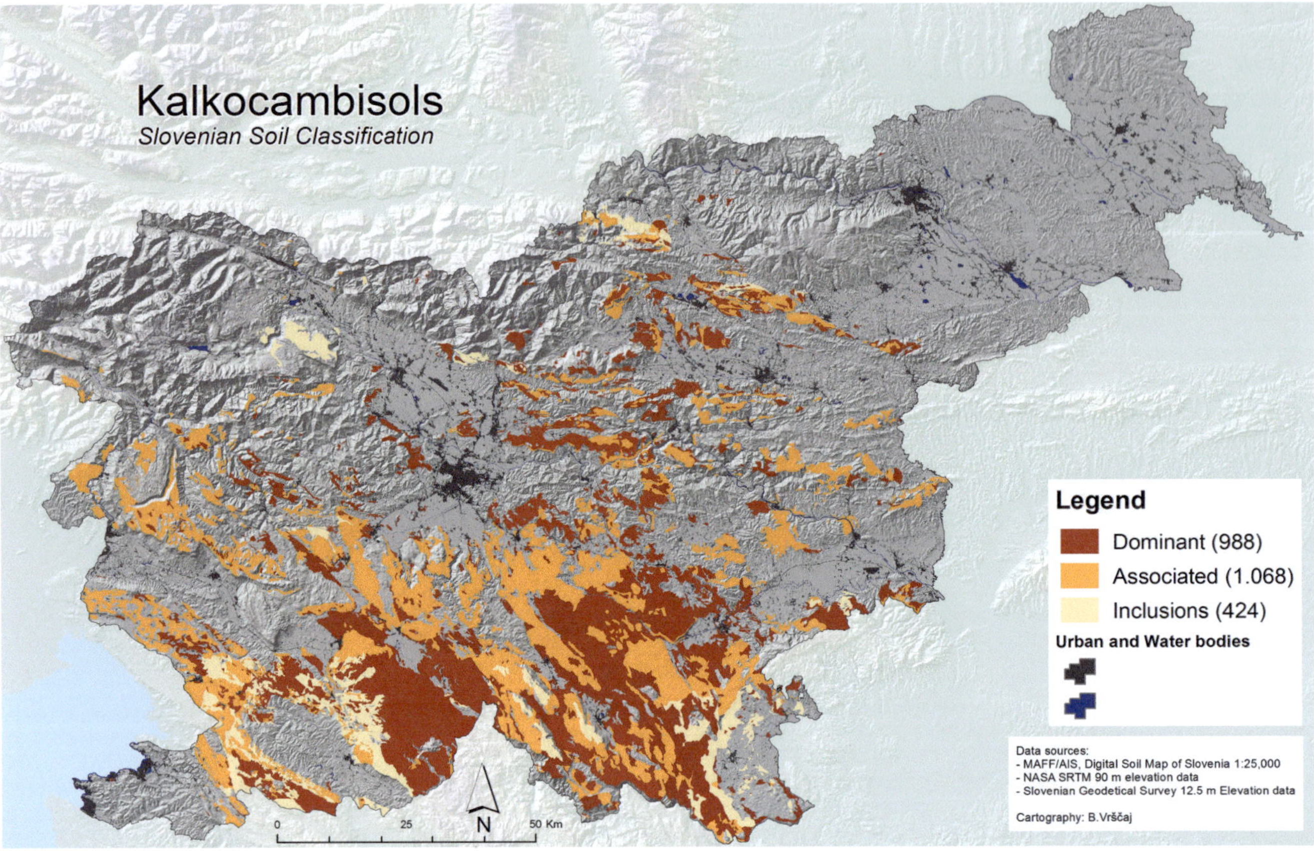

Fig. 5.33 The distribution of the Kalkocambisol

Fig. 5.32 Dolines in the Kalkocambisol area (*Photo* T. Kralj)

Fig. 5.34 Leached Kalkocambisol, Juršče, south-western Slovenia (*Photo* B. Vrščaj)

Fig. 5.35 Rock outcrops in a beech forest of the Kalkocambisol area (*Photo* B. Vrščaj)

Table 5.6 The data on a Kalkocambisol developed on dolomite, Dolenjska, south-eastern Slovenia (SP database)

	Depth		pH		Particle size distribution						P2O5	K_2O	N	C	*OS*	*C/N*
	From	To	KCl	$CaCl_2$	Fine sand (0.2–0.05 mm)	Coarse silt (0.05–0.02 mm)	Fine silt (0.02 - 0.002 mm)	Total silt (0.02–0.002 mm)	*Clay* (<0.002 mm)	Text. Class	Al extr. (plant-available)		Kyehl-dahl	Walkely Black		
Horizon	cm	cm	–	–	%	%	%	%	%	Class	mg/100 g		%	%	%	*Ratio*
Ap	0	18	6.9	–	11.4	14.0	39.7	*53.7*	34.9	SiClLo	10.8	42.3	6.4	–	3.7	–
A	18	29	7.0	–	14.0	11.3	37.2	*48.5*	37.5	SiClLo	6.1	38.2	5.5	–	3.2	–
Brzl	29	61	6.9	–	13.5	10.5	30.3	*40.8*	45.7	SiCl	–	–	2.3	–	1.3	–
Brz2	61	90	7.0	–	7.7	2.2	18.0	*20.2*	72.1	Cl	–	–	–	–	–	–

	Depth		Basic cations				Acidity	*Bases*	*CEC*	*Base saturation (%)*					
	From	To	Ca	Mg	K	Na	H	*Sum.*		*Total*	*Ca*	*Mg*	*K*	*Na*	*H*
Horizon	cm	cm	meq/100 g soil (1 M NH_4OAc, pH7)							%					
Ap	0	18	17.8	6.4	1.0	0.1	5.2	*25.3*	*30.5*	*83.0*	*58.3*	*21.1*	*3.3*	*0.3*	*17.0*
A12	18	29	17.1	6.4	0.8	0.1	5.2	*24.4*	*29.6*	*82.4*	*57.7*	*21.6*	*2.8*	*0.3*	*17.6*
Blrz	29	61	13.5	6.0	0.4	0.1	5.2	*20.0*	*25.1*	*79.5*	*53.8*	*23.7*	*1.6*	*0.4*	*20.5*
B2rz	61	90	18.9	8.6	0.4	0.1	5.9	*28.0*	*33.9*	*82.6*	*55.6*	*25.4*	*1.2*	*0.3*	*17.4*

5.1.2 Order H. Hydromorphic Soils

Hydromorphism is the basic characteristic of soils in this order. Such soil is saturated with groundwater or influenced by the surface water long enough to develop reducing conditions at various depths of the soil profile. As a result of the long-term presence of water, gleyic properties are expressed. The water fundamentally influences the soil properties, the suitability for different uses, and the productivity. When drained, some of the hydromorphic soil can be used for agriculture.

5.1.2.1 Class H2. Pseudogley Soils [A–E–Bg–C Sequence of Horizons]

Pseudogley

Genesis and Soil Profile Development The SSC classifies Pseudogley among the soils affected by stagnant rainwater in the upper part of the profile. The sequence of horizons is A-E-Bg or A-g-Bg. The basic criteria according to the SSC are as follows: stagnation of water in the upper part of the profile, which overlies a predominantly silty, loamy, or clayey, dense, and compact, very low to practically impermeable Bg horizon. Regarding the position in the relief, the SSC distinguishes between two sub-types: the Pseudogley formed on gentle slopes and the Pseudogley formed on flat, silty, clayey bottoms of valleys or Pleistocene-Pliocene silty-clayey alluvial terraces. In relation to the pH and base saturation, Pseudogley is classified as dystric or eutric. The primary pedogenetic processes are redox processes and the stagnation of water, seasonally drying out, and the wetting of the topsoil (Oset 1987; Radišek et al. 2014a, b; Stepančič et al. 1986a, 1984a; Stritar 1991) (Fig. 5.36).

Main Properties The topsoil of the Pseudogley is composed of an A horizon and an E or a g horizon. The E horizon features a coarser texture and a lighter, paler colour. Both the A and the E may be dense but are still crumbly and

Fig. 5.36 Pseudogley, Dystric, Savinja River valley, central Slovenia (*Photo* B. Vrščaj)

Fig. 5.37 A large concretion in Pseudogley (*Photo*: B. Vrščaj)

hard and significantly less dense than the underlying horizons. Pseudogley is largely characterized by the subsoil—the properties of the Bg horizon. It is primarily silty or silty-clayey, dense, compact, and practically impermeable to water and air. The stagnating water causes intensive mottles and the presence of Fe and Mn concretions. In general, mottles define the colour pattern of the horizon. In rarer cases, the mottles are less expressed and Fe and Mn concretions prevail. Concretions are of various size; typically 2–5 mm in diametre; concretions of 2 cm and bigger are not rare (Fig. 5.37).

Reductimorphic conditions are present primarily in cracks and macrospores between the aggregates while the inner parts of the aggregates are oxidized. Being too dense and impermeable to water percolation, the Bg at the same time prevents ascendant water flow. This is an additional reason why the topsoil dries out in summer months. It becomes hard and the cracks that appear due to shrinking tear the roots. The rooting depth extends to the upper boundary of the Bg horizon; therefore, the volume of the soil available to roots is limited. The aerobic and anaerobic processes alternate in the topsoil in accordance with the wet or dry seasons. In the wet season, the soil is too wet while in summer months it gets too dry for successful agricultural production. The time appropriate for tillage is limited. Pseudogley is a distinctly inconvenient soil type for agriculture (Figs. 5.38 and 5.39).

Representative Measured Profile Data on a Dystric Pseudogley on a gentle slope, Črni Les, north-eastern Slovenia (SP database) (Table 5.7).

Distribution Pseudogley is present in Slovenia on a total extent of 3.8% of the country, of which 2.4% on slopes and 1.4% on plains. The main Pseudogley areas are the borders and terraces of large basins and valleys. As a dominant soil type, it is present in the Savinja River valley, at the edges of the Murska plain, and all north-eastern Slovenia. It surrounds an area of Hypogley and Eutric Brown soils in the Krško basin, south-eastern Slovenia. To a minor extent, it is present in the Vipava valley in the west. The prevailing land use is hydrophilic grassland (Fig. 5.40).

Land Use/Land Cover Due to the favourable relief, Pseudogley was used for agriculture. The fields on slopes are ploughed downslope to prevent the stagnation of water. Orchards are a quite common land use on Pseudogley. Careful management and tillage is required to achieve satisfactory yields. The prevailing agricultural land use on Pseudogley is meadows.

Fig. 5.38 Driving over Pseudogley in its wet stage (*Photo* B. Vrščaj)

Fig. 5.39 Pseudogley is often used for orchards, Savinja River valley (*Photo* B. Vrščaj)

5.1.2.2 Class H3. Gleysol [A–G Sequence of Horizons]

Hypogley

Genesis and Soil Profile Development According to the SSC, Hypogley is characterized as a groundwater-affected soil. The gleyic horizons are located at the bottom of the profile, typically 50–90 cm deep. The idealized sequence of horizons is A-Go-Gr or A-Bv-Go-Gr. The reductimorphic Gr horizon that occurs at a depth of less than 100 cm is characterized by almost permanently wet conditions. The groundwater is present in the Gr horizon the entire year. The oxymorphic Go horizon is drained in the dry part of the year and is therefore seasonally aerated. In Slovenia, Hypogley develops on river sediments, on loamy-clayey Pleistocene/Pliocene terraces, and on valley bottoms where clays and loams accumulate. In general, Hypogley in Slovenia develops

Table 5.7 Data on a Dystric Pseudogley on a gentle slope, Črni Les, north-eastern Slovenia

	Depth		pH		Particle size distribution						P2O5	K_2O	N	C	*OS*	*C/N*
	From	To	KCl	$CaCl_2$	Fine sand (0.2–0.05 mm)	Coarse silt (0.05–0.02 mm)	Fine silt (0.02–0.002 mm)	Total silt (0.02 - 0.002 mm)	*Clay* (<0.002 mm)	Text. Class	Al extr. (plant-available)		Kyehl-dahl	Walkely Black		
Horizon	cm	cm	–	–	%	%	%	%	%	Class	mg/100 g		%	%	%	*Ratio*
Aa	0	18	4.1	–	14.3	31.7	38.9	*70.6*	15.1	SiLo	1.2	2.9	2.4	*1.1*	1.4	*1.3*
E	18	41	4.0	–	13.8	31.3	37.0	*68.3*	17.9	SiLo	–	2.4	1.4	*0.1*	0.8	*8.1*
g	41	55	3.9	–	11.9	32.1	35.6	*67.7*	20.4	SiLo	–	–	1.1	–	0.6	–
Bg1	55	80	4.0	–	11.0	32.1	37.5	*69.6*	19.4	SiLo	–	–	–	–	–	–

	Depth		Basic cations				Acidity	*Bases*	*CEC*	*Base saturation (%)*					
	From	To	Ca	Mg	K	Na	H	*Sum.*		*Total*	*Ca*	*Mg*	*K*	*Na*	*H*
Horizon	cm	cm	meq/100 g soil (1 M NH_4OAc, pH7)							%					
Aa	0	18	0.3	0.2	0.1	0.0	14.3	*0.5*	*14.8*	*3.6*	*1.8*	*1.1*	*0.5*	*0.2*	*96.4*
E	18	41	0.3	0.1	0.1	0.0	12.8	*0.5*	*13.3*	*3.7*	*2.0*	*0.9*	*0.5*	*0.3*	*96.3*
g	41	55	0.8	1.4	0.1	0.1	12.8	*2.4*	*15.2*	*15.7*	*5.5*	*9.2*	*0.5*	*0.4*	*84.3*
Bg1	55	80	2.9	3.5	0.1	0.1	10.9	*6.5*	*17.4*	*37.4*	*16.5*	*20.1*	*0.3*	*0.5*	*62.6*

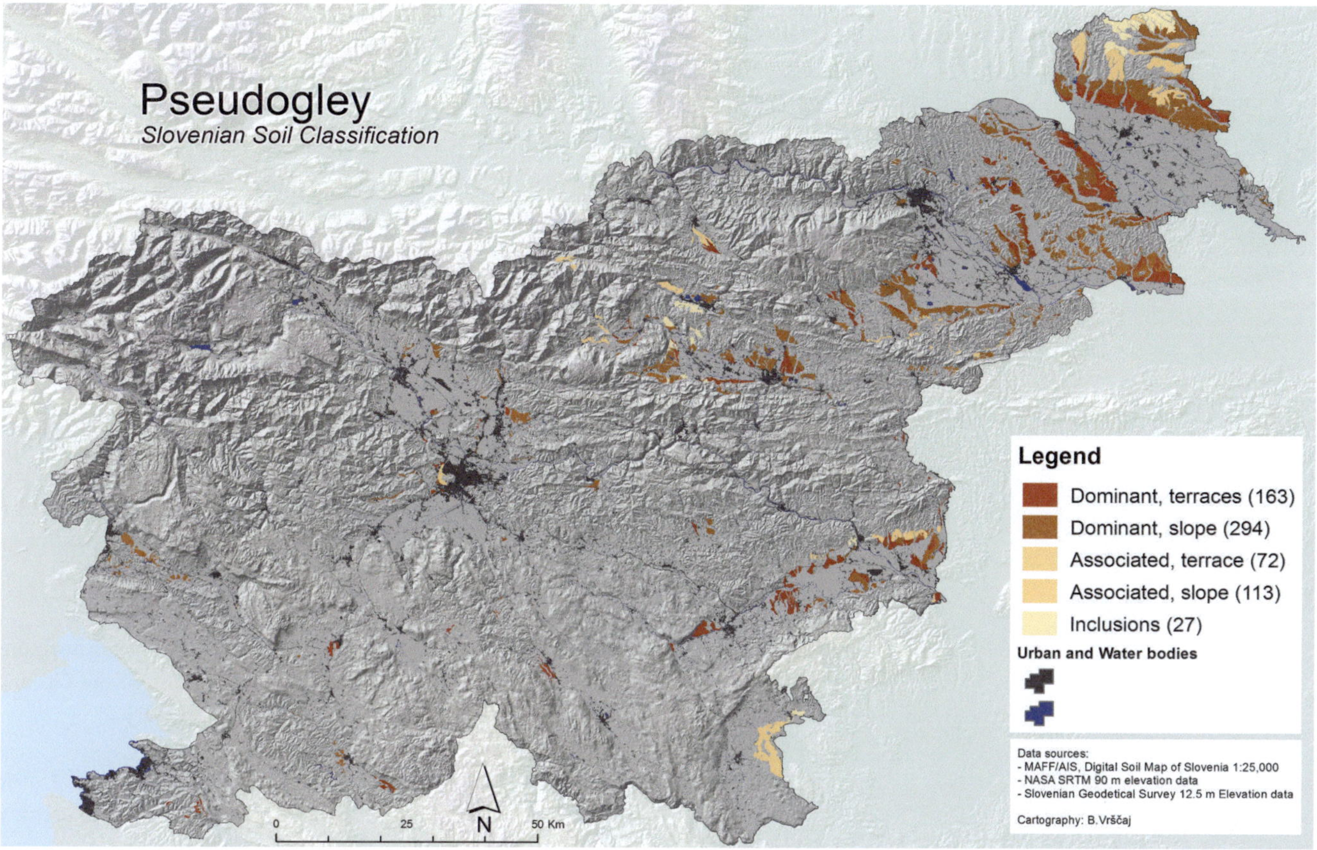

Fig. 5.40 The distribution of Pseudogley

on a heavy-textured material. Sandy Hypogley is present to a very limited extent (Lobnik et al. 1995b; Prus et al. 1986a, b, 1985a; Stepančič et al. 1987b, 1984a) (Fig. 5.41).

Main Properties Hypogley is a heavy-textured soil. The A horizon on the top may or may not show signs of hydromorphism in the soil, especially in the event Go and Gr are below 60 cm. The properties of the A horizon vary in relation to the specifics of the parent material and the land use. The Go and Gr horizons predominantly have a silty-clayey texture. In the Go horizon, the colour of the matrix reflects the oxidizing conditions. The grey areas occupy less than 50% of the exposed surface of the horizon. The moisture of the horizon largely depends on the time of the year. In the summer, it is fresh to moist, brittle, and, rarely, malleable and plastic. In the dry period of the year, it can be thick, a bit brittle, and less plastic. In the Gr horizon, blue-green-grey colours predominate. The horizon is wet or moist, plastic, malleable and can be sticky. In the dry period of the year, it can be dense, thick, less plastic, and a bit brittle. The reddish-brown areas are less present and are mainly situated around rare cracks and root tunnels.

The SSC distinguishes between two sub-types: Eutric Hypogley and Dystric Hypogley. Very similar as regards morphological properties, they vary as to chemical characteristics, mainly acidity and base saturation.

The median values for the pH(KCl), silt, clay, organic matter, and the base saturation for Dystric Hypogley are the following:

- A horizon: 3.9, 51, 24, 6.4, and 44%,
- Go horizon: 4.1, 53, 35, 2.1, and 60%, and
- Gr horizon: 5.8, 39, 36, 2.5, and 77%, respectively.

For Eutric Hypogley, the median values for the pH(KCl), silt, clay, organic matter, and the base saturation are:

- A horizon: 5.2, 59, 20, 8.1, and 64%,
- Go horizon: 5.9, 55, 26, 2.9, and 76%, and
- Gr horizon: 5.8, 39, 36, 2.5, and 77%, respectively.

Fig. 5.41 Eutric Hypogley, Letuš, Savinja River valley (*Photo* B. Vrščaj)

Representative Measured Profile Data on a Eutric Hypogley from the Pesnica River, north-eastern Slovenia (SP database) (Table 5.8).

Distribution Hypogley is dominant mainly in the river valleys and basins of the Murska Sobota plain, the Ščavnica and Pesnica Rivers, and south of the Dravsko-Ptujsko plain, all north-eastern Slovenia. Smaller Hypogley areas are scattered around the country, following the courses of rivers that flow through areas of heavy-textured parent material such as flysch, marl, and shale. Big Hypogley areas can be found on the bottoms of the intermittent Cerknica Lake and other karst poljes, and south of the city of Ilirska Bistrica, in southern Slovenia (Fig. 5.42).

As a dominant soil type, Hypogley covers 2.6% of the territory and 0.7% when it is associated. In total, it covers 3.3% of the territory Slovenia and ranks as the most common hydromorphic soil type in the country (Fig. 5.43).

Land Use/Land Cover The dominant land use is grassland, arable land, and forests. In years of droughts, Hypogley to a certain extent compensates for the decrease in yields on shallow and sandy Cambisols. Average to low yields in normal or wet years are higher in dry years.

Amphigley

Genesis and Soil Profile Development Amphigley is characterized by the excessive presence of surface and groundwater at the same time. Two reductimorphic Gr horizons are present in the soil, close to the top and at the bottom of the profile. The idealized sequence of horizons is A- Gr-Go-Gr. The A horizon might show traces of hydromorphism and in such case can be marked as Aa (Prus et al. 1986a, 1985a; Stepančič et al. 1987b; Stritar 1991).

Main Properties The properties of the Go and Gr horizons of the Amphigley are similar to the properties of the Go and Gr horizons of Hypogley. The soil is rarely dry.

Representative Measured Profile Data on a Eutric Amphigley from the intermittent Cerknica Lake, central Slovenia (SP database) (Table 5.9).

Distribution The extent of Amphigley is only 0.72% of the territory. As a dominant soil type, it is present in the area of Ljubljana moor, central Slovenia, Krakovski forest, west of Novo Mesto, and south of the cities Slovenska Bistrica and Lendava, north-eastern Slovenia. The dominant land uses are hydrophilic grassland, moor, and forest (Figs. 5.44, 5.45, 5.46 and 5.47).

Epigley

Genesis and Soil Profile Development According to the SSC, Epigley is characterized as a soil affected by surface waters. The gleyic horizons are located at the top of the soil profile. The sequence of horizons is A-Gr-Go. The reductimorphic Gr horizon occurs at a depth of between apx. 12 and 25 cm and is characterized by almost permanently wet conditions. Groundwater gleyization is not present. The A horizon frequently shows traces of hydromorphism and in such case is marked as Aa.

Main Properties The properties of Go and Gr horizons of Epigley are similar to the properties of the Go and Gr

Table 5.8 Data on a Eutric Hypogley from the Pesnica River, north-eastern Slovenia

	Depth		pH		Particle size distribution						P2O5	K_2O	N	C	*OS*	*C/N*
	From	To	KCl	CaCl2	Fine sand (0.2–0.05 mm)	Coarse silt (0.05–0.02 mm)	Fine silt (0.02–0.002 mm)	Total silt (0.02–0.002 mm)	*Clay* (<0.002 mm)	Text. Cl ass	Al extr. (plant-available)		Kyehl-dahl	Walkely Black		
Horizon	cm	cm	–	–	%	%	%	%	%	Cla ss	mg/100 g		%	%	%	*Ratio*
A1	0	6	5.1	–	18.6	15.0	38.9	*53.9*	27.5	SiClLo	4.0	10.8	9.7	*0.4*	5.6	*12.5*
Ago	6	28	4.8	–	16.5	6.3	42.4	*48.7*	34.8	SiClLo	–	4.6	3.9	*0.3*	2.3	*9.0*
Go1	28	56	5.3	–	14.0	11.1	38.0	*49.1*	36.9	SiClLo	–	3.8	2.2	*0.2*	1.3	*7.5*
Go2	56	70	5.7	–	15.3	17.8	20.0	*37.8*	46.9	–	–	–	–	–	–	–

	Depth		Basic cations				Acidity	*Bases*	*CEC*	*Base saturation (%)*					
	From	To	Ca	Mg	K	Na	H	*Sum.*		*Total*	*Ca*	*Mg*	*K*	*Na*	*H*
Horizon	cm	cm	meq/100 g soil (1 M NH_4OAc, pH7)							%					
A1	0	6	20.4	6.7	0.3	0.3	13.4	*27.7*	*41.1*	*67.4*	*49.7*	*16.4*	*0.6*	*0.6*	*32.6*
Ago	6	28	20.2	6.0	0.1	0.2	11.6	*26.5*	*38.1*	*69.6*	*53.1*	*15.7*	*0.3*	*0.5*	*30.4*
Go1	28	56	17.9	5.8	0.1	0.2	8.4	*24.0*	*32.4*	*73.9*	*55.2*	*17.9*	*0.3*	*0.5*	*26.1*
Go2	56	70	14.0	5.1	0.1	0.2	5.8	*19.4*	*25.2*	*76.8*	*55.6*	*20.4*	*0.3*	*0.6*	*23.2*

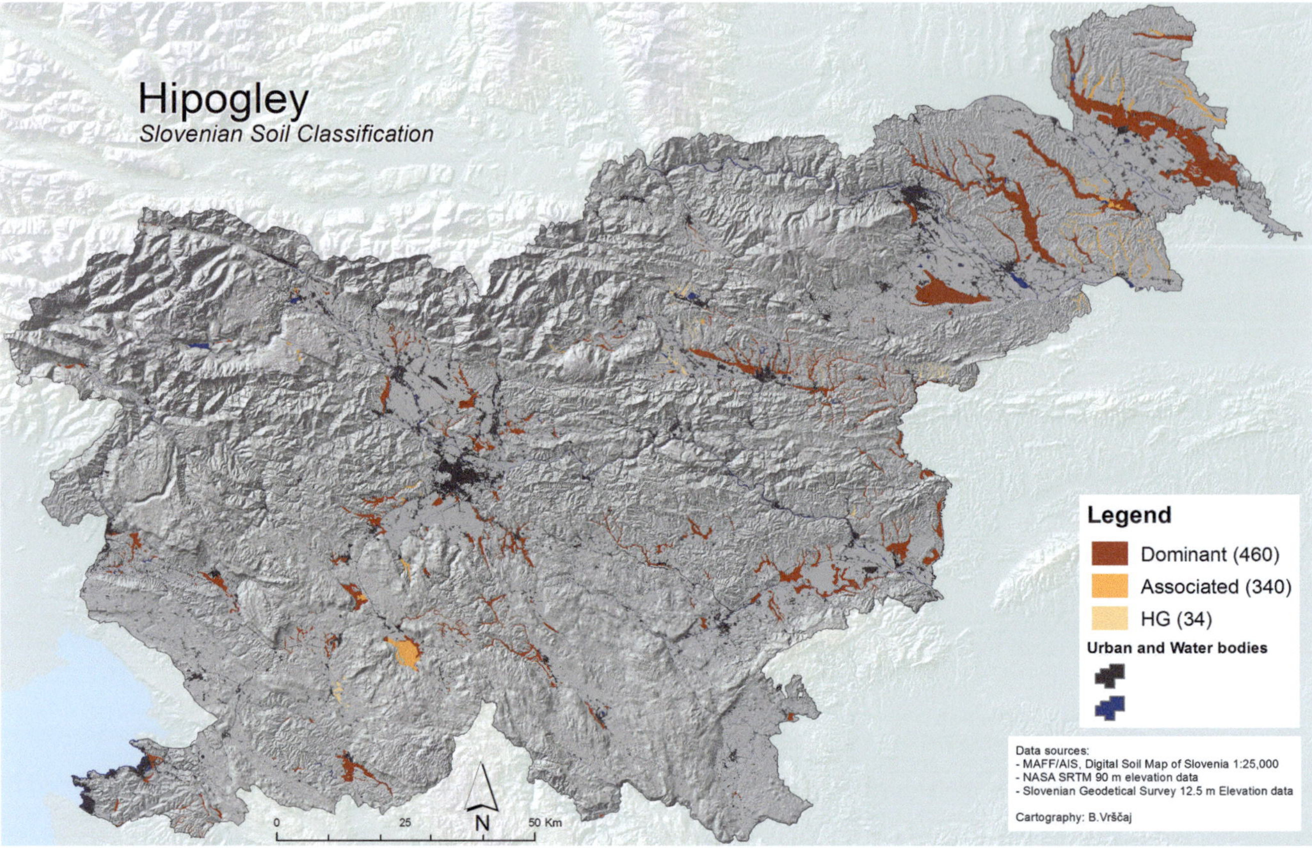

Fig. 5.42 The Hypogley areas in Slovenia

Fig. 5.43 A Hypogley area in the Vipava River valley, western Slovenia (*Photo* B. Vrščaj)

Table 5.9 Data on a Eutric Amphigley from the intermittent Cerknica Lake, central Slovenia

	Depth		pH		Particle size distribution						P2O5	K_2O	N	C	*OS*	*C/N*
	From	To	KCl	CaCl2	Fine sand (0.2--0.05 mm)	Coarse silt (0.05–0.02 mm)	Fine silt (0.02–0.002 mm)	Total silt (0.02–0.002 mm)	*Clay* (<0.002 mm)	Text. Cl ass	Al extr. (plant-available)		Kyehl-dahl	Walkely Black		
Horizon	cm	cm	–	–	%	%	%	%	%	Cla ss	mg/100 g		%	%	%	*Ratio*
Aa	0	12	–	6.7	8.2	8.5	42.7	*51.2*	40.6	–	2.5	15.8	16.1	*0.7*	9.3	*13.3*
GoGr	12	35	–	7.4	5.2	8.6	33.0	*41.6*	53.2	SiCl	0.3	12.6	4.4	*0.3*	2.5	*10.0*
Go	35	60	–	7.3	2.9	8.7	28.8	*37.5*	59.6	–	–	–	2.7	*0.1*	1.6	*13.3*
Gr	60	–	–	7.4	3.9	7.5	28.3	*35.8*	60.3	Cl	–	–	1.6	*0.1*	0.9	*10.0*

	Depth		Basic cations				Acidity	*Bases*	*CEC*	*Base saturation (%)*					
	From	To	Ca	Mg	K	Na	H	*Sum.*		*Total*	*Ca*	*Mg*	*K*	*Na*	*H*
Horizon	cm	cm	meq/100 g soil (1 M NH_4OAc, pH7)							%					
Aa	0	12	47.8	7.2	0.4	0.3	8.9	*55.6*	*64.4*	*86.2*	*74.1*	*11.1*	*0.6*	*0.4*	*13.8*
GoGr	12	35	47.1	4.3	0.3	0.2	5.4	*51.9*	*57.3*	*90.5*	*82.1*	*7.6*	*0.6*	*0.3*	*9.5*
Go	35	60	41.7	5.2	0.3	0.2	7.4	*47.4*	*54.8*	*86.4*	*76.0*	*9.5*	*0.6*	*0.3*	*13.6*
Gr	60	–	38.8	5.3	0.4	0.1	4.8	*44.7*	*49.5*	*90.3*	*78.5*	*10.7*	*0.8*	*0.3*	*9.7*

Fig. 5.44 A field in an Amphigley area. Ponding frequently occurs. In spite of the favourable topography, grassland predominates (*Photo* B. Vrščaj)

Fig. 5.45 An Amphigley field. The soil is saturated with water, but the surface is dried by strong spring sun and wind. The soil structure is unfavourable (*Photo* B. Vrščaj)

Fig. 5.46 An Amphigley field on the Murska Sobota plain, north-eastern Slovenia (*Photo* B. Vrščaj)

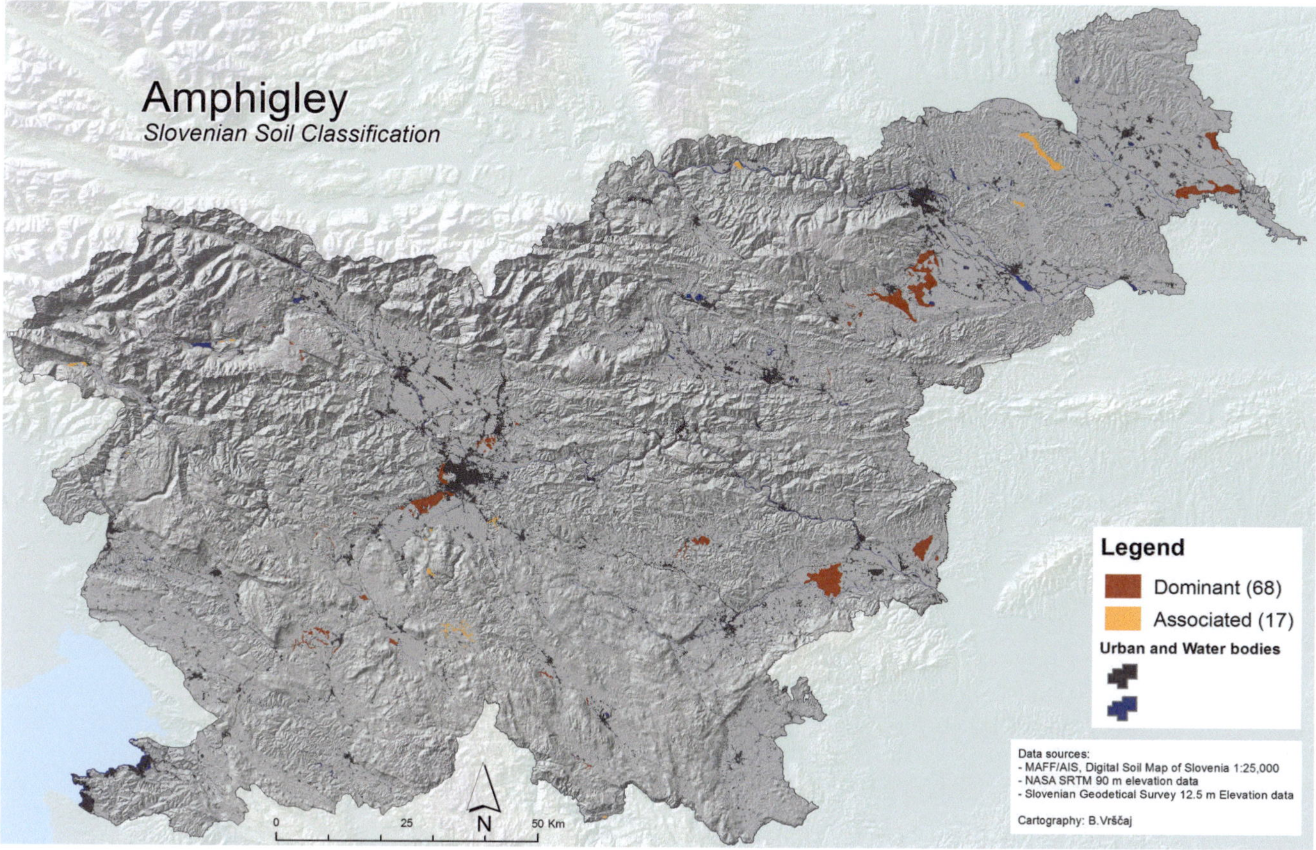

Fig. 5.47 The distribution of Amphigley in Slovenia

horizons of Hypogley. As a soil, it is wet. After precipitation, the water very slowly drains.

Representative Measured Profile Data on a Eutric Epigley, Petelinje Lake, north-eastern Slovenia (SP database) (Table 5.10).

Distribution and Land Use The Epigley is present in Slovenia to a marginal extent of 2.6 ha. Its appearance is related to depressions on riverbanks and geomorphological conditions that accumulate surface waters, as well as concave forms with heavy-textured soils such as karst fields. The prevailing land use is hydrophilic grassland.

5.2 Other Soils

In this chapter, the authors wish to present a number of soils that are important, interesting, and rare—as well as unexpected in Slovenia's climates. Another two soils that are more common to other climates and other parent materials were identified in Slovenia. The first is the podzolized soil of the cold Alpine Pokljuka plateau, and the second is a soil with Vertisol characteristics that had been neither expected nor detected in Slovenia but which has recently been identified in the Savinja River valley. Although marginal in terms of extent, all three soil types contribute to the diversity of Slovenian soil resources. Among local soil surveyors, they are regarded as interesting rarities that are worth being presented.

An important soil in the Slovenian context is Jerina, the soil of the Karst plateau. It develops on limestone, which besides calcite ($CaCO_3$), the predominant mineral, often contains large amounts of chert, fine-grained or amorphous sedimentary rock composed almost entirely of quartz (SiO_2). The Jerina soil type is frequently used to illustrate the impact of soil on agricultural production and agricultural products. Namely, the Refosco grape yields Refosco wine when grown on Eutric Cambisol, but Teran wine when grown on Jerina, a distinctly different soil. With regard to its organoleptic properties, colour, and chemical composition, a Teran wine differs significantly from a Refošk and is considered a Slovenian wine specialty protected by European legislation.

Vertisols and Vertisol-like types of soil are not present in the SSC. Additionally, the vertic classifier for soil horizons is not present. The climate conditions, parent material, and low content of expansive clays in Slovenian soils did not suggest soils with vertic properties at the time of the

Table 5.10 Data on a Eutric Epigley, Petelinje Lake, north-eastern Slovenia

	Depth		pH		Particle size distribution						P2O5	K_2O	N	C	*OS*	*C/N*
	From	To	KCl	CaCl2	Fine sand (0.2–0.05 mm)	Coarse silt (0.05–0.02 mm)	Fine silt (0.02–0.002 mm)	Total silt (0.02–0.002 mm)	*Clay* (<0.002 mm)	Text. Cl ass	Al extr. (plant-available)		Kyehl-dahl	Walkely Black		
Horizon	cm	cm	–	–	%	%	%	%	%	cla ss	mg/100 g		%	%	%	*Ratio*
Aa	0	13	–	7.0	6.2	15.9	44.3	*60.2*	33.6	SiClLo	0.3	9.1	9.3	*0.4*	5.4	*12.6*
Gr	13	25	–	7.3	7.0	7.7	29.4	*37.1*	55.9	Cl	0.2	7.0	5.1	*0.2*	3.0	*13.0*
GoGl	25	35	–	7.4	7.9	9.8	29.3	*39.1*	53.0	Cl	–	–	4.9	*0.2*	2.8	*18.7*
BC	35	–	–	–	–	–	–	–	–	–	–	–	–	–	–	–

	Depth		Basic cations				Acidity	*Bases*	*CEC*	*Base saturation (%)*					
	From	To	Ca	Mg	K	Na	H	*Sum.*		*Total*	*Ca*	*Mg*	*K*	*Na*	*H*
Horizon	cm	cm	meq/100 g soil (1 M NH_4OAc, pH7)							%					
Aa	0	13	46.9	0.7	0.3	0.1	4.4	*47.9*	*52.4*	*91.5*	*89.5*	*1.3*	*0.5*	*0.2*	*8.5*
Gr	13	25	59.0	1.0	0.2	0.1	7.3	*60.2*	*67.6*	*89.1*	*87.3*	*1.4*	*0.2*	*0.2*	*10.9*
GoGl	25	35	55.5	0.9	0.2	0.1	6.8	*56.7*	*63.5*	*89.2*	*87.4*	*1.4*	*0.2*	*0.2*	*10.8*
BC	35	–	–	–	–	–	–	–	–	–	–	–	–	–	–

elaboration of the SSC. This chapter presents Gleysol with typical vertic properties—slickensides as well as deep cracks, and a very gently undulating surface—found in central Slovenia.

Podzolized soils and Podzols developed on the limestone/dolomite moraine of the Pokljuka plateau are the third group of rare Slovenian soils. The combination of soil formation factors created an interesting juxtaposition of very different soil types within an area that at first glance appears to be uniform. Lithosol, Rendzina, Luvisol, Brown Podzolized Soil, and Podzol appear close together within an area of similar lithology and micromorphology and at the same altitude, typically at 1200–1300 m.a.s.l. This chapter presents two examples: a Luvisol developed on the limestone/dolomite moraine and Brown Podzolized Soil developed on beds of chert.

5.2.1 The Jerina Soil

The area of the elevated Triest-Komen plateau in south-east Slovenia is called *Kras* in Slovenian and the Karst in English. The area is considered to be a native karst due to its numerous and well-expressed karstic phenomena, which were the subject of scientific description back in 1689 when Baron Janez Vajkard Valvasor published his collection of books entitled *The Glory of the Duchy of Carniola* [*Die Ehre dess Hertzogthums Crain* in German]. In his fourth book, comprising 232 pages, Valvasor described the peculiarities of the Duchy such as fossils, the specificities of the animal and plant world, and especially the Karst and its subterranean world in Inner Carniola such as caves and intermittent lakes. This work was the reason he was elected a foreign member of the English Royal Society in 1687, which had been founded in 1660. Due to Valvasor's early description of karstic phenomena the Slovenian words *dolina* and *polje* became English geomorphological terms (Fig. 5.48). The Karst is something Slovenes are particularly proud of and many karstic features are associated with national attributes. These include the wine from the Karst and Karst ham wind-dried by the Bora. But it is not only the subterranean world and geomorphology that are important for Slovenians, the soils of the Trieste-Komen plateau are also well known in the national context.

Soil science professors in Slovenia often illustrate the impact of soil on agricultural production and agricultural products by presenting the Refošk vs. Teran wine example to students. It is well known that the Refošk (Refosco) vine variety grown on Eutric Cambisols, which forms on calcareous flysch formations, produces Refošk wine. However, not far away, the same variety of grape grown on a different soil called *Kremenica* in the area of the Trieste-Komen plateau produces Teran, a Slovenian wine specialty. With regard to its organoleptic properties, colour, and chemical composition, a Teran wine significantly differs from a Refošk. Due to its specific properties, which are largely linked to the soil and geographic area, the production of wine designated as Teran is protected by European legislation and must occur under these specific soil-geomorphology-climate conditions (Figs. 5.49, 5.51 and 5.52).

Jerina, the soil of the Karst plateau, develops on limestone, which besides calcite ($CaCO_3$), as the predominant mineral, often contains large amounts of chert, fine-grained, or amorphous sedimentary rock composed almost entirely of quartz (SiO_2). The typical limestone formations of the plateau (Tomaj limestone and shales of Komen) contain platy and nodulos forms of grey, reddish-brown, or black chert (Figs. 5.50 and 5.53).

In contrast to calcite, the chert is resistant to weathering so it gradually accumulates in the soil. In places where colluvial and erosion processes have been intensive chert may accumulate in large quantities and may form an important part of the total volume of the soil, typically 25% or even more (Figs. 5.50 and 5.53).

Detailed morphological and pedogenetic descriptions of the Karst soils were not very common in the past. Marija Kodrič, a soil scientist, defined two main soil types in a short paper (Kodrič 1958): *Ilovka*, a clayey-silty soil very low in skeleton content that forms on radiolite limestone, and *Kremenica*, a predominantly clayey soil rich in chert that forms on limestone (Slovene: *kremen*: quartz or chert in English). The profile of *Kremenica* features an abundance of residual chert skeleton on the surface and throughout the profile (Fig. 5.53). Professor Sušin described several different sub-species of *Jerina* in his doctoral thesis (Sušin 1964). He focused on *Ilovka*, the soil formed on lower Cretaceous grey and dark grey limestone, and on Terra Rossa or *Kremenica*, which forms on dark grey Upper Cretaceous limestone with chert. According to the WRB, Jerina is classified as a Chromic Cambisol (Skeletic). The predominant land use of *Ilovka* is arable land or grassland where fields are abandoned. *Kremenica* is a soil traditionally reserved for Teran vineyards.

Figures 5.51 and 5.53 present the tilled topsoil of a Teran vineyard. Nowadays, when machinery is available, an important share of the litter (i.e. vine plant residues: leafs, branches, grapes lying on the surface of the soil) is incorporated into the topsoil to a depth of apx. 25 cm. In the continuation, a description of a typical Kremenica soil profile is presented.

5.2.1.1 Description of a Representative Kremenica Profile

This soil profile is located on an old vineyard on the eastern outskirts of the village of Pliskovica in the centre of the Teran wine production area. It is a mild terraced hillside

Fig. 5.48 The Karst landscape in the central part of the Teran wine production area (*Photo* B. Vrščaj). The Karst is facing very strong land abandonment and intensive spontaneous afforestation. The agricultural production in the dolinas is changing from arable to grassland. Typical limestone walls transect the landscape and divide properties. Isolated vineyards are common in the remote areas, while in the vicinity of villages vineyards are the predominant land use

Fig. 5.49 A young Teran vineyard (*Photo* B. Vrščaj)

vineyard exposed to the south. *Refoško* (Refosco) grapes grow with the support of trellises (Fig. 5.55).

Description of the horizons The profile horizon structure is Ap-P-Brz-R.

Ap: 0–17 cm. The horizon is fresh to moist, friable, and almost loose. The structure is well-expressed nutty to polyhedral, aggregates are smaller and a bit sticky due to the moisture. The texture is clayey-loamy. The organic matter content is medium. The primary colour is an intense red-brown 7.5YR 4/3. The skeleton content is estimated to be 15%, up to 3 cm in size, mostly sharp-edged chert mixed with dark limestone. The horizon features signs of ploughing. The transition to the P horizon is level.

P: 17–40 cm. The horizon is moist, dense, and crumbly. The texture is clayey-loamy. The polyhedral structure is well expressed. It is poor in organic matter. The primary colour 2.5YR 5/3 passes into the secondary, which is a more intensive red 10R 3/4. The limestone skeleton is up to 4 cm in size, flat, apx. 10% of the total volume. The transition to the lower Brz horizon is gradual and irregular within an apx. 17-cm-thick layer. The horizon has barely visible indications of deep ploughing at the time of the vineyard's establishment. The boundary to the lower horizon is undulating and unclear. The transition to the lower horizon is wavy at an average depth of 40 cm.

Brz: 40–50 (68) cm. The horizon is moist, wet on the surface of well-developed, sharp-edged, polyhedral aggregates. The consistency is dense but still crumbly, partially compacted, and still brittle. Texture is clayey. It is mineral; organic

Fig. 5.50 Chert (*black*) and limestone (*white*) skeleton in a clayey Brz horizon (*Photo* B. Vrščaj)

Fig. 5.51 A typical Teran vineyard (*Photo* B. Vrščaj)

Fig. 5.52 An isolated Teran vineyard (*Photo* B. Vrščaj)

Fig. 5.53 Chert skeleton largely characterizes Kremenica (*Photo* B. Vrščaj)

Fig. 5.54 A Karst dolina (*Photo* B. Vrščaj)

Fig. 5.55 The Jerina profile in a Teran vineyard (*Photo* B. Vrščaj)

matter is present in rare root tunnels. The primary colour is 2.5YR 4/4, and the secondary 2.5YR 4/6. The predominant chert skeleton is sharp-edged, while the limestone skeleton is rounded due to weathering and platy, up to 3 cm big. The transition to the R is lithic and straight.

R: 50–70 cm. Compact dark limestone that shows intensive surface weathering.

The soil has a neutral acidity. The plant-available phosphorus content is optimal according Slovenian best-practice fertilization practices (the optimal C class: 13–25 mg P_2O_5/100 g). As expected, the concentration of plant-available potassium is higher than optimal. Regarding the high clay content, it is ranked in the D class (an excessive concentration of plant-available K: 34–45 mg K_2O/100 g). The content of the predominant chert skeleton is considered to be medium high. It constitutes apx. 25% of the total soil volume. Measured in the Ap, P, and Brz horizons, the skeleton content is 25.6, 26.0, and 22.9%, respectively (Table 5.11).

The soil is medium deep. In spite of the high clay content, the soil profile is well permeable due to the well-expressed polyhedral structure. The structure is stable due to the high clay and Ca^{2+} content and resists precipitation throughout the profile in spite of the low organic matter content in the clayey subsoil. The stable, almost permanent surfaces of the aggregates form a system of macropores that drains the soil profile in a short period of time. The high-clay-content soil shows no traces of hydromorphism. The workability of the soil is high. The good percolation and relatively rapid removal of sufficient water enable the tillage of the soil five to six hours after rain ceases. The high clay content suggests

Table 5.11 The set of analytical data of the Kremenica profile

	Depth		pH		Particle size distribution						P_2O_5	K_2O	N	C	OS	C/N
	From	To	KCl	$CaCl_2$	Fine sand (0.2–0.05 mm)	Coarse silt (0.05–0.02 mm)	Fine silt (0.02–0.002 mm)	Total silt (0.02–0.002 mm)	Clay (<0.002 mm)	Text. class	Al extr. (plant-available)		Kyehl-dahl	Walkely Black		
Horizon	cm	cm	–	–	%	%	%	%	%	Class	mg/100 g		%	%	%	Ratio
Ap	0	17	–	7.1	15.2	10.3	32.1	*42.4*	42.4	SiCl	49.0	48.0	0.3	*2.7*	4.7	*11.0*
P	17	40	–	7.3	15.8	8.6	26.3	*34.9*	49.3	Cl	12.0	23.0	0.2	*1.6*	2.8	*10.1*
Brz	40	50(68)	–	7.5	16.8	9.2	26.8	*36.0*	47.2	Cl	6.0	16.0	0.1	*1.4*	2.4	*10.0*
R	50	70+	–	–	–	–	–	–	–	–	–	–	–	–	–	–

	Depth		Basic cations				Acidity	*Bases*	*CEC*	*Base saturation (%)*					
	From	To	Ca	Mg	K	Na	H	*Sum.*		*Total*	*Ca*	*Mg*	*K*	*Na*	*H*
Horizon	cm	cm	meq/100 g soil (1 M NH_4OAc, pH7)							%					
Ap	0	17	25.2	2.9	1.0	0.1	5.0	*29.1*	*34.1*	*85.3*	*73.9*	*8.4*	*1.3*	*0.7*	*14.7*
P	17	40	26.3	3.0	0.5	0.1	5.0	*29.9*	*34.9*	*85.7*	*75.3*	*8.7*	*0.7*	*0.7*	*14.3*
Brz	40	50(68)	27.5	3.1	0.3	0.1	4.3	*31.0*	*35.3*	*87.9*	*77.9*	*8.9*	*0.4*	*0.6*	*12.1*
R	50	70+	–	–	–	–	–	–	–	–	–	–	–	–	–

Fig. 5.56 A piece of limestone rock that slowly dissolves in Brz horizon (*Photo* B. Vrščaj). The rock surface shows the embedded chert fragments (*left*). Brz is very clayey but shows no traces of hydromorphism (*right*)

a good water holding capacity, yet drought can be severe because of the numerous narrow and deep cracks that develop in relatively shallow soil (Vrščaj 2014a) (Fig. 5.56).

5.2.2 Amphigley with Vertic Properties

Vertisols and vertisol-like types of soil are not present in the SSC. The vertic classifier for soil horizons is also not present. The climate conditions, parent material, and low content of expansive clays in Slovenian soils did not suggest soils with vertic properties at the time of the elaboration of the SSC. A very recent soil survey in Slovenia discovered an interesting hydromorphic soil that features unexpected properties common to Vertisols. A number of surprising vertic features, i.e. slickensides as well as deep cracks, and a very gently undulated surface, were found in central Slovenia, at a location 46.3° north and 15.0° east. The mean annual precipitation is 1200 mm. The area where the soil appears is small in extent and situated on a very gently sloping plateau at the foot of gently undulating hills (Fig. 5.57). The parent materials of the hills to the north-west are fine-grained

Fig. 5.57 A gently sloping plateau with hills in the background (*Photo* B. Vrščaj)

Paleogene mudstone and Pliocene terrestrial sediments. The rock weathering material has the characteristics of smectitic clays.

The area was surveyed during a long dry season. During that season, the soil surface featured cracks that were frequently 5–12 cm wide and apx. 30 cm deep. At that depth, they were 1 cm wide.

5.2.2.1 A Description of the Profile of This Amphigley

At first glance, the soil is reminiscent of well-expressed Amphigley. The topsoil consists of two A horizons. Both of them, Aa_1 and Aa_2, are rich in organic matter (16.4 and 10.8%) and contain 30 and 40.7% clay, respectively. They show hydromorphic features such as mottles and small rusty concretions that are barely visible in a dark humus-rich matrix. The boundary between Aa_2 and the first gleyic Go_1 horizon is sharp and wavy; in some places, it remains in a tonguing pattern.

The clay content is very high in all horizons (30–70%). The permeability is very low and in wet periods close to nil, thus hydromorphic conditions occur. In microdepressions, ponding may occur. The transition between gleyic horizons is less sharp. Two clearly identifiable Gr horizons are present. Both of them are wetter than the Go horizons above, more plastic, malleable, and sticky. In the bottom of the profile, a GoC horizon of clay-coated gravel is present (Fig. 5.58).

Description of the Horizons The horizon sequence is Aa_1-Aa_2-Go_1-Gr_1-Go_2.Gr_2-GoC.

Aa_1, 0–8 cm. The horizon is friable, dry to fresh; the structure is crumbly granular to polyhedral. It is very rich in organic matter and is similar to mulch. Rusty spots 2–3 mm wide are present.

Aa_2, 8–14 cm. The horizon is dense, friable, and slightly malleable. Moist to wet. The structure is weakly expressed blocky to polyhedral. It is rich in organic matter and abundant on thin grass roots. Rusty spots 2–3 mm wide are present throughout the horizon.

Go_1, 14–30 cm. The horizon is dense, malleable, and plastic. Moist. The structure is angular blocky, numerous sharp-edged prisms are well expressed. The surface of the prisms is covered with thin clayey organo-mineral coatings.

Gr_1, 30–45 cm. The horizon is dense, sticky, and plastic. Moist to wet. The structure is angular blocky, prisms are weakly expressed. Mottles and Mn Fe concretions are rare. A few single fine roots are present.

Go_2, 45–65 cm. The horizon is dense, plastic, and malleable. Moist. The horizon is brittle and disintegrates on the vertical fine cracks that separate angular blocky peds. The surfaces of the cracks are covered with thin, organic matter-rich clayey coatings.

Fig. 5.58 The Amphigley profile. Slickensides can be observed at a depth of 90 cm (*Photo* B. Vrščaj)

Gr_2, 65–90 cm. The horizon is dense, malleable, plastic, and sticky. Moist. Well-expressed, polished surfaces, i.e. slickensides, are present at a depth between 80 and 90 cm. The angle of the slickenside surfaces is 25–30°.

GoC, 90–110 cm+. The horizon is dense, plastic, and malleable. Wet. It consists of clay-coated gravel 0.5–3 cm in diametre (Table 5.12).

5.2.2.2 The Vertisol-like Features

Close examination of the soil profile and the area reveal typical features that are characteristic of Vertisol. The main and unexpected characteristics are sheared planar structures that are known as *slickensides*, polished surfaces that extend in the direction of shear. They appear at a depth between 80 and 100 cm and extend across the entire width of the profile (Fig. 5.59).

Table 5.12 The Amphigley set of analytical data

	Depth		pH		Particle size distribution						P2O5	K_2O	N	C	*OS*	*C/N*
	From	To	KCl	$CaCl_2$	Fine sand (0.2–0.05 mm)	Coarse silt (0.05–0.02 mm)	Fine silt (0.02–0.002 mm)	Total silt (0.02–0.002 mm)	*Clay* (<0.002 mm)	Text. Class	Al extr. (plant-available)		Kyehl-dahl	Walkely Black		
Horizon	cm	cm	–	–	%	%	%	%	%	Class	mg/100 g		%	%	%	*Ratio*
Aa1	0	8	–	5.6	14.3	11.0	44.7	*55.7*	30.0	SiClLo	13.0	23.0	1.1	*9.5*	*16.4*	*8.7*
Aa2	8	14	–	6.1	14.3	9.1	35.9	*45.0*	40.7	SiClLo/SiCl	6.6	17.0	0.7	*6.3*	*10.8*	*8.9*
Go1	14	30	–	6.5	6.8	6.4	33.8	*40.2*	53.0	SiCl	3.4	20.0	0.2	*1.7*	*3.0*	*8.6*
Gr1	30	45	–	6.6	4.6	5.2	35.6	*40.8*	54.6	SiCl	4.3	23.0	0.2	1.4	*2.4*	*9.1*
Go2	45	65	–	6.6	8.4	10.9	34.2	*45.1*	46.5	SiCl	3.5	18.0	0.1	*0.9*	*1.6*	*10.0*
Gr2	65	90	–	6.7	5.6	3	21.5	*24.5*	69.9	Cl	5.5	27.0	0.1	0.6	*1.0*	*8.4*

	Depth		Basic cations				Acidity	*Bases*	*CEC*	*Base saturation (%)*					
	From	To	Ca	Mg	K	Na	H	*Sum.*		*Total*	*Ca*	*Mg*	*K*	*Na*	*H*
Horizon	cm	cm	meq/100 g soil (1 M NH_4OAc, pH7)							%					
Aa1	0	8	42.7	9.9	0.5	0.4	16.5	*53.5*	*70.0*	*76.4*	*61.0*	*14.2*	*0.9*	*2.5*	*23.6*
Aa2	8	14	45.6	10.8	0.4	0.4	12.8	*57.1*	*69.9*	*81.8*	*65.2*	*15.4*	*0.6*	*2.6*	*18.2*
Go1	14	30	33.4	9.3	0.5	0.4	8.8	*43.5*	*52.3*	*83.3*	*63.9*	*17.7*	*0.8*	*2.1*	*16.7*
Gr1	30	45	29.3	8.5	0.5	0.3	8.8	*38.5*	*47.3*	*81.5*	*61.9*	*17.9*	*0.8*	*1.6*	*18.5*
Go2	45	65	25.6	7.5	0.4	0.2	7.8	*33.7*	*41.5*	*81.3*	*61.8*	*18.1*	*0.6*	*1.2*	*18.7*
Gr2	65	90	33.2	10.8	0.6	0.3	7.8	*44.8*	*52.6*	*85.3*	*63.1*	*20.5*	*0.9*	*1.3*	*14.7*

Fig. 5.59 Slickensides. They extend across the entire width of the profile (*Photo* B. Vrščaj)

In G horizons, angular blocky peds are present throughout the gleyic horizons. In Slovenian Gleysols, the presence of numerous well-expressed sharp-edged prisms, or in other words a big angular blocky structure, had not hitherto been detected. Such structures are common in Vertisol as a result of intersecting shear planes.

Cracks are another feature characteristic of Vertisol that are present in the area. The cracks in the surface show a different pattern than is usually present in Gleyic and other Slovenian soils. In general, they are long, much wider, and deeper. It is expected that they are open less than 60 consecutive days during the summer. High precipitation in the winter and frequent snow cover suggest that the cracks close during the winter. Not every summer is dry enough for the cracks to open wide. The cracks are open wide a short time and very little surface soil rich in organic matter falls into them (Figs. 5.60 and 5.61). Self-mulching is not present and the soil does not exhibit homogenization. It is marginally expressed in topsoil in rare tongue-like transitions between A and G horizons. Numerous narrow vertical cracks are frequent and coated in clay and organic matter. They allow the surface water to drain into a certain depth of the subsoil, to the Gr_2 in the case of this profile. The swelling pressures are present because of the differential wetting in this depth and slickensides occur. The soil has a vertic structure, but the grade of the structure is weaker.

Fig. 5.60 Cracks in the surface (*Photo* B. Vrščaj)

Fig. 5.61 Fine cracks close to a wide crack (*right*) and the $CaCO_3$-covered surface (*Photo* B. Vrščaj). Figure shows the topsoil rich in organic matter. The granular structure of the soil is covered with $CaCO_3$ accumulations. Numerous thin cracks are connected to long typically 5–10-cm-wide cracks that are up to 30 cm deep.

The soil surface within a short distance of the examined profile shows no uneven topography. The area is almost flat, and very gentle mounds and microdepressions are present in the area nearby sharing the same slope.

According to the SSC, the presented soil profile is classified as Amphigley. The features that are present allow the introduction of the *vertic* qualifier. Hitherto, vertic characteristics were not expected to be present in Slovenian soils. The parent material, the topography, and the climatic conditions in some areas of Slovenia, especially in the Mediterranean part as well as dryer and hotter north-eastern Slovenia, suggest that the SSC needs to be updated with a vertisol classifier and, potentially, a Vertisol soil type (Radišek et al. 2014a).

5.2.3 The Luvisol and Brown Podzolized Soil of the Pokljuka Plateau

Pokljuka is a plateau at altitudes in general between 900 and 1300 m.a.s.l. The parent material is primarily massive Carnian Triassic dolomite with chert, Triassic limestone with chert, older consolidated carbonate moraine, and unconsolidated limestone-dolomite moraine. The mean annual precipitation is between 2000 and 2600 mm and 2600–3200 mm at altitudes above 1400 m.a.s.l. The mean annual temperature is 2–1 °C at altitudes above 1200 and 0–1 °C above 1400 m.a.s.l.

The combination of soil forming factors created an interesting juxtaposition of very different soil types within an area that at first glance appears to be uniform. Lithosol, Rendzina, Luvisol, Brown Podzolized soil, and Podzol appear close together within an area of similar lithology and micromorphology and at the same altitude, typically at 1200–1300 m.a.s.l.

5.2.3.1 Luvisol Developed on the Limestone/Dolomite Moraine

In the spruce forest area presented in Fig. 5.62, very different soil types develop within a short distance, typically less than 50 m.

Figure 5.63 shows a profile of a shallow Luvisol developed on typical limestone/dolomite moraine. The sequence of horizons is Ol-Of-Oh/Ah-AE-E-Bv-Bt. The transitional AE and the eluvial E horizons are coarser-textured. The clay and organic matter and partly the iron oxides leach into the illuvial Bt horizon. Figure 5.64 shows the sidewalls of the same Luvisol profile. The coarser, bleached material interfingers into the Bv and/or Bt horizon. The irregular distributed areas of albic material and the accumulation of iron oxides is a result of the surface micromorphology. The microdepressions in the moraine accumulate the rainwater and the melting snow. The concentrated water flow increasingly translocates the clay, humic substances, and free iron oxides through narrow preferential percolation areas and channels. In some parts of the profile, reducing conditions appear and stagnic properties develop (Fig. 5.64).

At a close distance, Lithosol (Oh-R, or R) develops on top of elevated areas of small ridges, boulders, and rock

Fig. 5.62 Spruce forest on the juxtaposition of different soil types, the Pokljuka plateau (*Photo* B. Vrščaj)

Fig. 5.63 A shallow Luvisol that developed on carbonate moraine, the Pokljuka plateau, north-western Slovenia (*Photo* B. Vrščaj)

Fig. 5.64 The side walls of a shallow Luvisol profile (*Photo* B. Vrščaj)

outcrops. The Rendzina (Ol-Oh-Ah-C-R profile) is characteristic of short slopes. On lower slopes and level surfaces, Luvisol develops (Ol-Oh-Ah-E-R). In the small microdepression areas, Brown Podzolized soil can be found in various stage of development. Figure 5.64 presents a soil profile in an early podsolization stage, while Fig. 5.65 features a medium-deep Brown Podzolized soil that developed on carbonate moraine.

5.2.3.2 Brown Podzolized Soil Developed on Beds of Chert

The Pokljuka plateau features a karstic morphology. The dolinas that have a discharge in the karstic underground system are frequent. After the dissolution/weathering of the limestone and/or dolomite rich in chert, the erosional processes accumulate the remaining quartz material on the bottom of the dolinas in beds of various thicknesses and granulation. In such areas, which are limited in extent, Brown Podzolized soil develops. The soil is acidic (pH_{CaCl2} of the A horizon is 4.2). The acidophilus vegetation of such area (*Luzula luzuloides, Nardus stricta*, *Vaccinium myrtilus*) contrasts with the vegetation in areas with Renzina (Fig. 5.66) and contributes to the diversity of the vegetation cover. Figure 5.67 shows a soil profile that has developed on such accumulation of coarse quartz/chert sand and rubble on the bottom of a small dolina. The sequence of horizons is Ol-Of-Oh-A-CE-Bfe-Bv. The transitional CE horizon of irregular thickness is bleached and consists of 80% chert and

Fig. 5.65 Brown Podzolic soil, the Pokljuka plateau, north-western Slovenia (*Photo* T. Kralj)

Fig. 5.66 An acidophilus grassland in a depression, the Pokljuka plateau (*Photo* B. Vrščaj)

Fig. 5.67 Brown Podzolic soil, acidophilus grassland, the Pokljuka plateau (*Photo* B. Vrščaj)

20% bleached eluvial material of finer texture. The underlying Bfe horizon is loamy, has a polyhedron structure, and contains apx. 15% chert skeleton.

The profile represents a transitional stage of soil development towards a Podzol. In Slovenia, it is characteristic for small accumulations of quartz material within the limestone/dolomite area on the cold and humid Pokljuka plateau.

6 Soil Distribution

Borut Vrščaj

Abstract
The Digital Soil Map of Slovenia was completed in 1999, revised, and delivered to the Ministry of Agriculture, Food, and Forestry in 2001 using the Slovenian Soil Classification (SSC). Subsequently, the legend was translated in accordance with the revised legend of the Soil Map of the World (FAO and UNESCO 1997) in spite of the fact that the FAO map was not intended for mapping soils on a local scale but merely at continental scales. The translation was unofficial and unrevised and thus temporary. For this book, the SSC soil types were checked, and the primary and secondary qualifiers were determined for the appropriate translation of the SSC soil types into the WRB soil classification. The WRB translation was linked to the geometry of the Digital Soil Map of Slovenia 1:25,000, and a map of WRB reference soil groups found on Slovenian territory was elaborated. This chapter presents the distribution and extent of individual WRB RSGs: Cambisols, Leptosols, Gleysols, Stagnosols, Luvisols, and Histosols. Maps are accompanied by statistics on the area the RSGs occupy and a description of the relevant landscapes and are illustrated with photographs of characteristic landscapes that belong to the main WRB RSGs.

6.1 Mapping Slovenian Soils Using the WRB Classification

The Digital Soil Map of Slovenia was completed in 1999, revised, and delivered to the Ministry of Agriculture, Food, and Forestry in 2001 using the Slovenian Soil Classification (SSC). Later, the legend was translated in accordance with the revised legend of the Soil Map of the World (FAO and UNESCO 1997) in spite of the fact that the FAO map was not intended for mapping soils on a local scale but merely at continental scales. The translation was unofficial and unrevised and thus temporary. The Slovenian Soil Classification (SSC) was translated into the WRB (FAO 2006b) by Kralj (2008), Čuden (2009), and Kralj and Grčman (2010, 2009a) in 2010 within the framework of a national soil research project focused on soil informatics and pedotransfer functions (Vrščaj 2010).

The SSC soil types were revised for the purposes of this publication. For each soil type, the appropriate WRB 2014 legend (FAO IUSS 2014) reference soil group was identified and linked to the SSC legend. Where possible, the primary and secondary qualifiers were determined for the appropriate translation of the SSC soil types to the WRB soil classification. In order to visualize the maps and to identify the extent and the location of individual soil types, the WRB translation was linked to the geometry of the Digital Soil Map of Slovenia 1:25,000 (DSM25) (Fig. 6.1).

6.1.1 The Digital Soil Map of Slovenia

The Digital Soil Map of Slovenia 1:25,000 (DSM25), a vector polygon database, is the most detailed soil map that covers the complete territory of the country. Each polygon is a soil mapping unit (SMU) that can be composed of up to three soil typological units (i.e. soil types and STUs).

B. Vrščaj (✉)
Department of Agricultural Ecology and Natural Resources, Centre for Soil and Environment Research, Agricultural Institute of Slovenia, Ljubljana, Slovenia
e-mail: Borut.Vrscaj@kis.si

B. Vrščaj et al., *The Soils of Slovenia*, World Soils Book Series,
DOI 10.1007/978-94-017-8585-3_6

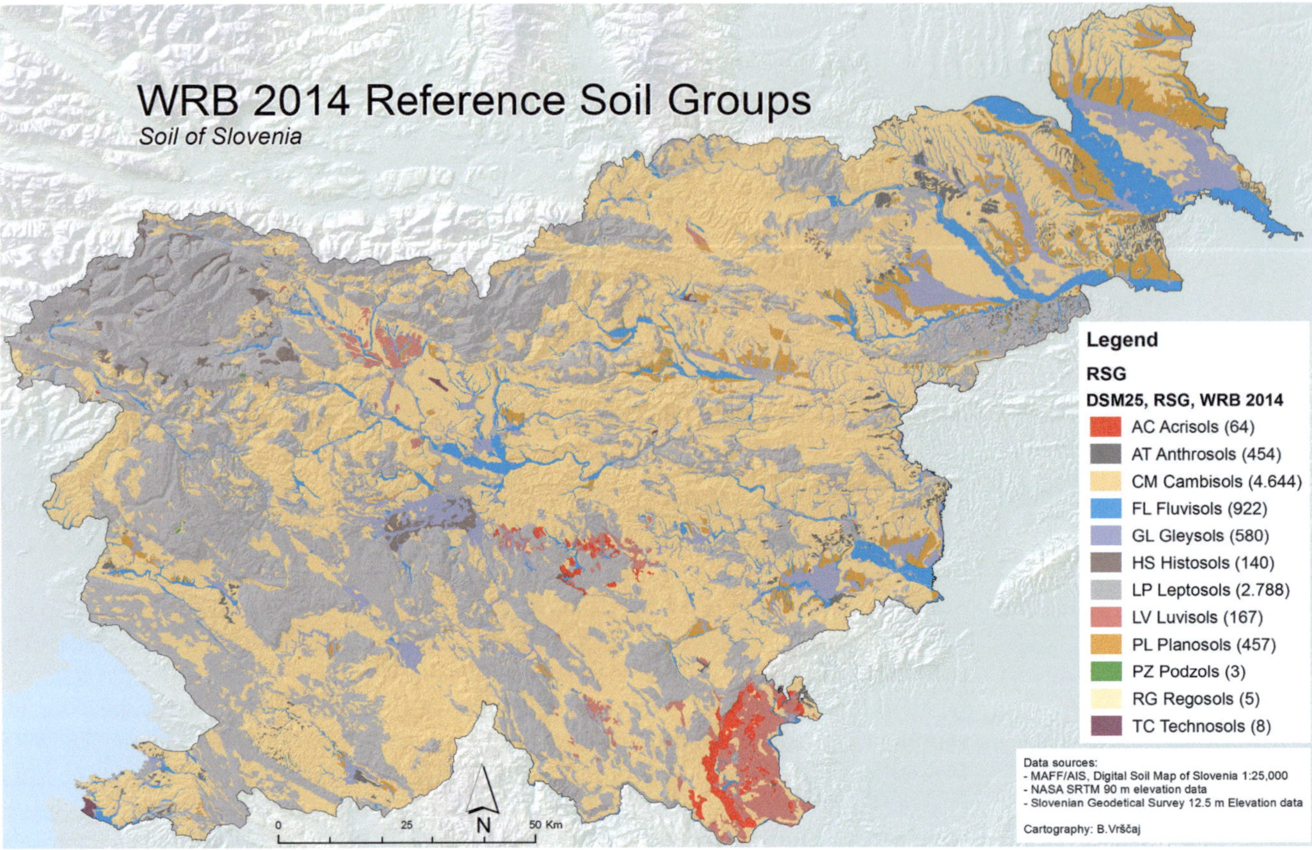

Fig. 6.1 Map of WRB reference soil groups derived from the Digital Soil Map of Slovenia (DSM25)

An additional STU that occurs to a marginal extent within the SMU polygon can be attributed to the polygon. The spatial distribution of an individual STU within the SMU is not defined. The attribute DSM25 data set holds information on the percentage of each STU present in the SMU area.

6.1.1.1 Spatial Representation of Reference Soil Groups According to the WRB

The basic statistics on the extent and distribution of soil types show that two reference soil groups (RSGs) significantly predominate in Slovenia. Almost one-half of the country is covered by a large variety of Cambisols (48.2%) and an additional third of the country by Leptosols in different forms (33%). In the rest of the territory (18%), four RSGs are mainly present: Fluvisols (5.0%), Gleysols (4.4%), Stagnosols (3.8%), and Luvisols (1.7%). Besides urban areas and land under the traffic infrastructure (1.6%), the total area of the six other RSGs does not exceed 1% of the Slovenian territory (Fig. 6.2).

6.1.2 The Distribution and Extent of Individual WRB Reference Soil Groups

6.1.2.1 Cambisols

Following the WRB, the Slovenian Cambisols are divided into three main sub-groups: Eutric, Dystric, and Chromic Cambisols. The Eutric Cambisols and Calcaric Eutric Cambisols in the Vipava valley and the coastal region are formed on carbonate rock, mainly Eocene flysch and carbonate sands. In central Slovenia, the south, and the south-western part of Slovenia, Eutric Cambisols are combined with Luvisols and Leptosols. The dominant parent materials of these areas are almost exclusively Triassic, Jurassic, and Cretaceous limestone and dolomite. An important Eutric Cambisol area is situated in the north-west hilly area of Slovenske Gorice. Sandy marl and sandstones are the predominant parent material there. The hills are intersected by parallel narrow valleys, where Eutric Gleysols and Eutric Gleyic Fluvisols dominate. In general, the Eutric Cambisols are deep, and very frequently, they are of fine texture; they have high base saturation and high water-holding capacity. These Cambisols and the climate make these areas in particular suitable for agricultural production. Being very suitable for fruit and vines, the Cambisols of these areas largely contribute to these areas being important fruit and especially wine regions of Slovenia (Fig. 6.3).

In southern Slovenia in the area of Ilirska Bistrica, Eutric Cambisols interlace with Dystric Cambisols and several dystric-eutric transitional forms of Cambisols. They have formed on decalcified mudstone, marl, and in some areas on carbonate or quartz sandstone. In the large basins of the

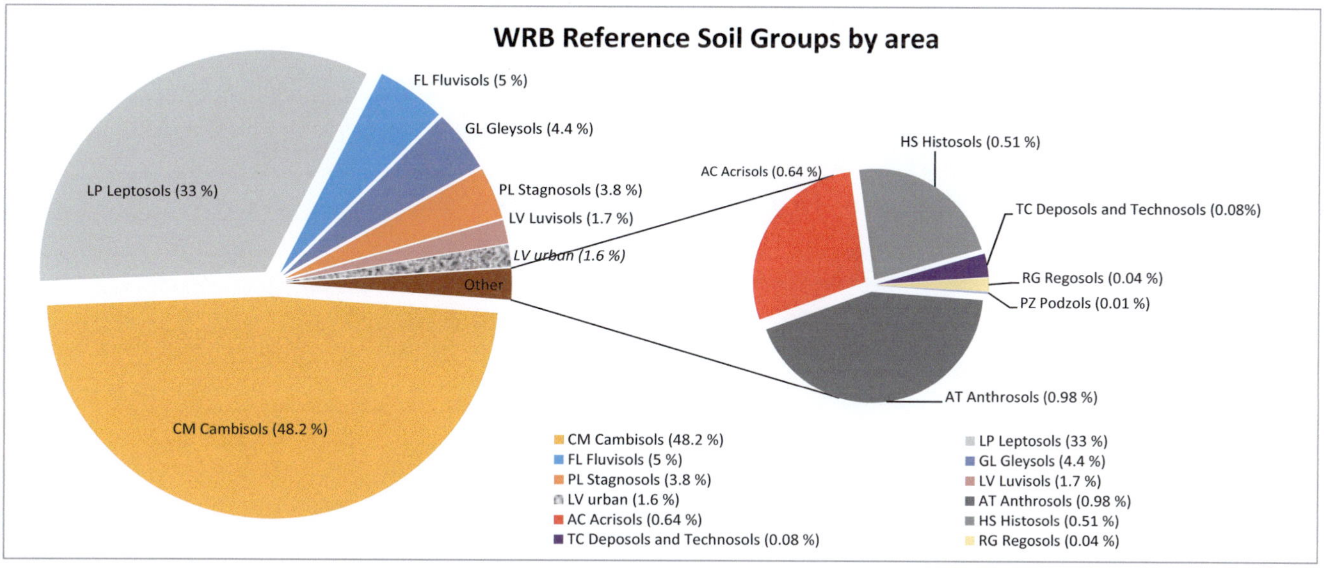

Fig. 6.2 Most important soil types of Slovenia by area. *Source* Digital Soil map 1:25,000

Fig. 6.3 Vineyards in an area of Eutric Cambisols (*Photo* T. Kralj)

Savinja River east of the city of Celje (Oset 1987; Radišek et al. 2014a) and north of Ljubljana, shallow (Stepančič et al. 1985), sandy, and consequently highly permeable Eutric Cambisols have formed on carbonate Quaternary glacio-fluvial sand and gravel deposits (Fig. 6.4). Accompanied by Hyperskeletic and Calcaric Fluvisols and/or Gleysols, they overlie important groundwater protected areas—sources of drinking water for a significant part of the Slovenian population. Being flat, these areas have traditionally been used for agriculture; thus, conflict related to nitrate leaching and groundwater quality is entrenched.

In the past, the Eutric Cambisols of this area were ranked among the high-quality agricultural soils of Slovenia, but severe and frequent agricultural droughts between the years 2000 and 2013 exposed their main weakness: their low, or better stated, very low water-holding capacity. The yields on these shallow and sandy Cambisols were dramatically affected. The damages were high, and farmers were not

Fig. 6.4 An aerial view of an area of shallow and permeable Eutric Cambisols and Leptosols formed on carbonate sands and gravel deposits of the Savinja valley during the 2012 drought (*Photo* B. Vrščaj). The picture clearly shows the typical pattern of an interleaved mixture of deep and shallow Cambisols and shallow and sandy Leptosols that have developed on alluvial carbonate sands and gravel

easily compensated by the state. Many stable yields in the neighbouring Gleysol areas have caused a change in the general opinion regarding the suitability and agricultural quality of the Eutric Cambisols in these basins.

Very different Eutric Cambisols combined with Chromic Cambisols are the main soil types of hilly Notranjska, Dolenjska, and the Kras, all of them situated in the southern part of Slovenia (Stepančič et al. 1987b). The dominant parent material in this area is Triassic, Jurassic, and Cretaceous limestone spotted with areas of accumulated loamy Quaternary deposits and rarely and to a minor extent the remains of silicate rocks. The surface of the limestone rock (R horizon) is very undulating, and soil depth is difficult to predict in spite of the even ground surface. The profile of these Cambisols can significantly vary from very shallow forms (typically 30–40 cm) to pockets of accumulated loamy clay several metres deep within very short distances, frequently less than 20 m. Shallow Rendzic Leptosols are the main associated soils that vary with Cambisols and, very frequently, rock outcrops. The relief, soil depth, and rock outcrops make such areas less suitable or even marginal for modern agriculture based on the use of heavy machinery. Pastures, the primary land use in the past, are nowadays abandoned and overgrown by transitional vegetation, or they are already under forests. Land abandonment is one of the significant agricultural problems in these areas in terms of food safety (Fig. 6.5).

Dystric Cambisols in Slovenia are formed on a variety of non-carbonate parent materials, frequently on Middle Triassic volcanic rocks, igneous, or metamorphic parent material. According to the area, igneous grandiorite and cizlakite and metamorphic schist prevail in the Pohorje massif. The small extent of keratophyre and porphyry is present in the pre-Alpine hills of central Slovenia (Prus et al. 1985b), while clastic Upper Palaeozoic rocks can be found east of Ljubljana. The soils formed on these parent materials have a low pH, typically between 5.5 and 4.2 measured in $CaCl_2$ and a low base saturation, frequently below 10% in the Bv horizon. These areas are hilly, and the predominant land use is forest with patches of agricultural land, mainly individual farms oriented to diary production. Shallow and sandy Dystric Cambisols cover two large plains—basins filled with Quaternary quartz sands and gravel of Drava and Mura River deposits. These flat areas of the Dravsko-Ptujsko plain (Stepančič et al. 1986a) and the Prekmurje plain (Stepančič et al. 1984a) are the two most agriculturally important areas of Slovenia. Dystric Cambisols here are associated with Dystric Leptosols over a very short distance following old river beds in a similar way as is presented in Fig. 6.6. Along the rivers, shallow and sandy Dystric Cambisols are associated with or border Fluvisols (Lobnik et al. 1996a). Shallow depth, coarse texture, and a significant share of gravel quartz skeleton in the soil profile make the soils less fertile and sensitive to droughts (Table 6.1).

6.1.2.2 Leptosols

Leptosols are the second most dominant WRB reference soil group, covering more than 33% of Slovenia. Rendzic Leptosols cover by far the most significant part of the country (almost 21%). The shallow forms of Leptosols that have developed on compact limestone and/or dolomite dominate. The main reason for that is the rough topography. The steep slopes and consequently permanent erosion, the altitude, and the parent material, which decomposes slowly, keep the soils

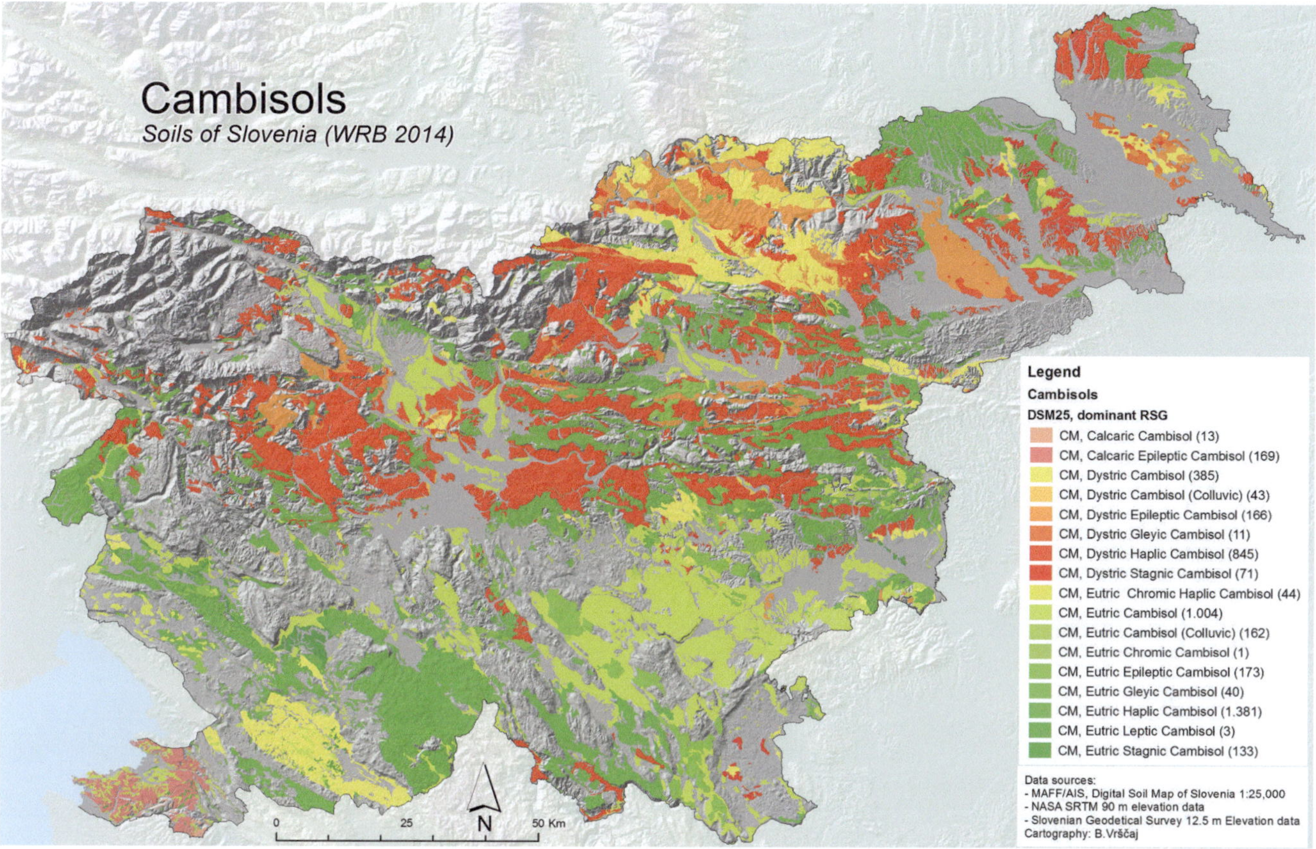

Fig. 6.5 The distribution of Cambisols in Slovenia

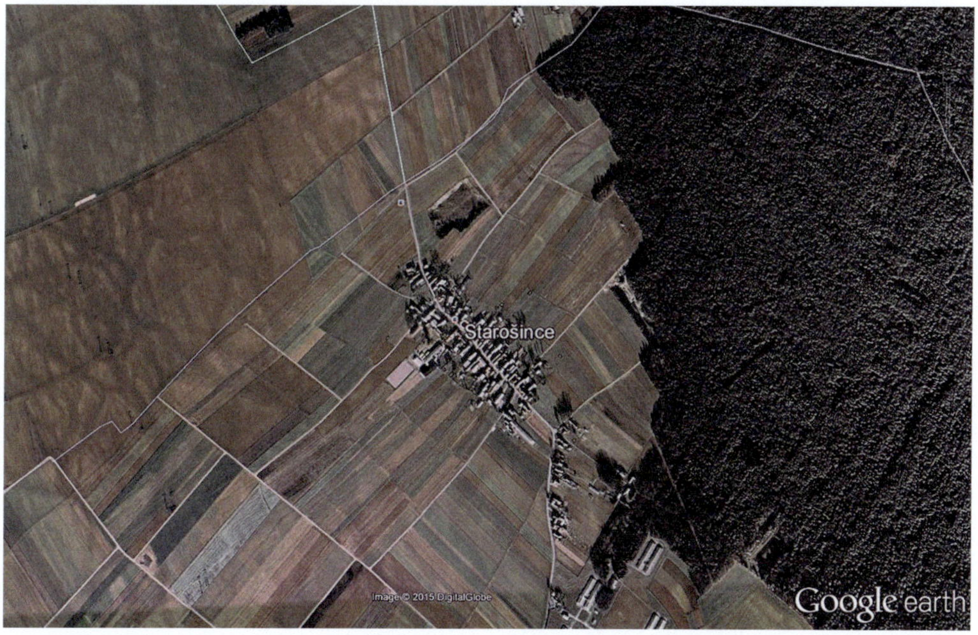

Fig. 6.6 Interlacing Dystric Cambisols and Leptosols on quartz sand and gravel deposits of the Dravsko-Ptujsko plain, north-east Slovenia (*Image* Google Earth). In years of average or high precipitation, the old river beds cannot easily be detected from the ground. A very clear picture of river beds and soil depth is presented by satellite imagery or aerial photography

in young developmental stages. Nudilithic Leptosols are frequently and mainly present in the Alps, especially at altitudes above 2000 m.a.s.l. or on northern slopes above 1800 m.a.s.l. (i.e. above the forest line) (Stepančič et al. 1986b 1987a). Rendzic Lithic Leptosols dominate in the area of Notranjska and Dolenjska. Formed on compact limestone, they vary with Chromic Cambisols and less frequently with Eutric Cambisols and Luvisols. In general, their appearance

Table 6.1 Cambisols by area, according to the DSM25

Cambisols by area	
%	*Soil type*
25.81	Eutric Cambisol
1.72	Eutric Leptic and Epileptic Cambisol
0.51	Eutric Stagnic Cambisol
0.22	Eutric Chromic Cambisol
0.18	Eutric Gleyic Cambisol
15.65	Dystric Cambisol
2.60	Dystric Epileptic Cambisol
0.76	Dystric Stagnic Cambisol
0.24	Dystric Gleyic Cambisol
0.48	Calcaric Cambisol
48.19	Total % of the territory

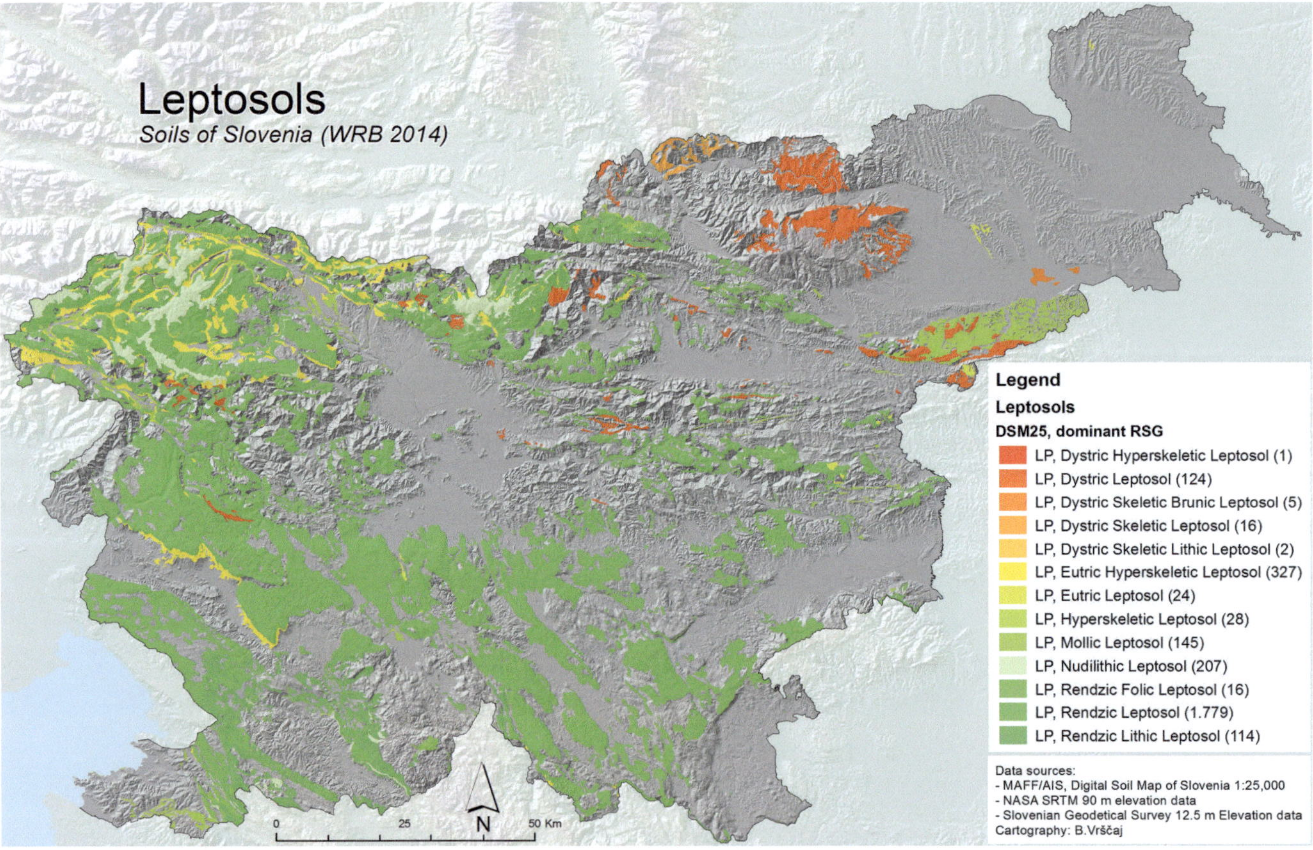

Fig. 6.7 The distribution of Leptosols in Slovenia

is mainly related to the topography and the typical weathering—slow dissolving of relatively nearly pure limestone. Leptosols dominate on steep slopes and rugged relief forms. The contact is frequently lithic; the A–R profile is very common on the limestone and A–C on the dolomite.

The proportion of Dystric Leptosols (4.9%) is significantly smaller than different forms of Eutric or Rendzic Leptosols (the sum of the latter two is 28.9% in total). The occurrence of Dystric Leptosols in Slovenia is mainly linked to non-carbonate parent material. Besides some patches of Triassic volcanic rocks in central–western Slovenia, various quartz, silicate, or non-calcareous parent material is mainly present in north-eastern Slovenia. The metamorphic rocks (schists) and grandiorite are present in the Pohorje massif and Kobansko, where Dystric Leptosols are associated with the Dystric Cambisols. The andesite, tonalite, and to a larger extent Tertiary volcaniclastic or coherent parent materials dominate in the area of Smrekovec between the city of

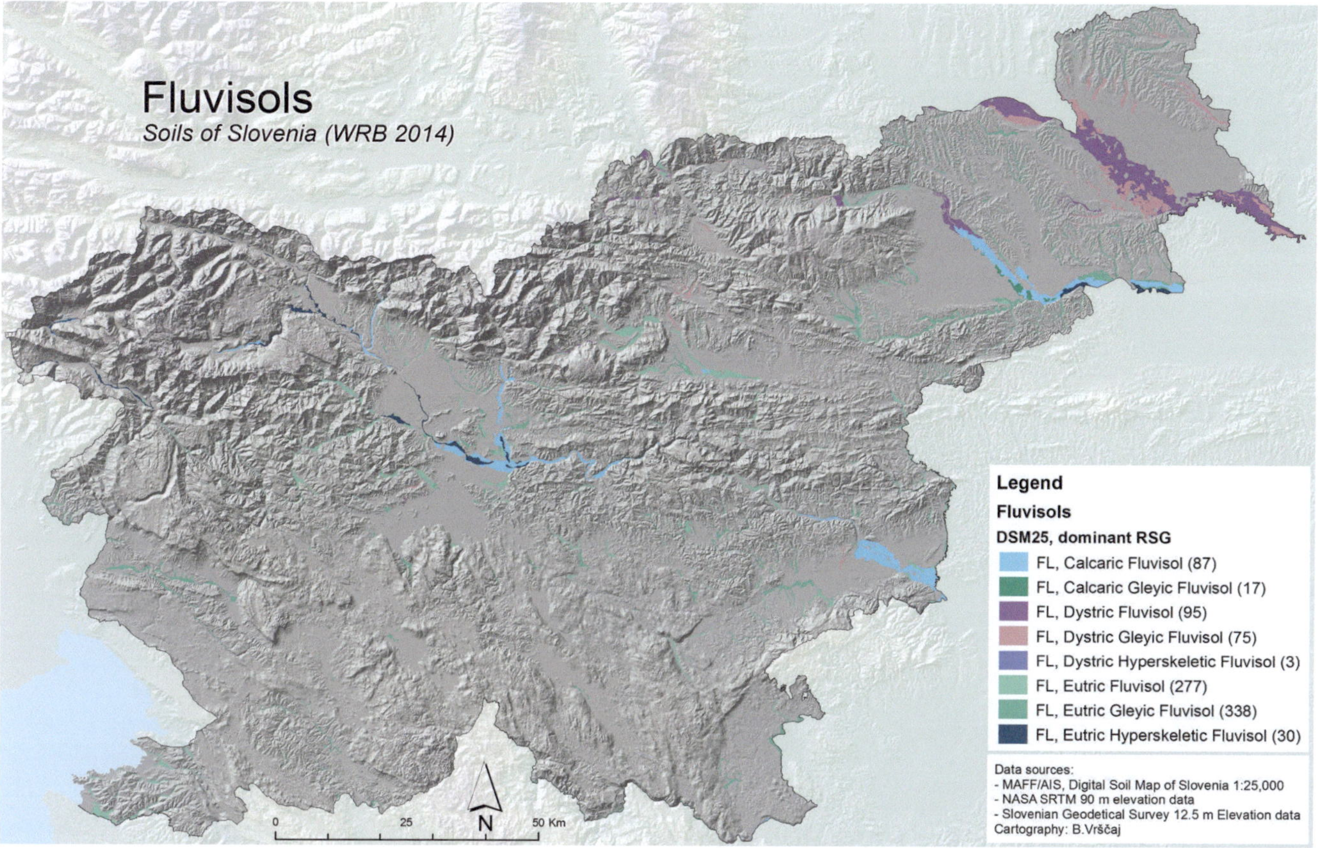

Fig. 6.8 The distribution of different types of Fluvisols

Velenje and the border with Austria. The relief is mountainous and hilly but not as rugged as the limestone/dolomite landscape; the relief forms are rounded, and the slopes are gentler. Large areas of Dystric Leptosols are rare; they are subordinate to Dystric Cambisols and predominate only at higher altitudes or on mountain ridges and steep slopes. Frequently, they develop on compact parent material more resistant to weathering such as andesite, tonalite, and grandiorite (Fig. 6.7, Table 6.2).

6.1.2.3 Fluvisols

The Sava and Savinja in central Slovenia, the Drava and Mura in the north-east (Stepančič et al. 1984a), as well as the Soča and Vipava in the west are the main rivers that cross the Slovenian areas with a large extent of Fluvisols. Fluvisols associated with smaller rivers and creeks are scattered all across the country, but the limestone or karst areas do not contribute significantly to the total of 5% of the Slovenian territory that belongs to this soil group. Eutric Fluvisols (2.5%) are present in the watersheds of the Savinja, Soča, and Vipava rivers (Lobnik et al. 1996b, 1989a; Oset 1987; Stepančič et al. 1986b) and especially along the entire course of the Sava River (Stepančič et al. 1987a, 1985). Large Eutric Fluvisol areas can be found in the Ljubljana basin and south of Krško, a city close to the border with Croatia. Dystric Fluvisols (1.6%) have been identified in north-eastern Slovenia mainly on silicate sandy deposits of the Drava and Mura rivers (Stepančič et al. 1984a). In general, the Fluvisols along the mentioned rivers, including the Idrijca River and the creeks of the pre-Alpine hills of north-western Slovenia, are predominantly sandy (Prus et al. 1985b). Particles of carbonate-rich parent material dominate the composition of the gravely, sandy, and silty, less frequently clayey, material that is deposited along the rivers. Clayey strata in Fluvisol profiles are rare. This is not valid for smaller Fluvisols areas of southern Slovenia, mainly the Dolenjska and Notranjska regions, as well as the Vipava River valley (Lobnik et al. 1996b). The texture of Fluvisols in these areas is predominantly loamy due to the flysch and/or limestone or dolomite origin of the deposited material (Prus et al. 1986b) (Fig. 6.8, Table 6.3).

Fluvisols in Slovenia are frequently under agricultural land use and less commonly under forests. The very different development of the soil profile in terms of the thickness and texture of the strata, the degree of hydromorphism, and the frequency of floods determine the agricultural quality and suitability. In the past, the Fluvisols were considered to be flooding area soils; their important functions are to buffer

Table 6.2 Leptosols by area, according to the DSM25

Leptosols by area		
%	*Soil type*	ha
20.87	Rendzic Leptosol	423,097.44
2.20	Rendzic Lithic Leptosol	44,636.86
1.99	Nudilithic Leptosol	40,386.27
1.68	Eutric Hyperskeletic Leptosol	34,094.80
0.40	Eutric Leptosol	8172.10
1.56	Mollic Leptosol	31,715.18
0.11	Rendzic Folic Leptosol	2131.50
0.08	Hyperskeletic Leptosol	1651.64
0.03	Hyperskeletic Rendzic Leptosol (colluvic)	682.16
3.80	Dystric Leptosol	77,021.42
0.24	Dystric Skeletic Leptosol	4947.71
0.04	Dystric Skeletic Brunic Leptosol	840.42
0.01	Dystric Skeletic Lithic Leptosol	123.53
0.00	Dystric Hyperskeletic Leptosol	89.91
33.03	Total % of the territory	669,590.93

Fig. 6.9 Fluvisols near the Mura River, north-west Slovenia (*Photo* T. Kralj)

Table 6.3 Fluvisols by area, according to the DSM25

Fluvisols by area	
%	*Soil type*
2.5	Eutric Gleyic Fluvisol
1.6	Dystric Fluvisol
0.9	Calcaric Fluvisol
5.0	Total % of the country

and mitigate floods. These functions were to a certain extent lost in the general perception. In comparison with other soil groups, Fluvisols were disproportionally urbanized due to their attractiveness for housing (Fig. 6.10).

6.1.2.4 Gleysols

Gleysols occupy 4.4% of Slovenian territory, an extent similar to the frequently associated Fluvisols (Vrščaj and Kralj 2014). A large area of predominantly Dystric Gleysols in the Mura plain in north-east Slovenia has developed on Holocene and clayey and loamy sediments (Fig. 6.11). The parent material originates in Pliocene clay, sand, and gravel sediments from Goričko to the north (Stepančič et al. 1984a). Predominantly, Dystric Gleysols are combined to a minor extent with Eutric Gleysols and transitional forms such as eutric/dystric qualifiers. The pH of the Go horizon of Dystric Gleysols is typically 4.3 (determined by 1 m KCl or $CaCl_2$), while the base saturation is frequently around 48% (measured by 1 m NH_4OAc, pH 7) ("Talni profili, digitalna baza podatkov (TP) [soil profile; digital database (SP)]", 2001). The Go horizon of Eutric Gleysols has a pH of approximately 5.9 and a typical base saturation of approximately 70%. In both cases, a Go horizon overlays a G horizon at a depth of between 25 and 55 cm.

Fig. 6.10 Urbanized Fluvisols along the Savinja River, central Slovenia (*Photo* B. Vrščaj). The photograph above shows the pattern of urbanization. On the *left side* of the picture are Eutric Cambisols and Leptosols on the *upper terrace*. The *thin line* of trees in the centre separates the *upper terrace* from the *lower terrace* covered by Fluvisols. The new settlements emerged along the riverbank (*right*)

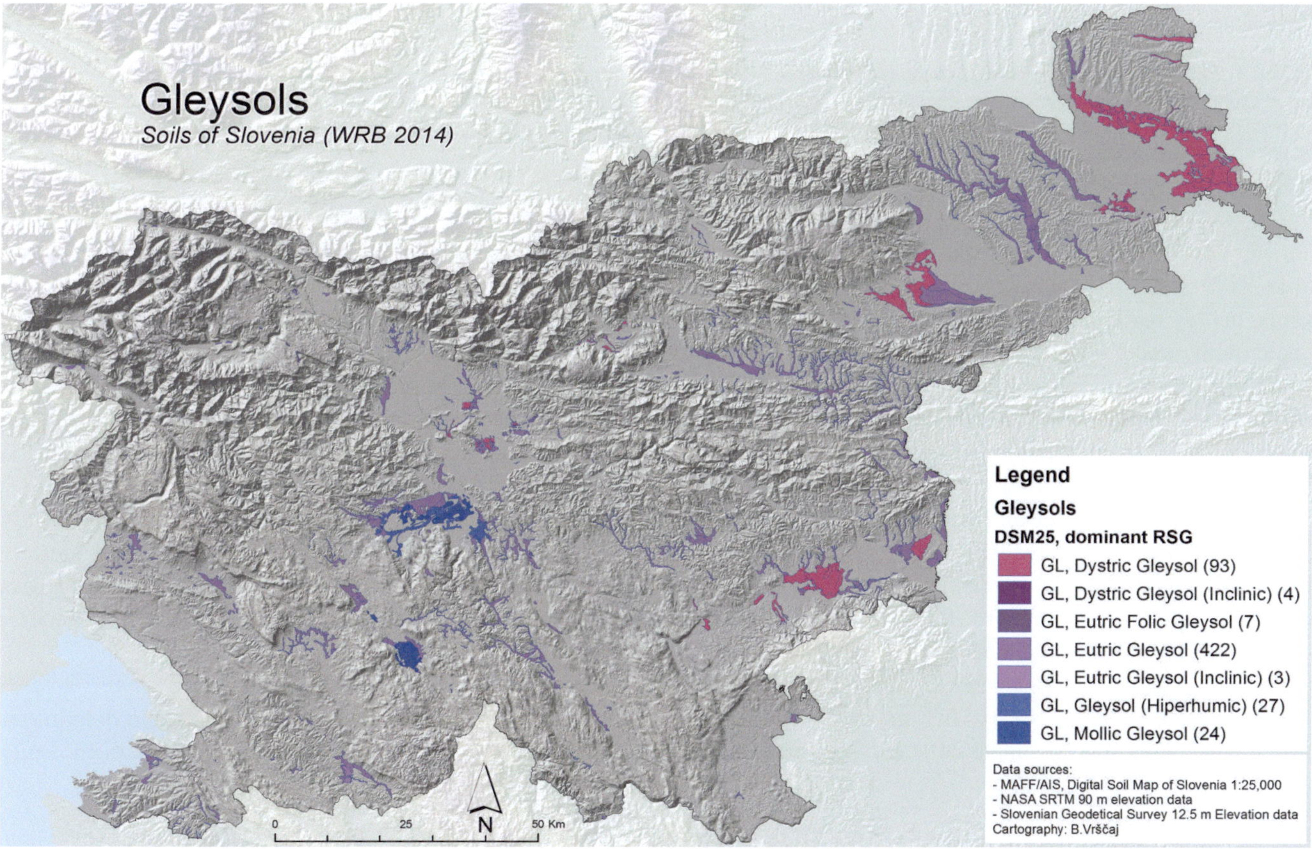

Fig. 6.11 The distribution of different types of Gleysols

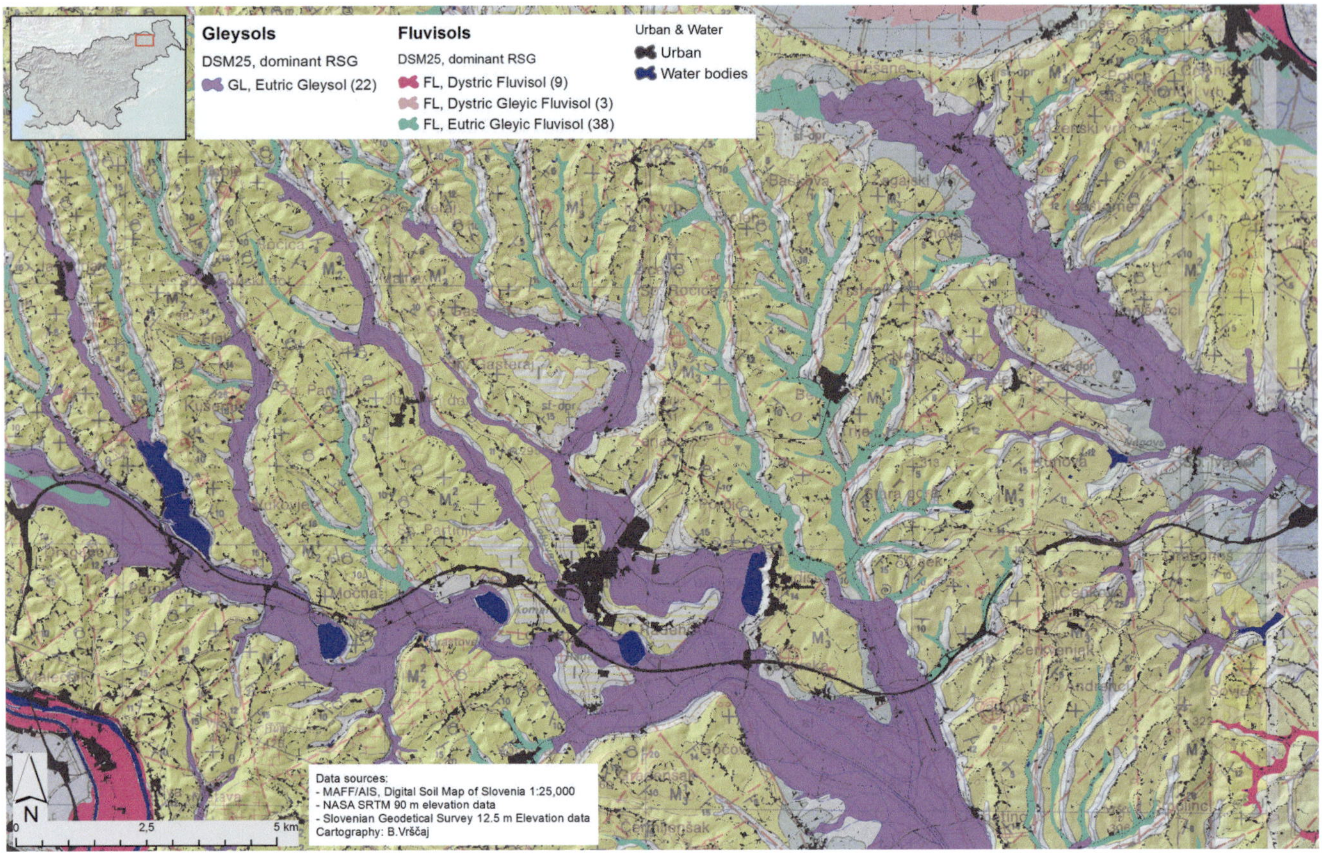

Fig. 6.12 Eutric Gleysols and Fluvisols in the valleys of the Slovenske Gorice area, north-east Slovenia. The extent of Gleysols within the wide valleys of the Pesnica (*centre left*) and Ščavnica rivers (*upper right*) and tributaries. Fluvisols cover the *bottoms* of the *narrow valleys* that intersect the hilly region of Slovenske Gorice, which is famous for good vineyards

The parent material of the large hilly area of Slovenske Gorice, which stretches from the Austrian border in the north-west to Croatia in the south-east, are Miocene sediments, mainly marl and sandy marl. Both of them are locally mixed with predominantly quartz gravel. Washed off the slopes, the fine-textured materials accumulate at the bottoms of numerous valleys (Prus et al. 1985a, 1986a, b; Zupan et al. 2002b). This constitutes the main mineral fraction of the material on which Eutric Gleysols have developed (Fig. 6.12). A large extent of Gleysols can be found in the Ledava and Ščavnica River valleys. At selected locations, the gravely parent material is cemented with Fe and Mn and together with large concretions forms a hard layer with petroplinthic properties (Stepančič et al. 1984a) Table 6.4).

A large area with Eutric and Dystric Gleysols is the Dravsko-Ptujsko plain south of the Pohorje massif (Fig. 6.14). The parent materials are clayey alluvial deposits from the neighbouring area of Plio-Quaternary clayey and sandy sediments and gneiss from the Pohorje massif.

Mollic and Eutric Gleysols can be found south of the city of Ljubljana in the area of the Ljubljana Moor. Both soil types are frequently qualified as hyperhumic and combined with Histosols, the main soil type of the Ljubljana Moor area, which has an approximate size of 135 km^2 (Fig. 6.15).

In the coastal region north-east to south-east of the city of Koper Eutric, Gleysols are present on the bottoms of a few wide and many narrow river valleys that intersect areas of Eutric Calcic Cambisols formed of Eocene flysch. Heavy loamy-clayey texture, a high base saturation, the frequent presence of carbonates, and a high groundwater level define the agricultural quality of these Gleysols. In the area of Jurassic, Triassic, and Cretaceous limestone and dolomite of the Dolenjska and Notranjska regions in southern Slovenia,

Fig. 6.13 Vineyards of the Slovenske Gorice area, north-east Slovenia (*Photo* T. Kralj)

Table 6.4 Gleysols by area, according to the DSM25

Gleysols by area	
%	*Soil type*
2.5	Eutric Gleysol
1.5	Dystric Gleysol
0.2	Mollic Gleysol
0.2	Gleysol (hyperhumic)
4.4	Total % of the territory

Eutric Gleysols, as well as Cambisols, Leptosols, and Luvisols, formed (Lobnik et al. 1987; Prus et al. 1986a, b) (Fig. 6.16).

6.1.2.5 Stagnosols

Stagnosols stretch over 3.8% of the Slovenian territory. Occurrence is predominantly localized to the Pleistocene, Pliocene, and Plio-Quaternary clayey and silty-clayey terraces at the borders of wide plains in Prekmurje in the north-east and Krško in the south-east, the wide Savinja valley in central Slovenia, the Vipava valley in the west, and north of Ormož in the east (Stepančič et al. 1984a, 1986a) (Fig. 6.17). According to the DSM25, the total extent of Stagnosols in Slovenia is approximately 76,500 ha (Vrščaj and Kralj 2014) (Table 6.5).

Depending on the source of the parent material, Stagnosols are classified as eutric or dystric. The typical pH of the Bg horizon of Dystric Stagnosols is between 4.5 and 5.0 (measured in 1 m KCl), while the base saturation is approximately 30% (by 1 m NH4OAc, pH 7). In Eutric Stagnosols, the pH and base saturation of Bg and G horizons of Eutric Stagnosols can vary significantly, with a pH between 5.5 and 7.1, while the base saturation is between 60 and 90% ("Talni profili, digitalna baza podatkov (TP) [soil profile; digital database (SP)]", 2001). In both sub-types of Stagnosols, the characteristic Bg horizon is very compact and dense, hard in the dry period and plastic in the wet period of the year.

The distribution of Stagnosols in the Savinja valley is rather typical for the occurrence of Stagnosols in basins and large valleys. Pliocene sediments, mainly clays and loams in some areas mixed with predominantly quartz gravel, have covered the valley in the past. In the Holocene, the heavy-textured material in the central part of the valley was eroded frequently to the underlying rocks and filled with carbonate gravel and sands. Nowadays, the altitude of the remaining Pleistocene terraces with Stagnosols is 15–18 m above the gravely sandy valley bottom, where Eutric Cambisols and Fluvisols predominate (Oset 1987; Radišek et al. 2014a, b). In the case of the Savinja valley, Stagnosols surround the Cambisols in the central part of the valley

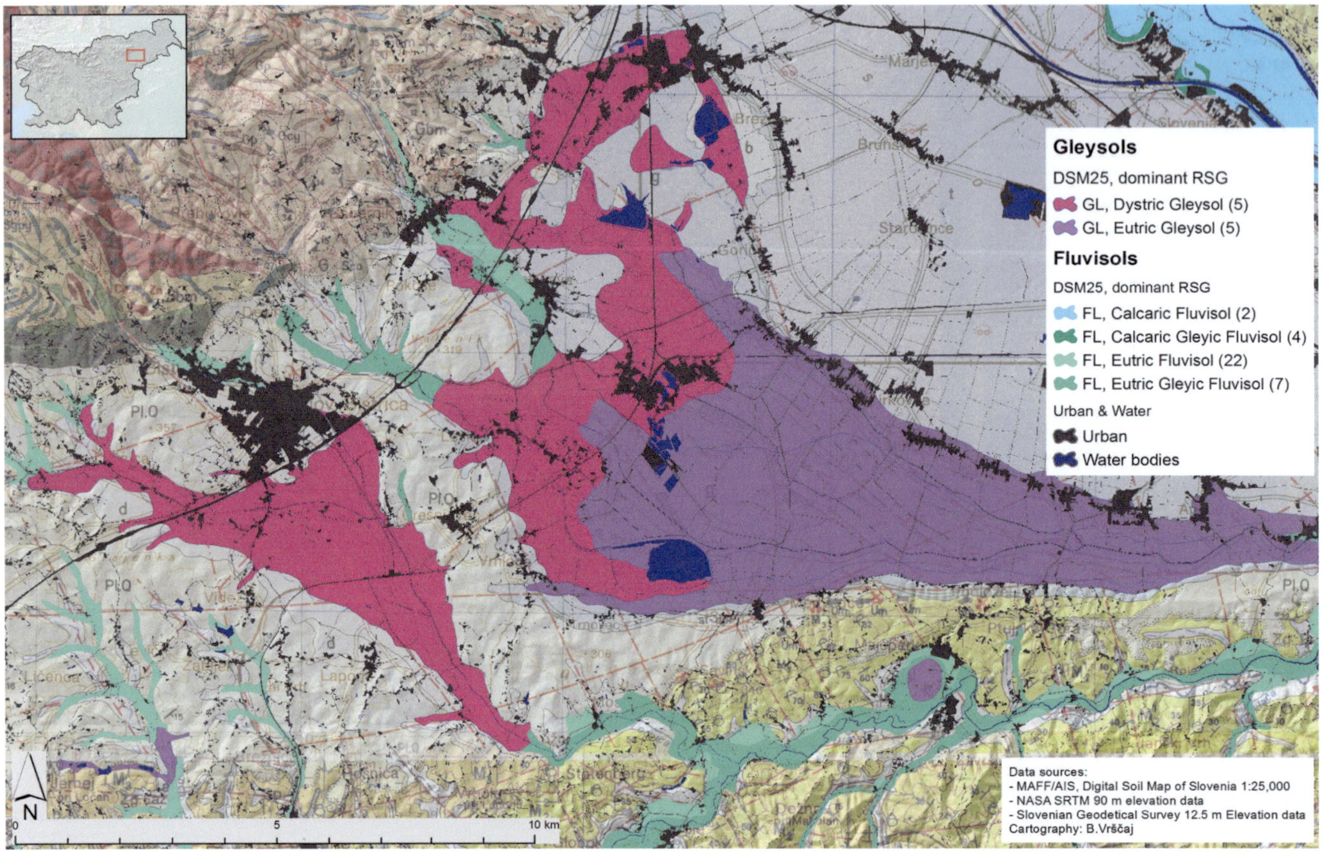

Fig. 6.14 Eutric and Dystric Gleysols south of the Pohorje massif, north-east Slovenia

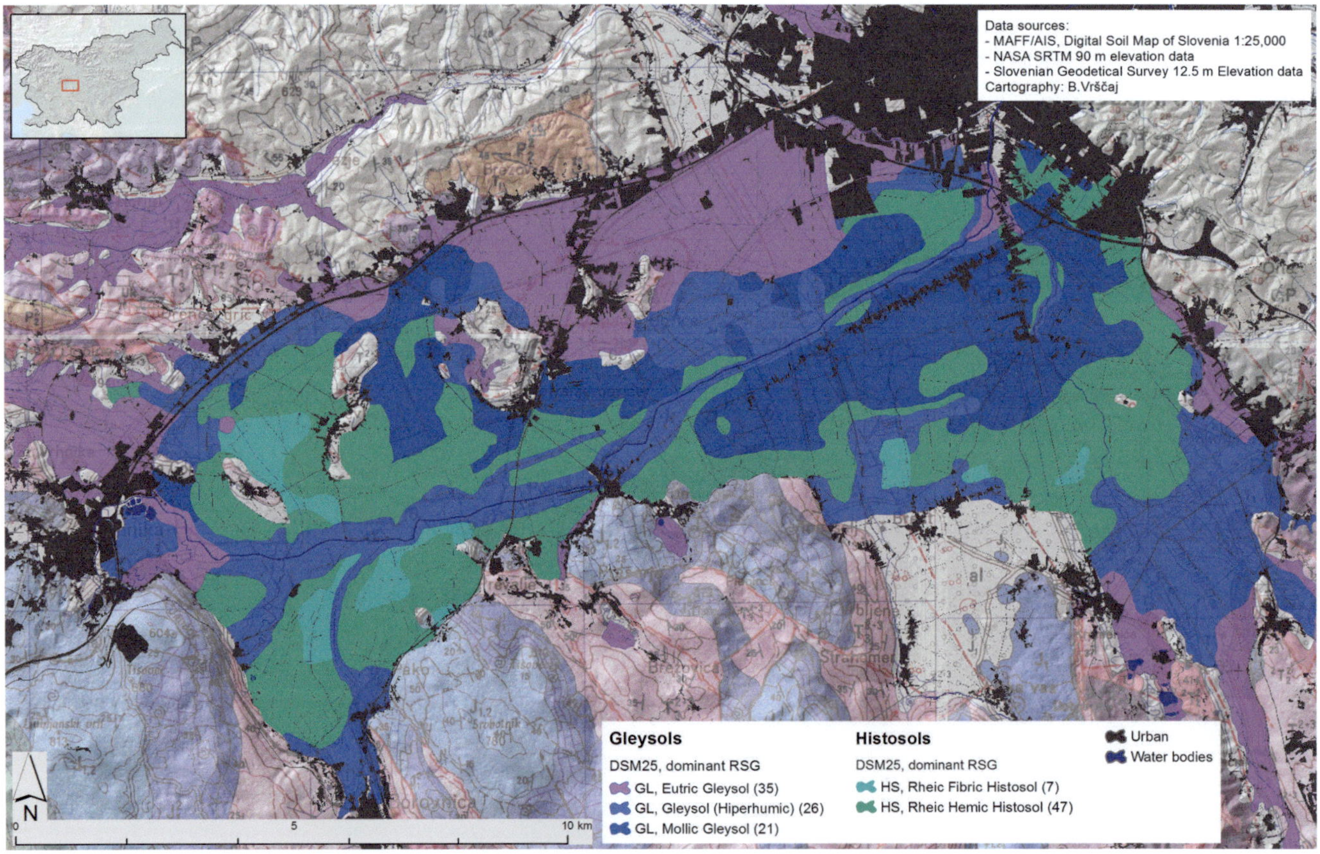

Fig. 6.15 Eutric Gleysols and Histosols at the Ljubljana Moor, central Slovenia

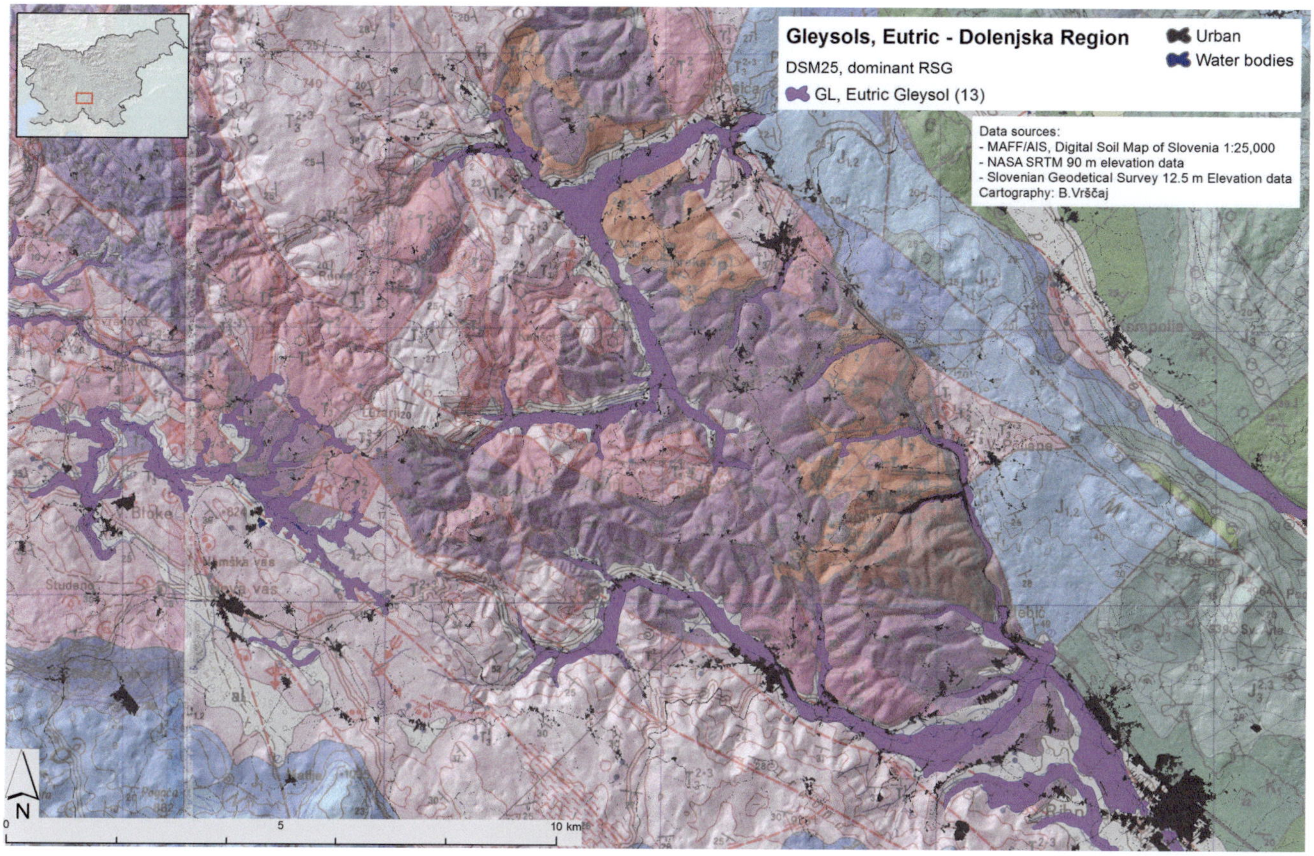

Fig. 6.16 Eutric Gleysol areas intersecting Eutric Cambisols and Luvisols in the areas of Jurassic, Triassic, and Cretaceous limestone and dolomite of the Dolenjska Region

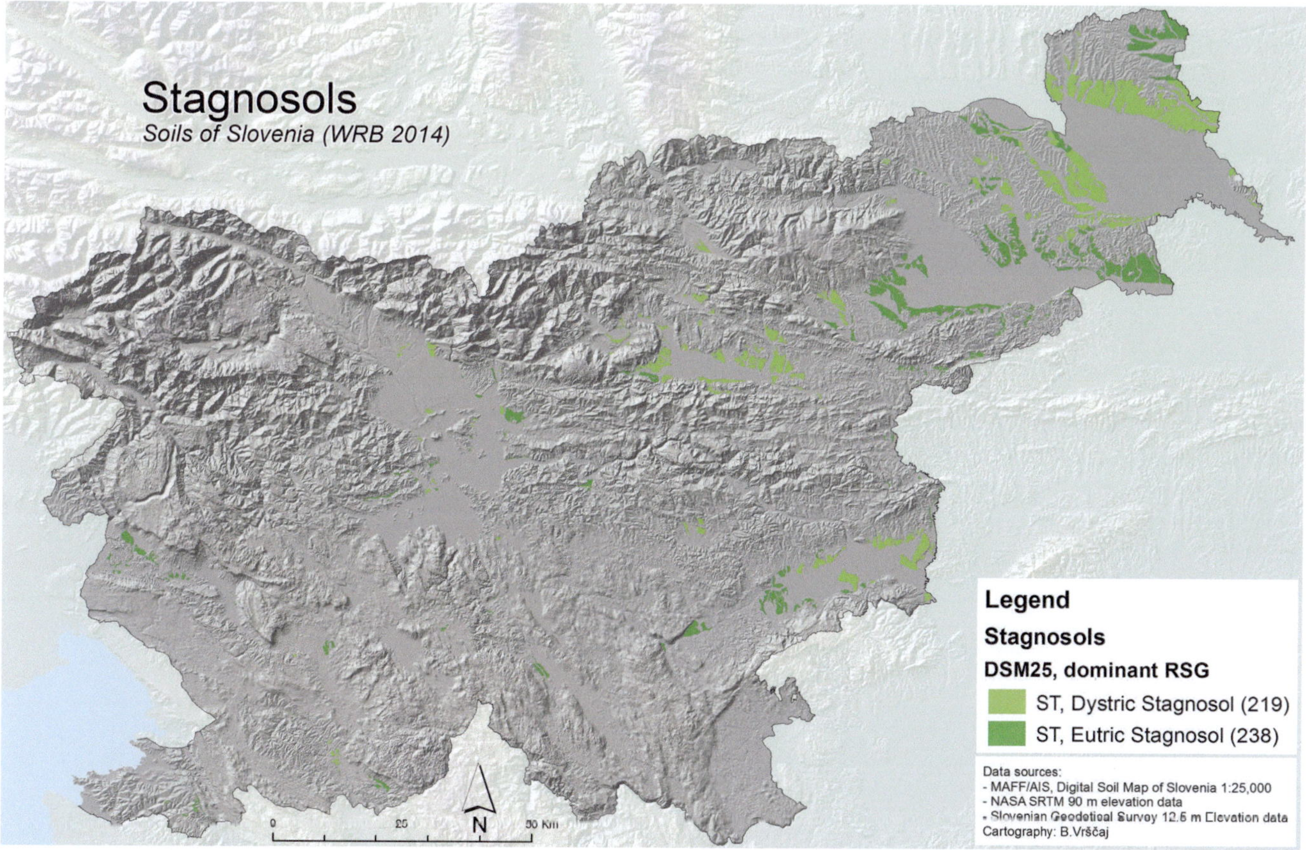

Fig. 6.17 The distribution of Stagnosols in Slovenia

Table 6.5 Stagnosols by area, according to the DSM25

Stagnosols by area	
%	*Soil type*
2.2	Dystric Stagnosol
1.6	Eutric Stagnosol
3.8	Total % of the territory

(Fig. 6.18). A similar situation can be identified in the Prekmurje plain (Fig. 6.19).

6.1.2.6 Luvisols

The extent of Luvisols in Slovenia is very limited. In total, they occupy only approximately 1.7% of the country (Vrščaj and Kralj 2014). Two main areas of Luvisols were identified, the upper Sava valley and the region of Bela Krajina (Fig. 6.20, Table 6.6).

The Luvisols in the upper Sava valley formed on glacial outwash terraces (Fig. 6.22). The Luvisols from that area are very old. According to determination by the Be10 method, the age of two minor and younger areas is 32 and 44 thousand years (i.e. ka), respectively, while the age of the largest homogenous Luvisol areas has been determined to be approximately 62 ka. Three old Luvisols that are situated on terraces of different age north-west and south-east of the city of Kranj have been determined to be 450, 960, and 980 ka, respectively. By far, the oldest are Luvisols on the small remaining terraces distributed in a patchy pattern north and north-west of Kranj, which have been determined to be between 1–1.8 Ma (Jaecks Vidic 1994; Vidic et al. 1991; Vidic and Lobnik 1997).

The other Luvisol area important in terms of extent is Bela Krajina (Fig. 6.23). The age of Luvisols in this area has not been investigated, but in some opinions, it is much younger, approximately 30–40 ka. Luvisols there have developed on Cretaceous dolomite, Jurassic limestone, and Pleistocene clayey deposits. They are combined with Acrisols, with Cambisols, and to a minor extent with Stagnosols. The specific land use is birch forests. The name of the region

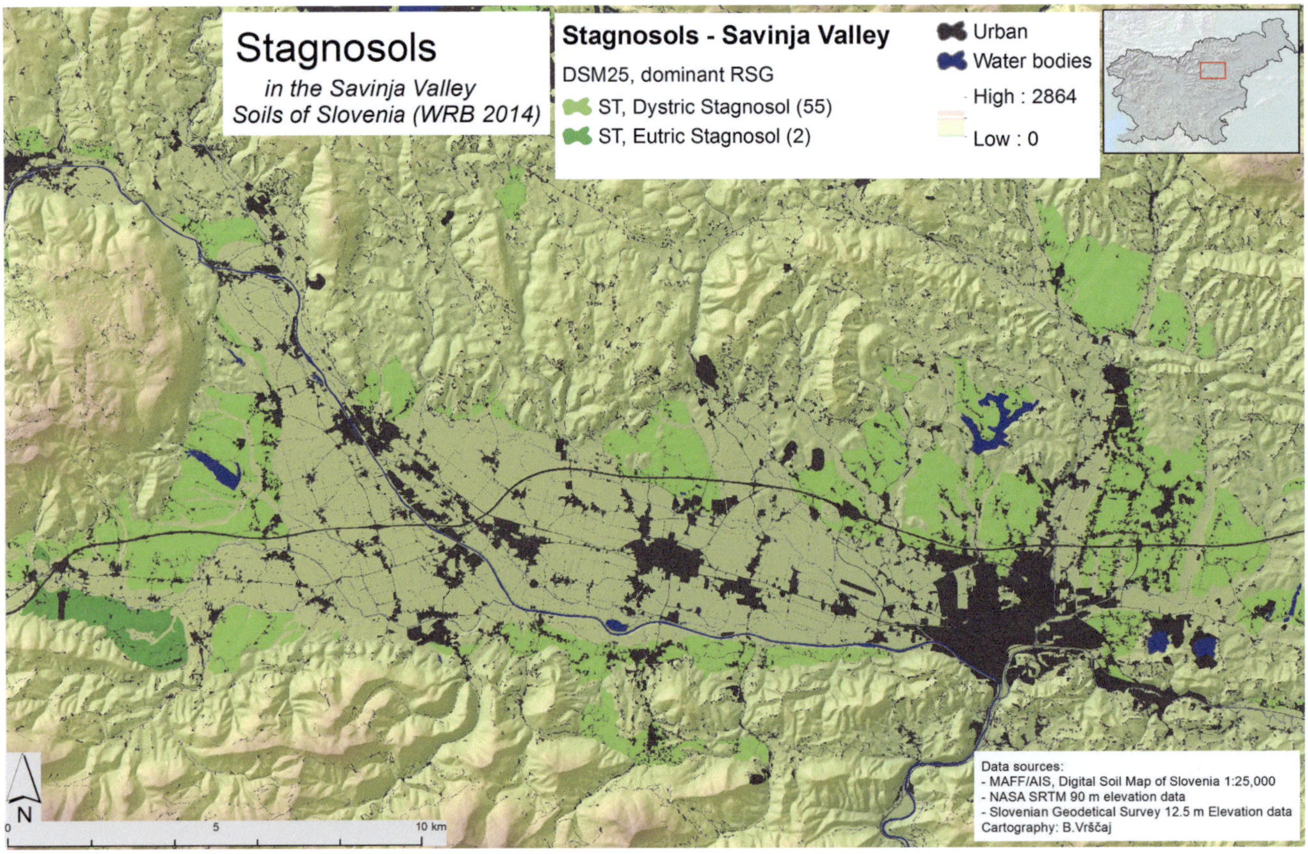

Fig. 6.18 The distribution of Stagnosols in the Savinja valley, Central Slovenia

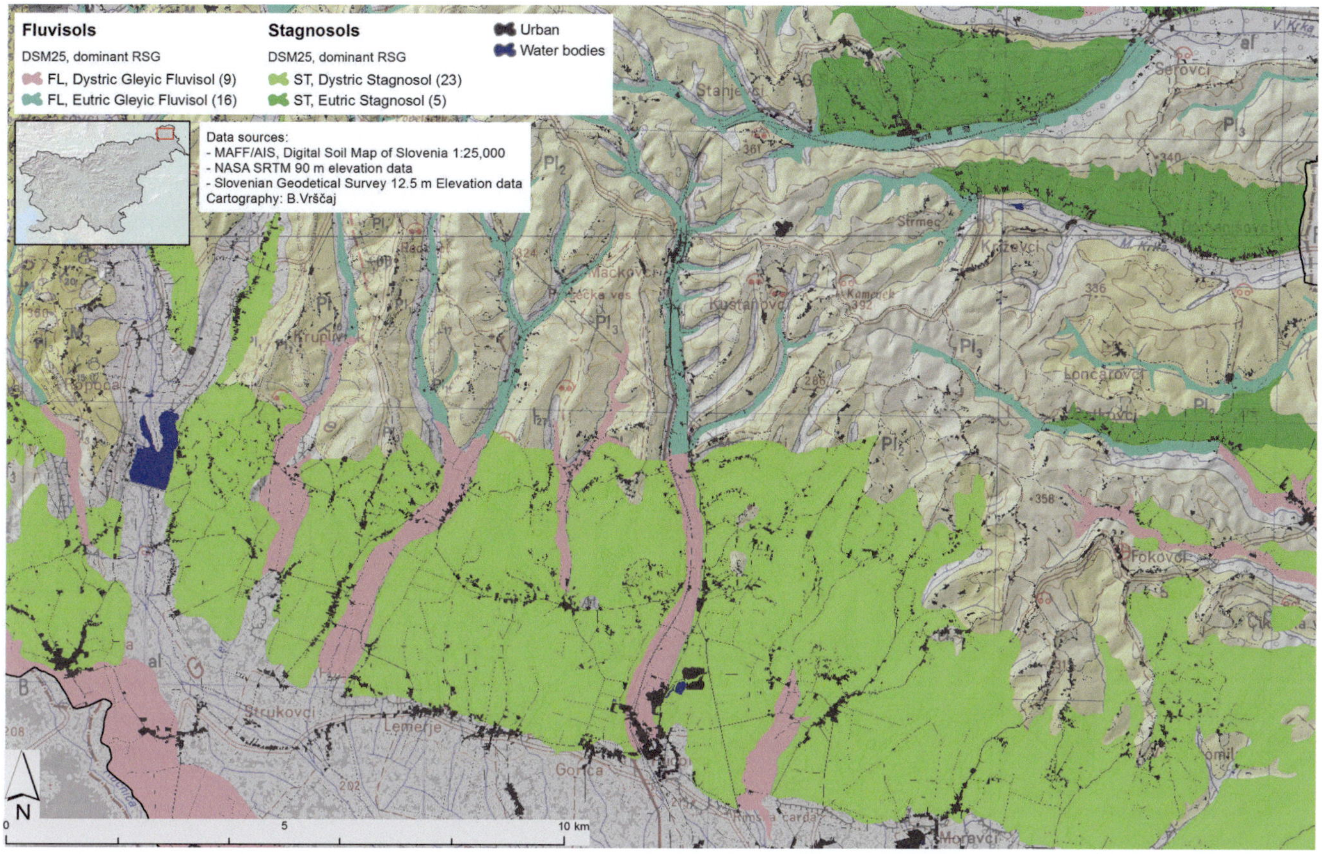

Fig. 6.19 The distribution of Stagnosols on the Pleistocene terraces of the northern border of the Prekmurje plain, north-east Slovenia

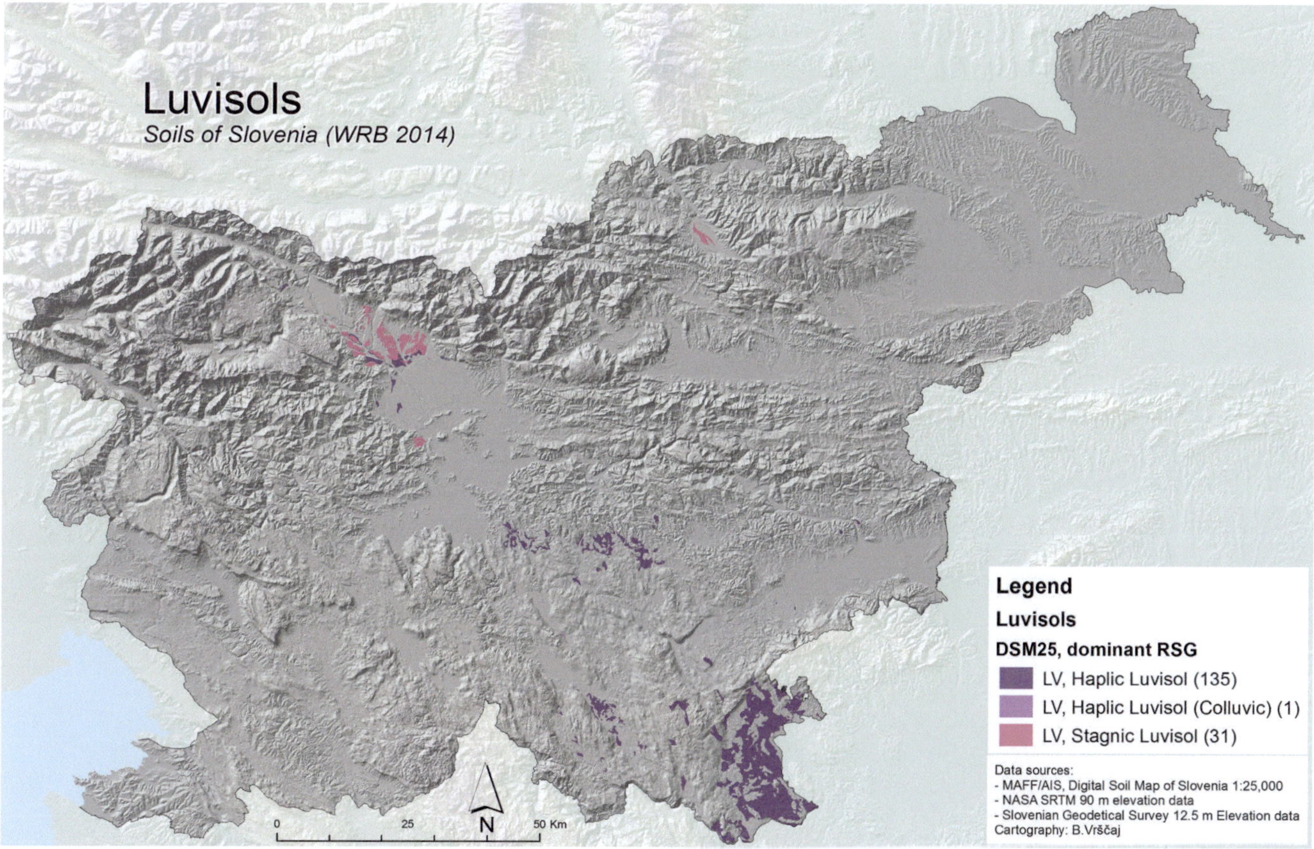

Fig. 6.20 The distribution of Luvisols in Slovenia

Table 6.6 Luvisols by area, according to the DSM25

Luvisols by area	
%	*Soil type*
1.46	Haplic Luvisol
0.17	Stagnic Luvisol
0.04	Gleyic Luvisol
0.02	Haplic Luvisol (Colluvic)
1.69	Total % of the territory

Bela Krajina stems from the white-coloured trunks of birches ("*bela*"—white in Slovenian). According to Stritar (1990b), agricultural practices in the past have significantly contributed to the properties of Luvisols. Namely, by harvesting ferns in sparse birch forests, farmers have significantly contributed to the removal of nutrients and calcium from the forests to the fields and thus additionally acidified the soils under forest (Fig. 6.21).

6.1.2.7 Histosols

The total area of soils that belong to the Histosol soil group is very limited in Slovenia. Approximately 10,400 ha amounts to only 0.5% of the country (Fig. 6.26, Table 6.7).

According to the WRB, 52% of soils that can be classified as shallow Dystric Folic Histosols are located within small areas scattered in Alpine valleys, high plateaus, and predominantly western and northern slopes of the Julian Alps in north-west Slovenia, the Pohorje massif in the north-east, the Savinja Alps in the north, as well as in some dolinas and slopes at high altitudes of Mt. Snežnik (1796 m. a.s.l.) in the south. The parent material is mainly limestone, dolomite, calcareous moraine, and scree (Fig. 6.26).

Rheic Hemic Histosols and Rheic Fibric Histosols are detected in the area of the Ljubljana Moor, a lowland south of the capital city Ljubljana at an average altitude of 290 m.a.s.l. (Fig. 6.15). The area is one of the most southern peat moors that was drained and turned into arable land by a decree issued in 1769 by the Empress Maria Theresa. Following her orders, an inventory of the Ljubljana Moor had to be made, a plan for its reclamation prepared, and finally, the area had to be turned into land suitable for farming. The basin was slowly drained after opening the Gruber channel in 1792. Almost the entire area was turned into arable land and grassland (Figs. 6.28 and 6.29). In the twenty-first century, only minor plots remained under semi-natural birch and alder forests (Fig. 6.27), which are now protected. Hemic Histosols

Fig. 6.21 The birch forests in Bela Krajina, south-east Slovenia (*Photo* T. Kralj)

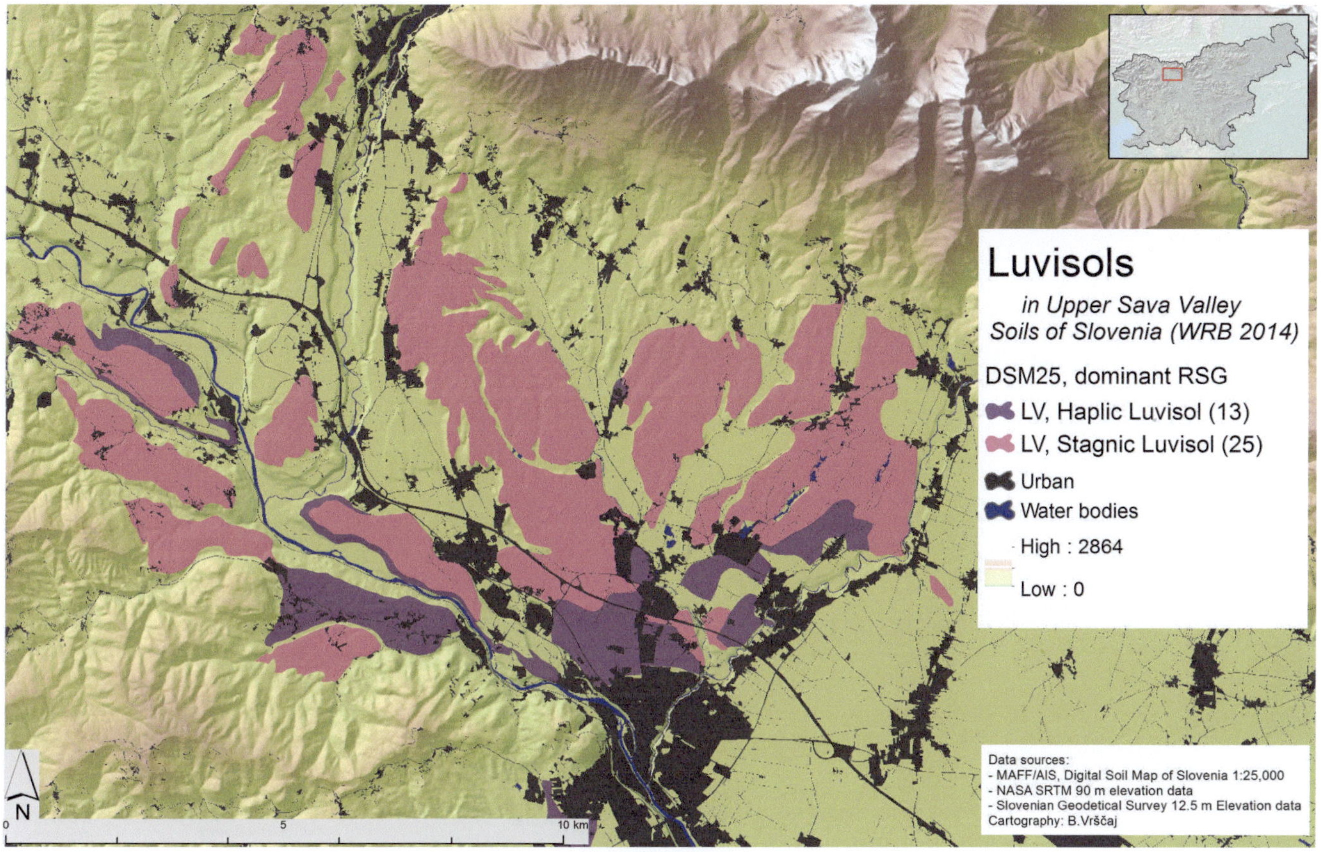

Fig. 6.22 The distribution of Luvisols in the Upper Sava Valley, Gorenjska region, north-west Slovenia

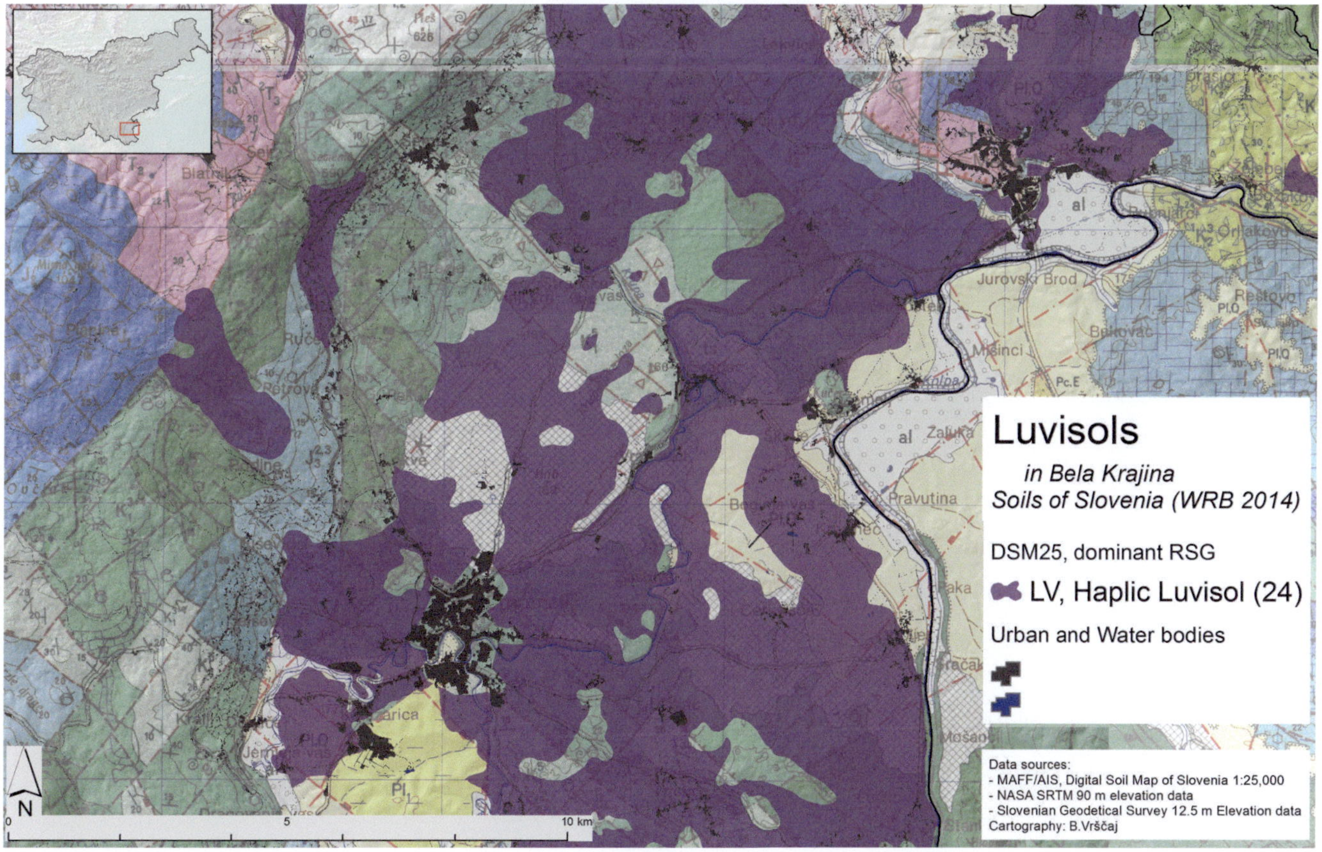

Fig. 6.23 The distribution of Luvisols in Bela Krajina, south-east Slovenia

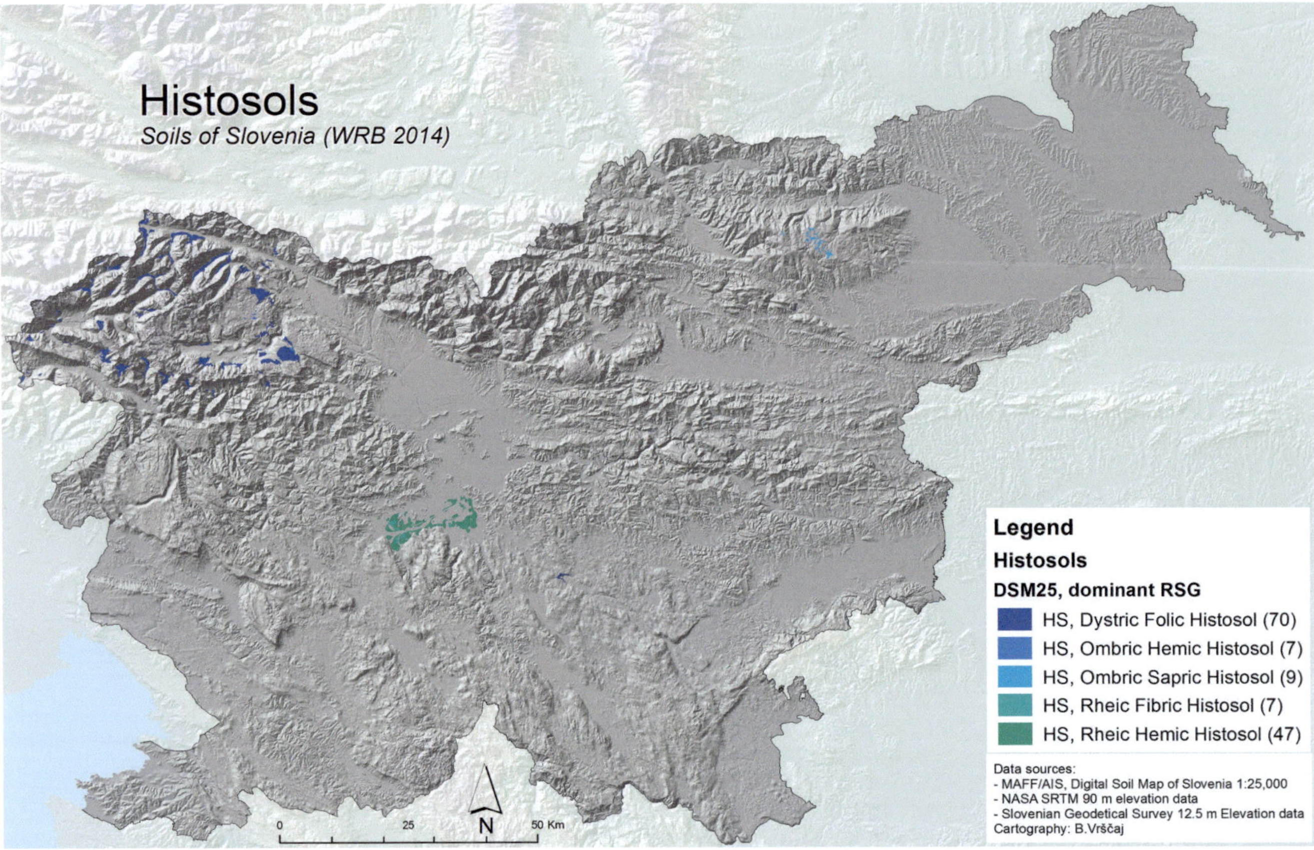

Fig. 6.24 The distribution of Histosols in Slovenia

Table 6.7 Histosols by area, according to the DSM25

Histosols by area	
%	*Soil type*
0.27	Dystric Folic Histosol
0.19	Rheic Hemic Histosol
0.02	Rheic Fibric Histosol
0.03	Ombric Sapric Histosol
0.01	Ombric Hemic Histosol
0.51	Total % of the territory

Fig. 6.25 The Ljubljana Moor, central Slovenia (*Photo* T. Kralj)

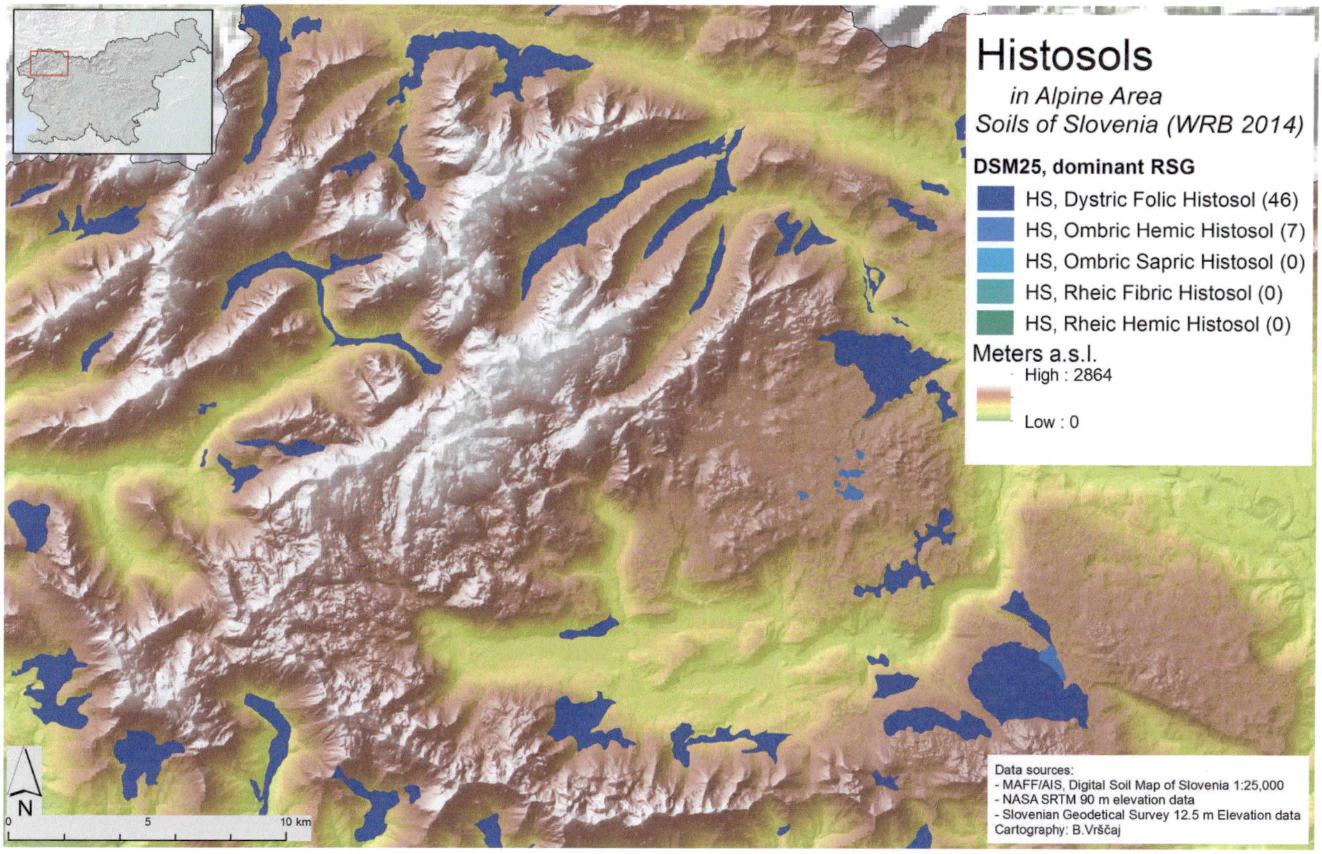

Fig. 6.26 The extent of Histosols in valleys, plateaus, and the slopes of the Julian Alps, north-west Slovenia

Fig. 6.27 The semi-natural vegetation of the Rheic Fibric Histosols of the Ljubljana Moor (*Photo* B. Vrščaj)

Fig. 6.28 An area of anthropogenized Rheic Hemic Histosols of the Ljubljana Moor under extensively used grasslands (*Photo* B. Vrščaj)

Fig. 6.29 Rheic Hemic Histosol on former arable land turned to grassland, Ljubljana Moor (*Photo* B. Vrščaj)

prevail under the arable land, while Fibric forms are the most common under the forests.

The extent of Ombric Histosols is limited to very marginal areas of raised bog on the Pokljuka Plateau (1200 m.a.s.l.), on top of the Pohorje massif (1500 m.a.s.l.), as well as very small areas in the Ljubljana Moor. Being very rare in Slovenia, all these areas are protected for nature conservation purposes.

7 Soil Information

Borut Vrščaj

Abstract

The first Slovenian digital soil information was collected in the late 1970s and early 1980s within the development of the federal Yugoslav Soil Information System. At that time, soil maps were kept separately in paper form, while the alphanumeric soil profile information was coded and integrated into the database maintained on mainframe computers. The rapid development of computer hardware and software in the early 1980s triggered the collection, processing, and subsequent mapping of soil information using desktop computers utilizing UNIX and Windows operating systems. Several software packages were developed to collect, process, store, and present alphanumeric data on physical and chemical soil properties—soil profile datasets. The digitalization of soil maps began in 1987 and was completed in 1999 at a scale of 1:25,000. The Slovene Soil Information System was designed in 1995 and completed in 2000. Nowadays, digital soil information is distributed at several institutions and widely used. A significant proportion thereof is available through contemporary web-GIS services available to the general public, researchers, and students.

7.1 The History of Digital Soil Information

7.1.1 The Yugoslav Soil Information System

The first Slovenian digital soil information was collected in late 1970s and early 1980s. The design of databases and attribute definitions were standardized as a part of the *Zemljišni informacioni sistem Jugoslavije* (*The Soil Information System of Yugoslavia*) (*Zemljišni informacioni sistem Jugoslavije—kodni sistem* [*Soil Information System of Yugoslavia—The Coding System*], 1978). The development of the system was a federal Yugoslav project coordinated by representatives of the soil science societies of individual Yugoslav Republics and thirty-three institutes or universities. Slovenia was represented by the Biotechnical Faculty, the Institute for Forest and Wood Management, and the Agricultural Institute of Slovenia (AIS). The entire process of data collection and interpretation as well as the soil survey was the responsibility of the Republics. The computer system was alphanumeric since the graphic capabilities of the hardware at that time were far too meagre to store soil information in the form of digital maps. Consequently, the soil maps were kept separately in paper form while the alphanumeric soil profile information was coded and integrated into the database. The Soil Information System coding book defined the extensive dataset of attributes in terms of four groups. The soil profile dataset was composed of fifteen attributes related to location, geomorphology, parent material, groundwater, vegetation community, and land use. This important set was related to twenty-one productive properties of soils, e.g., soil depth, stoniness, rockiness, rooting depth, erosion, and frequency of floods (Fig. 7.1). The morphology of soil horizons was described by twenty-three attributes, while the analytical table was populated by thirteen measurements, each of them with mandatory information on the method used. For that time, the structure of the digital attribute tables was contemporary and would, if completed, represent a modern and well-organized digital soil database of measured soil properties.

B. Vrščaj (✉)
Department of Agricultural Ecology and Natural Resources, Centre for Soil and Environment Research, Agricultural Institute of Slovenia, Ljubljana, Slovenia
e-mail: Borut.Vrscaj@kis.si

B. Vrščaj et al., *The Soils of Slovenia*, World Soils Book Series,
DOI 10.1007/978-94-017-8585-3_7

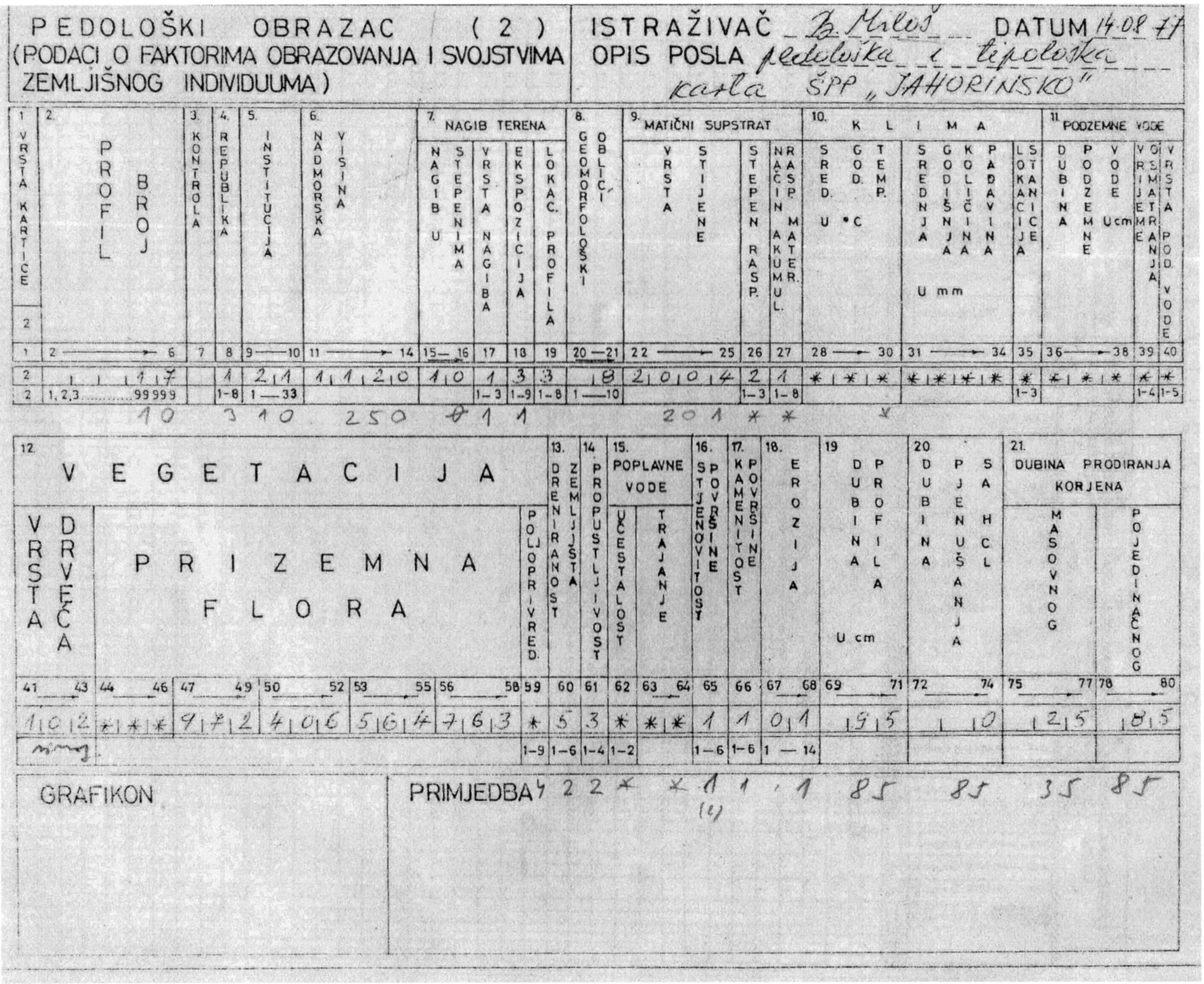

Fig. 7.1 The soil Information system of Yugoslavia—page from the soil coding manual

The data were collected and centrally maintained on mainframe computers located at the UPIM computer centre in Sarajevo. In Slovenia, the Centre for Soil Science and Environment of the Biotechnical Faculty, University of Ljubljana (CSSE) assumed responsibility for the coordination of the project and collected and stored the soil profile information in the VAX mainframe computer located in Ljubljana. In the next stage, the Slovenian soil profile information was sent to Sarajevo to be appended to the seamless Yugoslav soil dataset. The fate of these databases is unknown, but it is very likely that they were lost in the turmoil during the war in Bosnia and Herzegovina in the early 1990s (Čustović 2009). The majority of Slovenian soil profile data collected at the VAX mainframe in Ljubljana remained preserved, and after being exported to comma-delimited files, it became a part of the archives of the CSSE. Parallel to the development of the centralized alphanumeric soil profile database, the soil scientists in the various Republics carried out a soil survey to elaborate the soil maps with a scale of 1:50,000. A common methodology was prepared (Ćirić 1984b) and followed to elaborate the soil maps. In contrast to other Republics where 1:50,000 topographic maps were used, the Slovenian soil surveyors used contemporary military topographic maps with a scale of 1:25,000. The soil survey in individual Republics had been completed to various degrees at the time of the breakup of Federal Yugoslavia in 1991. By the end of 1992, approximately 65% of Slovenia had been mapped.

7.1.2 Early Computer Hardware

In the early 1980s reduced the cost of IT equipment and started the era of desktop computers and then that of

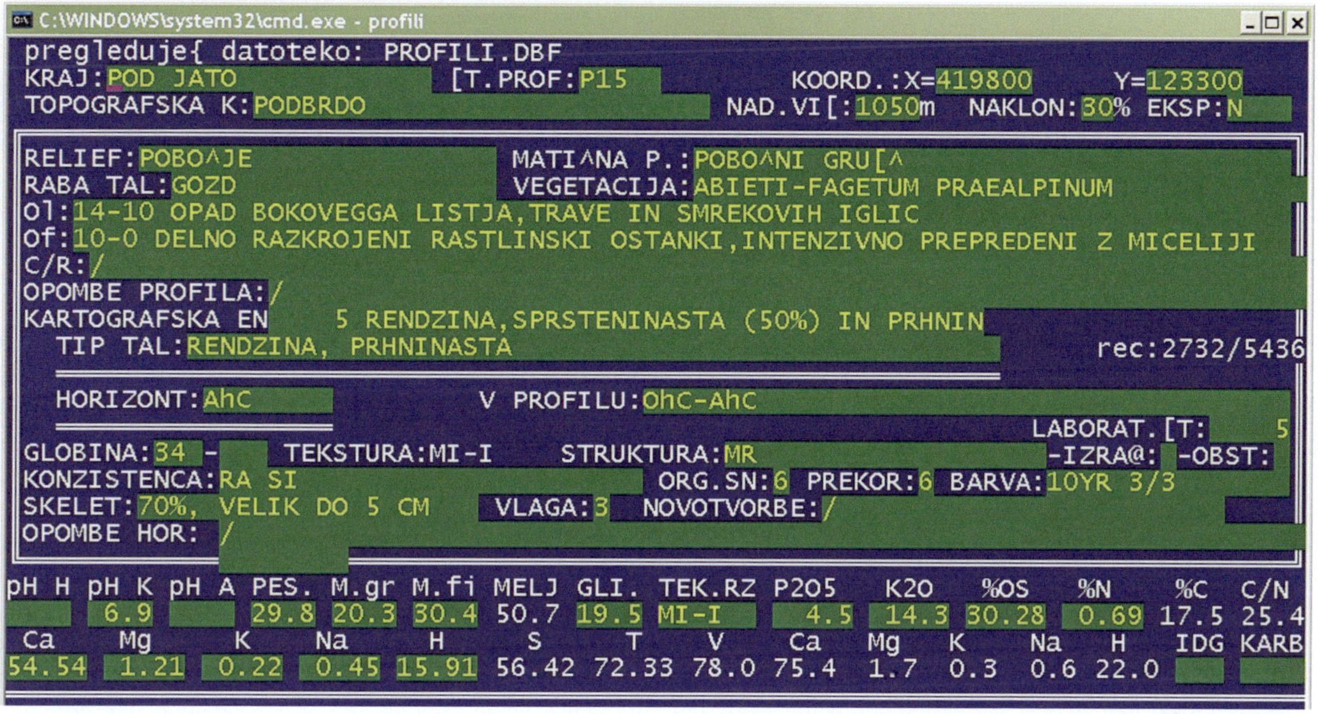

Fig. 7.2 A Profili.exe computer screen

personal computers (PCs). In 1983, the soil profile data archived by the CSSE were transferred to the first desktop computer available to the Slovenian soil community, a Commodore 64, a small and very popular 8-bit home computer launched on the market in January 1982. Despite the modest hardware capabilities, the complete digital soil information was entered into the C64 with great enthusiasm. Additional profiles were added, and the first Slovenian digital soil database was developed. The data were kept on cassettes and subsequently on floppy disks and regularly backed up to prevent loss of data. The C64 did not have a true computer monitor; a small portable black and white TV was used instead. A big—and for soil scientists important—step forward was the purchase of an IBM ProPrinter dot matrix printer. After costly reprogramming of the printer's EPROM to utilize the Slovenian character set, the first soil profile printouts were available to soil scientists—a valuable complement to their research reports. By such, the era of wide typewriters came to an end. The C64 was replaced in 1985 with a Commodore 128. Besides the usual Commodore Basic and due to its Zilog Z80 CPU, the C128 was able to run the CP/M operating system, the same OS as was used by the superior and expensive Triglav computer produced in 1984 by Iskra, the national computer company. The present author remembers the enthusiastic statement of one of his colleagues that this computer was about to store and process all Slovenian soil data. Actually, the Commodore era was very short. In 1985, the first genuine IBM XT computer ousted the Commodores and was quickly replaced by a computing "monster"—an AT 286 processor-based computer equipped with 4 MB of RAM, a 20 MB 80 ms average access time hard drive and Hercules graphics. From that point on, the hardware was frequently upgraded to 386 and 486 processor-based computers, which in 1991 were replaced by a then superior Silicon Graphics Iris (SGI) Indigo workstation equipped with one of the most sophisticated graphics capabilities of its era. The Indigo was designed to run IRIX, a SGI version of UNIX. The rapid development and affordability of Microsoft-based PCs and the relatively high cost of SGI maintenance and especially the expensive GIS software licenses were the main reasons to replace the SGI Indigo in 1995 with the first Windows 95-based workstation.

The computer peripherals that accompanied the computers were continuously upgraded between 1985 and 1990. A GTCO DIGI-PAD digitizer with a 16-button cursor was used to digitize the soil maps (Fig. 7.3), and an A0 size HP-7550A pen plotter was used for plotting soil maps. One interesting development was a system for plotting vector soil maps directly onto 1:25,000 scale topographic maps (Vrščaj et al. 1994a). The procedure included the automatized plotting of the map legend on the backside. Through continuous upgrades and improvements, the hardware slowly declined in importance. Over the years, the work changed from focusing on fighting hardware capability and software capacity problems to focusing on the database structure, standards, and data improvement. The frequent hardware changes required constant adaptation and improvement of the soil database structures, versioning of soil data processing software, and the development of data entry and data processing routines. In comparison to other Yugoslav republics and the neighbouring countries, Slovenian soil science relatively quickly adopted IT technology in order to capture, process, and reproduce soil data.

7.1.3 The Early Soil Software

The first computer program for soil data processing was developed for the Commodore 64 in 1983 (Vrščaj 1983). It was written using the Commodore Basic computer language and shortly upgraded with new functions using the improved Simons' Basic. The data input forms were programmed, and in the end, the software was capable of storing, organizing, and printing out the soil profile information. The first instructions were prepared for students employed to enter the soil data. In addition to the profile data, short programs were developed such that students of agriculture could learn and test their knowledge of soil science. The students were asked to select the correct answer from the database, and the system calculated the percentage of correct answers and gave a grade. The C64/C128 software was relatively simple. However it required a lot of typing, corrections were not easy, and most significantly the loading time and data backup procedures were long. All in all, it was an important learning experience for would-be programmers to learn the fundamentals of computer programming on the way to developing more serious soil data software.

7.1.3.1 The Soil Profile Data Processing Software

The DOS operating system available with the purchase of the first IBM PC required the adaptation of the C64/C128 software to GW-BASIC. Soon, the poor software capacities related to data management were identified, and a decision to move towards a better database platform seemed reasonable.

The first version of the IBM PC-based Profili.exe (Profiles.exe) soil profile data processing software was written using the Clipper 87 programming language (Vrščaj 1987). A dBASE III® database format was utilized to develop a simple relational database holding the profile location data, horizon field description data, and the analytical information. Figure 7.2 shows a Profilie.exe data page. Along with the development of the computer program, the soil profile data

Fig. 7.3 Digitizing the 1:25,000 soil map of Slovenia in 1989 (*Photo* B. Vrščaj)

coding system was developed, as well as the field soil profile description forms. Profili.exe was later adapted to the MS Windows® environment and used until the late 1990s, when it was replaced by similar Windows software.

7.1.3.2 The Software for Soil Fertilization Requirements

Storing and printing soil profile information was not the only task. The Gnojenje.exe (Fertilization.exe) software was developed in 1991 using the Clipper 87 programming language. The aim of the software development was to elaborate a tool for the agricultural extension service to accelerate and automatize the interpretation of soil nutrients, and the calculation and elaboration of plant nutrition recommendations for farmers (Vrščaj 1992, 1994). The program linked five databases: the soil analytical data, the mineral and organic fertilizer database on nutrient content, the crop nutrient requirement, and the crop standard rotation dataset. Users were able to evaluate and print out the soil nutrient data, select a predefined crop rotation or create a new one for up to five consecutive years, select mineral and/or organic fertilizers, and print out the complete fertilizing instructions —a recipe for the farmer at issue.

7.1.3.3 The Software for the Soil Laboratory

The soil laboratory of the Centre for Soil Science and Environmental Protection (CSES) lacked a computerized system to store and maintain data on soil analyses. The data were numerous, from the simple and limited selection of analytical figures on soil nutrients, to the standard soil analyses dataset. The LabTla.exe (LabSoil.exe) software was written for CSES laboratory staff to efficiently store, process, and print out the soil analytical results. To a certain degree, in later releases, the program was upgraded with data warning routines.

7.1.4 The GIS Software

Until 1987, the soil information of Slovenia was stored in alphanumeric form. The digitalization of spatial soil

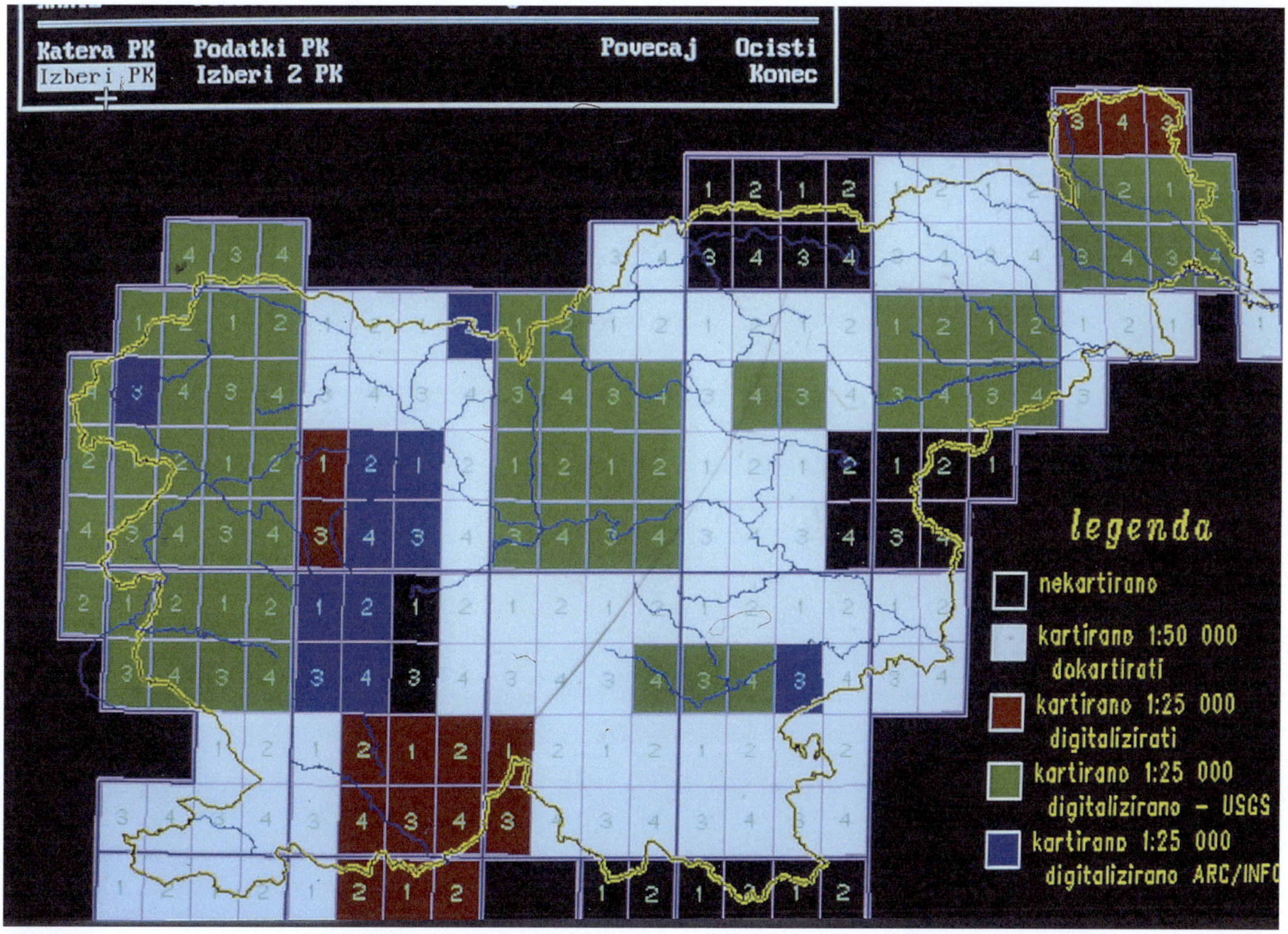

Fig. 7.4 Computer screen presenting the menu-driven application for the digitalization of the 1:25,000 soil maps developed in Arc Info 3.4d (*Photo* B. Vrščaj)

information in the form of maps, not to mention the spatial modelling of the soil data, was not possible due to the poor capacities of the computer hardware and software in the early 1980s and the availability and cost of GIS software in the following years. In 1987, Dr. Ronald Tidball, a USGS geologist and geochemist from Denver, Colorado, visited the Centre for Soil Science and Environment Protection and presented the USGS suite of soil mapping and modelling software, the GS MAP, GS DRAW, and GS PLOT, written by Gary I. Selner, and Richard B. Taylor. The software was used for the digitalization of the vector soil maps on a scale of 1:25,000. In spite of the quality of the USGS software, the requirements soon surpassed the capacities of the USGS software. In 1989, an ESRI PC Arc Info 3.4d software package was purchased and used on a 386-based PC and a GTCO DIGI-PAD digitizer to digitize the soil maps, while an HP A0 size pen plotter was used to plot the soil maps. Using the Arc Info 3.4d SML computer language, a menu-driven application for the digitalization and graphic editing of the 1:25,000 soil maps was developed (Figs. 7.4 and 7.5). ESRI products were used to digitize all 200 soil 1:25,000 maps that cover the Slovenian territory and to create a seamless polygon-based soil database. Subsequently, an ESRI Arc Info Workstation was used on SGI Indigo and Windows NT computers (Figs. 7.6 and 7.7). Nowadays, ESRI Arc GIS still predominates in Slovenian soil science, but SAGA, MapInfo, as well as QGIS, GRASS, and R are gaining in importance.

7.1.5 The Digitalization of the Soil Maps

Activities related to the digitalization of the Soil Map of Slovenia on a scale of 1:25,000 began in 1987. This effort was initiated by CSSE and financed by the Ministry of Science and Technology through various research funds. After Slovenia achieved independence in 1991, it was decided that the work on mapping soils, the digitalization of soil data, and the establishment of digital soil databases should be accelerated.

In 1992, a new contract was concluded between the Ministry of Agriculture, Food, and Forestry of the Republic

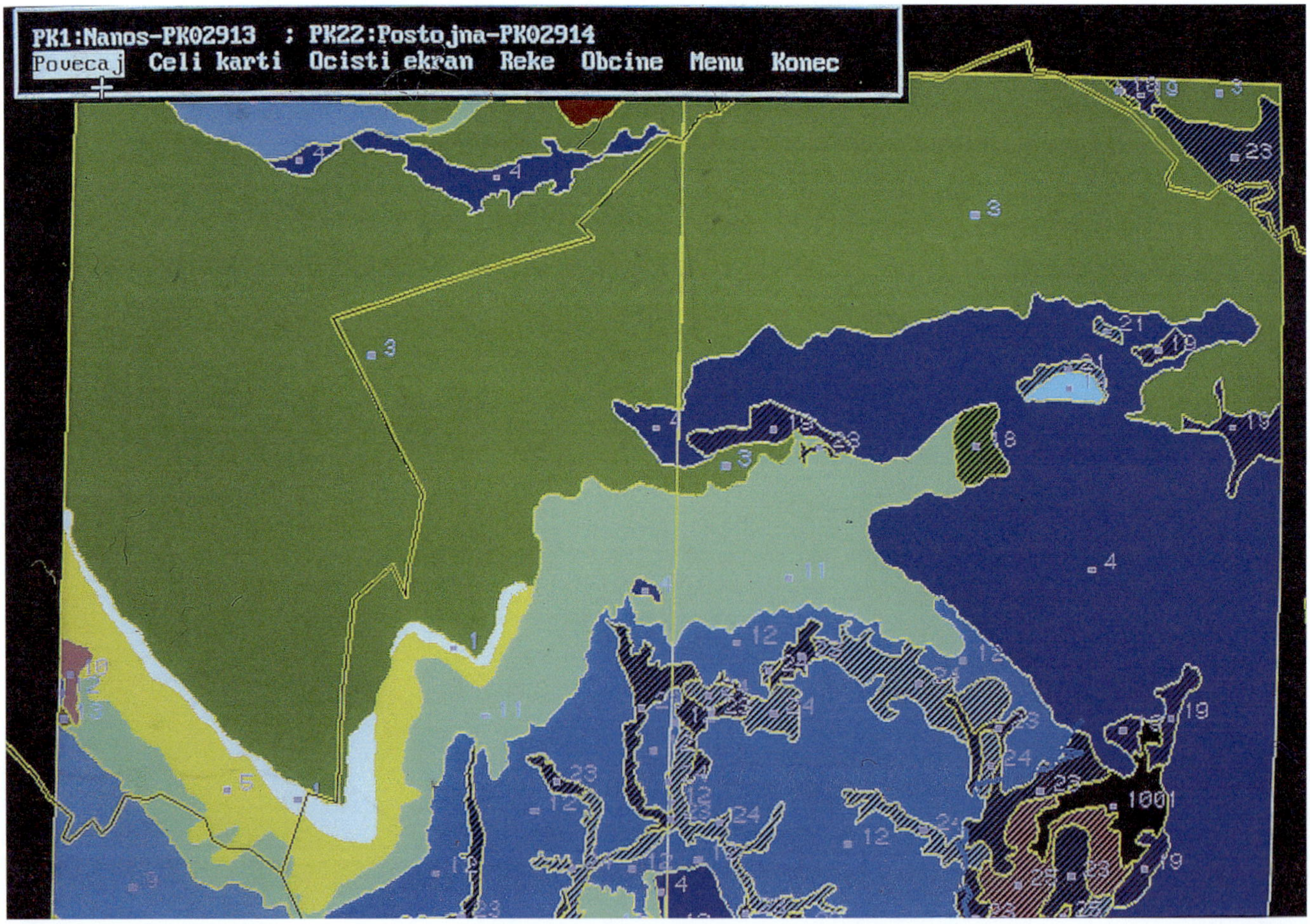

Fig. 7.5 The computer screen showing the edge-matching stage in digital soil map processing (*Photo* B. Vrščaj)

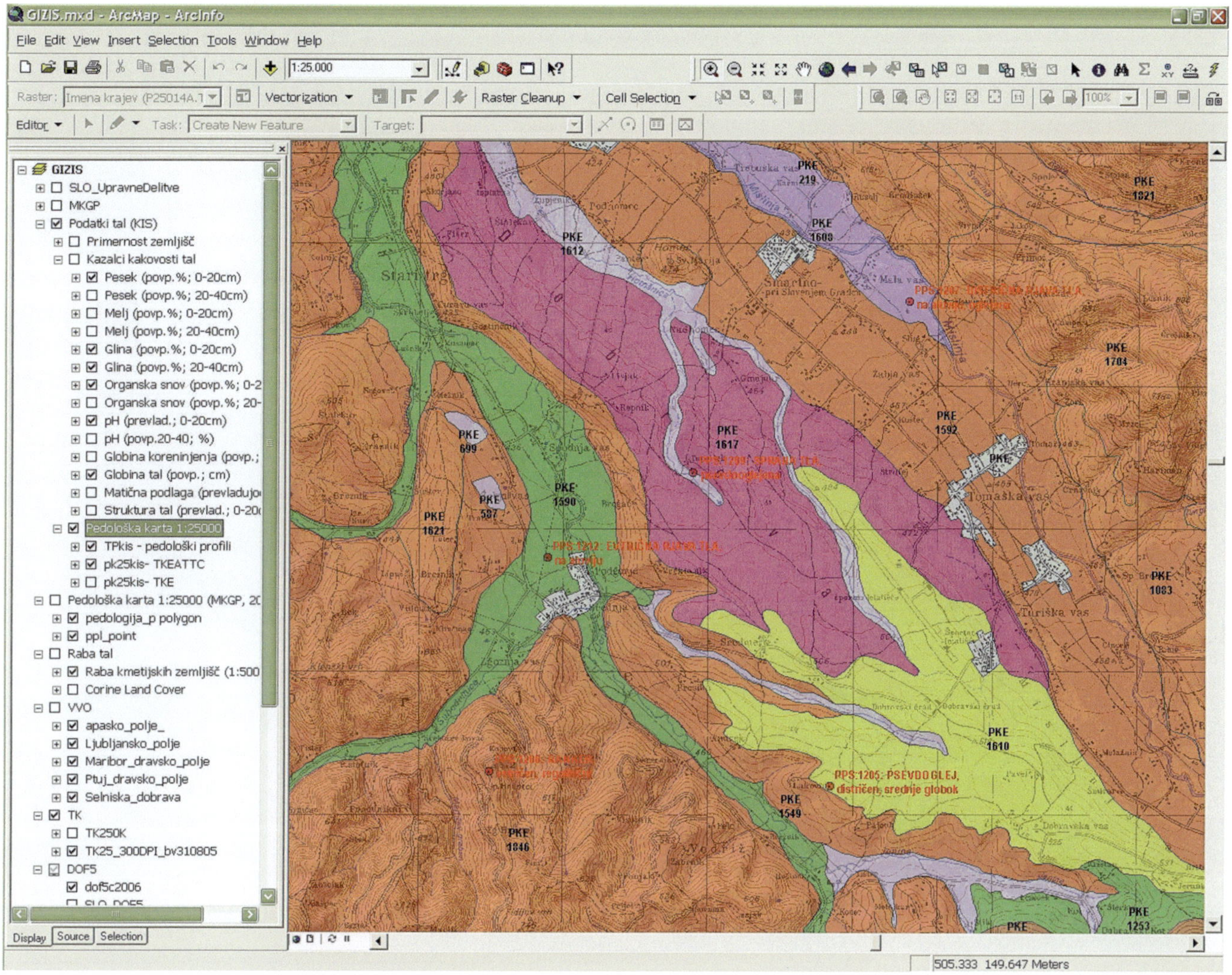

Fig. 7.6 A section of the 1:25,000 Digital Soil Map of Slovenia in an early version of Arc GIS

of Slovenia (MAFFS) and CSSE. The project had four major goals: to complete the soil mapping on a scale of 1:25,000 and to edge-match separate soil maps; to elaborate the seamless digital soil map; and to collect, verify, and enter soil profile data into the databases. Soil mapping was divided into 200 sheets of 1:25,000 topographic maps. Each topographic map covers apx. 13,000 ha. Between 1992 and 1998, additional soil survey fieldwork was carried out simultaneously with the digitalization of data. The project was completed in January 1999 when a unified, seamless Digital Soil Map on a scale of 1:25,000 (DSM25) was available for the entire territory of Slovenia.

7.1.6 Soil Profile Data

Currently, a dataset comprising approximately 1860 soil profiles is available. The work is still not yet finished, there are still some areas for which measured soil data are not available. A soil profile is a vertical cross-section of a soil type from the surface to the parent material. It is representative of a soil type (SSU), and it is geographically oriented by *x*, *y*, and *z* coordinates. A soil profile point layer (SP) contains data on the physical and chemical soil properties obtained by standard soil analyses. Thus, a soil profile data set contains the attribute data for an SSU in a certain location (Fig. 7.7).

The SP layer contains the point data of concentrations of the most important inorganic and organic contaminants present or expected in Slovenian soils. The soil-sampling pattern performed on predefined sampling locations is standardized and defined by legislation (Official Gazette RS, 1997). One sampling point represents the centre of a 100 m circle with six sub-sampling locations. From each of them, samples are taken at three different depths (0–5 cm, 5–20 cm, and 20–30 cm) and combined into three representative samples. In addition to soil samples, the indicator

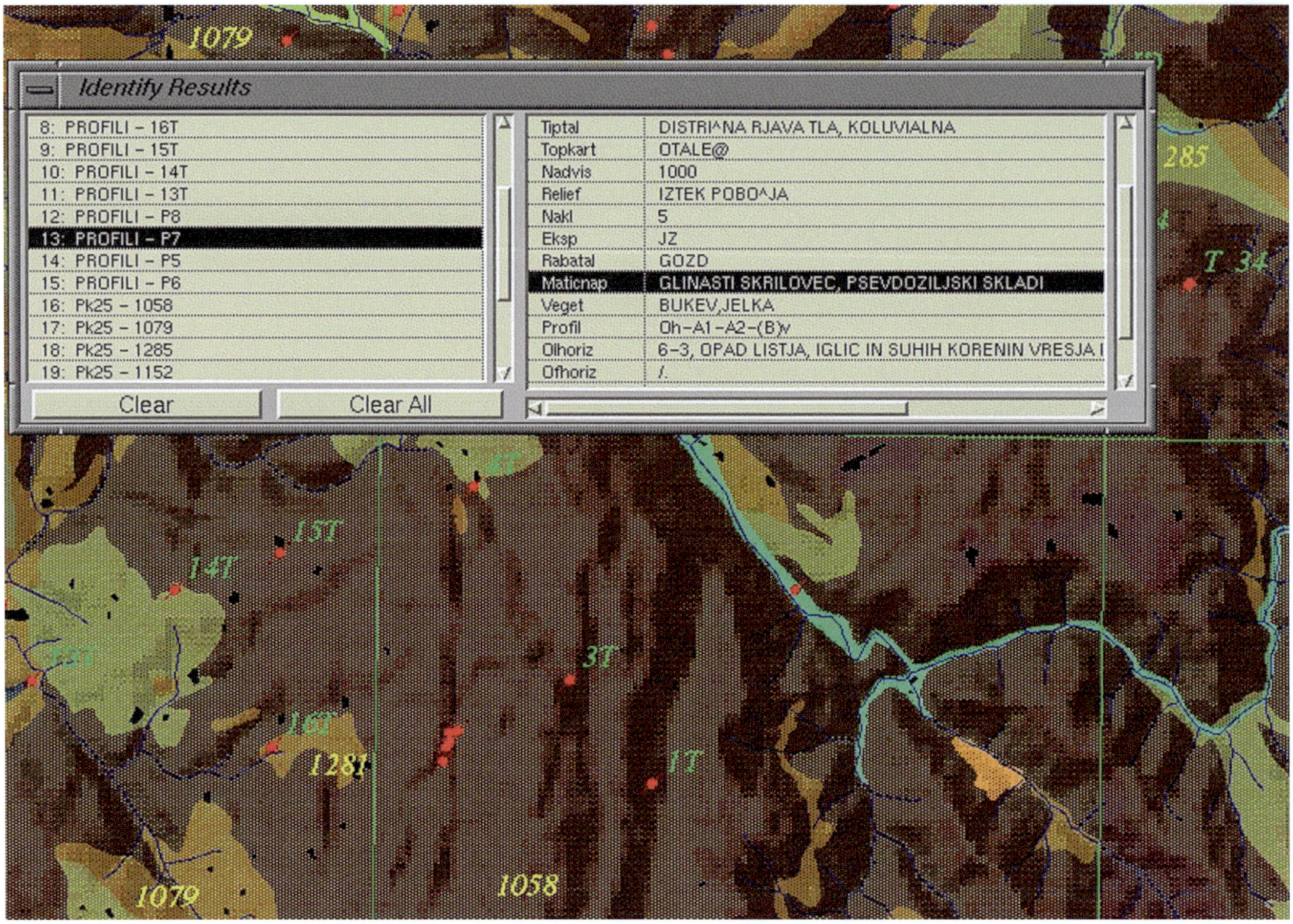

Fig. 7.7 Soil profile locations marked with *red dots* overlaying the digital soil map with shaded relief—an SGI Indigo screen copy

plant narrow leaf plantain (*Plantago lanceolata)* was sampled in the circle and analysed (Hudnik et al. 1994). Predefined sampling in a regular grid covers the entire Slovenian territory. The density of sampling points is 2 × 2 km in agricultural areas, while forests and high-elevation areas are covered with a 4 × 4 km grid. Denser 1 × 1 grid sampling was performed in the late 1980s and early 1990s in areas where a high degree of soil contamination with heavy metals and soil pollution with pesticides was anticipated (the city of Celje, selected areas of the city of Ljubljana, Jesenice, Dravsko-Ptujsko plain, Krško plain, and the hinterland of Koper) (Figs. 7.8 and 7.9).

Since July 2002, soil pollution data on 367 sampling points have been stored in the soil information system and interpreted according to legislation (Official Gazette RS 1996). The Soil Pollution Monitoring Project is being continued by the CSSE and the reports are available for download (Zupan et al. 2008).

7.2 Joining Soil Data—The Development of the Soil Information System of Slovenia

In addition to soil mapping, several other research projects concerning soil and soil pollution have been carried out in Slovenia. Altogether, they have produced a substantial amount of data that had to be organized and used in the most effective way; the data had to be stored in an adequate geographic soil information system. Several studies on digital soil information (Vrščaj et al. 1989) led to the decision to harmonize, structure, and unite available digital Slovenian soil data in the Soil Information System of Slovenia (SIS). The SIS activities were initiated, developed (Vrščaj 1996; Vrščaj et al. 1994b), and maintained at CSSE (Vrščaj 1999; Vrščaj et al. 2005; Vrščaj and Lobnik 1999a, b, 1997). The financing of the SIS is not directly supported by the state; it is realized through ongoing research grants. The SIS is a

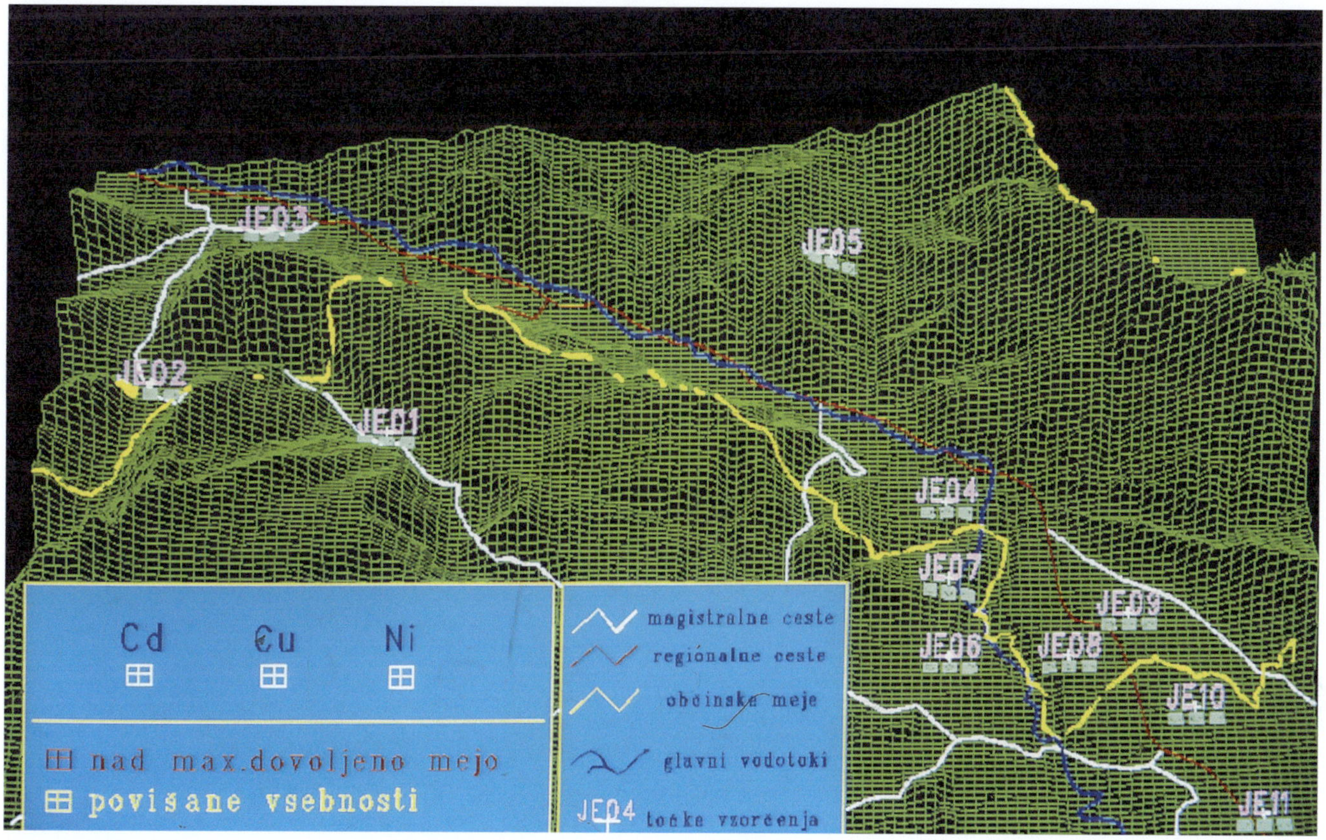

Fig. 7.8 3D visualization of Cd, Cu, and Ni concentrations in the topsoil of the area of Jesenice from late 1989 (*Photo* B. Vrščaj)

source of soil data, verified by and comparable to laboratory analyses, procedures, and spatial accuracy.

The SIS database comprises the three main data layers:

- the Soil Profile Data (SP);
- the Digital Soil Map on a Scale of 1:25,000 (DSM25);
- the Data on Soil Pollution Monitoring (SPM).

In 2005, the development of the SIS was completed and well documented (Tič and Vrščaj 2001a, b, c, d, e, 2002; Vrščaj and Tič 2001a, b). The data reports were predefined and intensively used.

The SIS represents the main source soil information. Unfortunately, it is currently not supported by the Slovenian authorities and represents an unofficial dataset accessible at the CSSE and the AIS.

7.3 The Use of Digital Soil Information

The digital soil information has been intensively used in various research projects, which in general were ordered by the Slovenian authorities. Here, we present selected examples of projects carried out at the CSSE and the Agricultural Institute of Slovenia.

The **drought -affected areas** in the year 2000 were identified with the help of the GIS. A model was designed to elaborate a raster database (100 m cell size) of soils affected by severe drought. Besides the DSM25 attribute data, the basic model inputs were soil type, expert classification of the water retention capacity for soil types and precipitation data, DEM, land use, and derived relief data (Lobnik et al. 1994). The problem of nitrate leaching in Slovenia is being evaluated in an ongoing project financed by the Ministry of the Environment and Spatial Planning. The soil properties derived from the DSM25 and data on actual agricultural practices were used as the key input data. The results of the study identified the areas with permeable soils and the locations where good agricultural practices should be strictly followed to avoid ground water contamination with nitrates. Severe droughts intensified the research at the Agricultural Institute of Slovenian using the digital soil information and the GIS. The irrigation areas in Slovenia were analysed and classified according to 6 soil properties for the **National Irrigation Program**. As a result, 65 maps were plotted where the soil suitability for irrigation was shown in 5

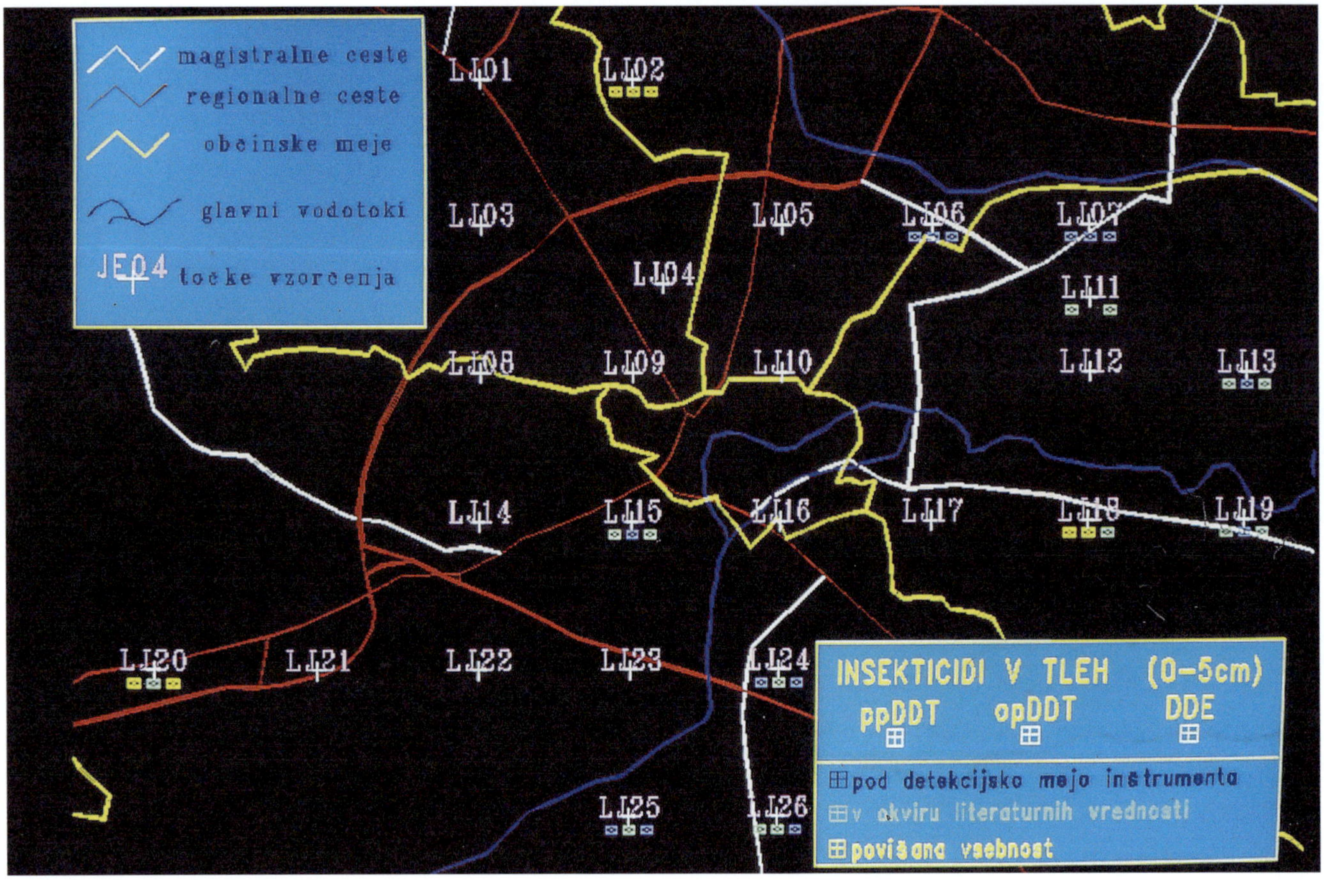

Fig. 7.9 Visualization of the concentrations of DDT and derivates in the topsoil of the area of Ljubljana (*Photo* B. Vrščaj)

categories (Lobnik et al. 1994; Vrščaj et al. 1994c). The **motorway construction** programme in Slovenia between 1995 and 2003 required several optimization and land assessment studies in which soil information was intensively used. Proposed routes for several motorway sections were evaluated: Arja vas—Vransko, Zadobrova—Blagovica, Blagovica—Šentjakob, Cogetinci—Radmožanci, Kastelec—Sermin, and the Vipava valley route. The losses of soils in intensive agricultural areas, environmental impact, and the loss of agricultural production were estimated (Lobnik et al. 1995a, b, c, d; 1996a, b; Prus and Vrščaj 1997). In order to assess the **land suitability** for six main types of agricultural production (arable land, vineyards, orchards, hops, grassland, and olive growing), a methodology based on raster modelling was developed (Vrščaj et al., 2003). The main input data were the DSM25, digital topography, and climate data. The technical parameters related to agricultural machinery were integrated into separate sub-models for individual types of agricultural production. The results were six raster databases in 100 m resolution presenting the suitability of the agricultural land for the particular agricultural use for the entire country.

At the Agricultural Institute of Slovenia, the DSM25 and soil profile information are improved, amended, and used for various purposes ranging from military and agricultural to environmental protection-oriented projects. The off-road trafficability of Slovenian soils for military use was assessed for different military vehicles. A GIS model that employed soil parameters and linked to the DSM25 was designed, and trafficability raster databases were created (Vrščaj and Godeša 2008) for path modelling (Šinigoj et al. 2008). Being the main crop in Slovenia, the maize production potentials of Slovenian soils were assessed by the AIS (Čergan et al. 2008). The land's suitability for trufficulture was assessed, and a web-GIS portal elaborated for model verification and data dissemination was created (Bergant and Vrščaj 2014). The coastal region in Slovenia is considered to be one of the most northern olive production areas. The land suitability for olive groves was assessed within a cross-border EU project (Vrščaj 2014b). Slovenia is facing a

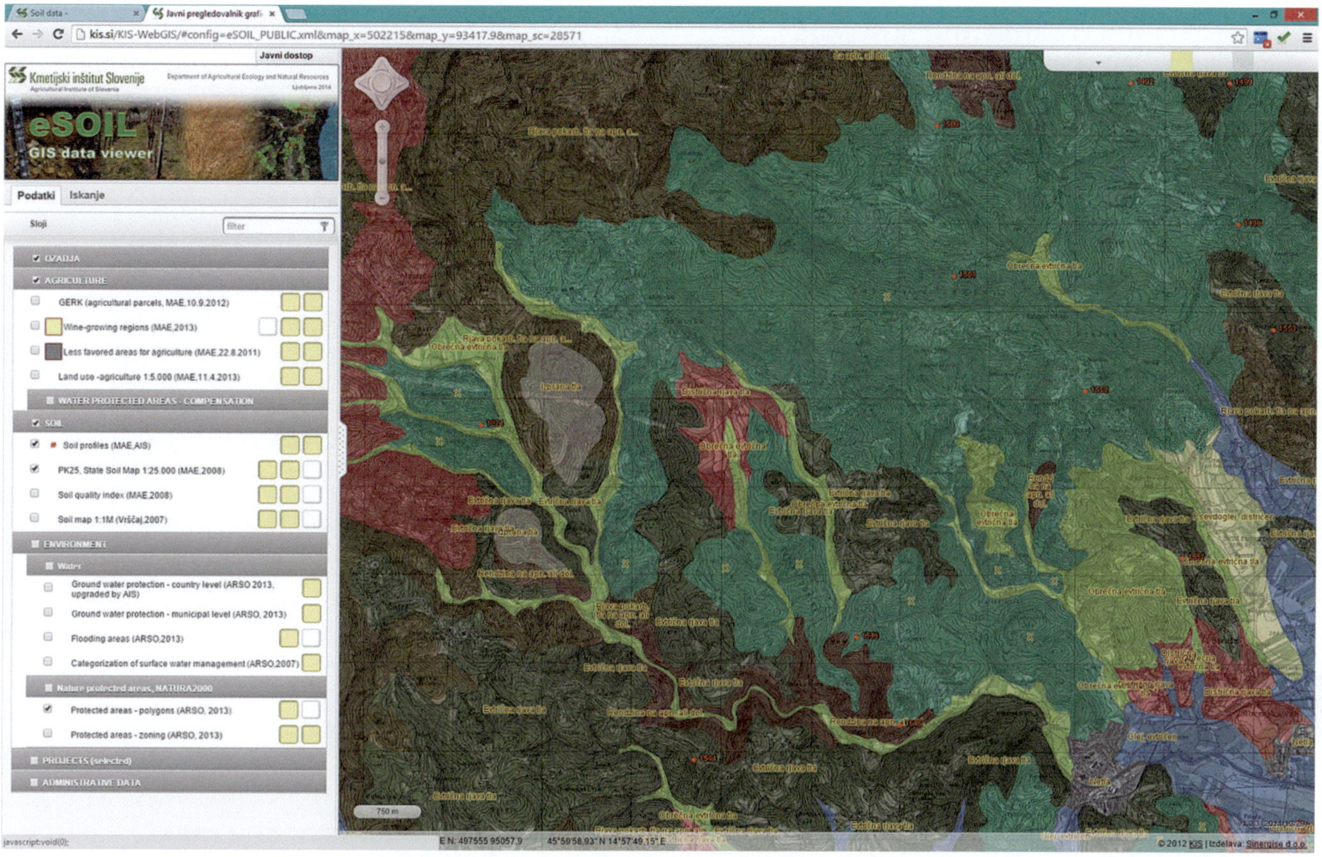

Fig. 7.10 Web GIS soil data viewer—English version, Agricultural Institute of Slovenia 2014

high soil sealing rate. The limited soil resources had to be protected, which is why soil information was employed to elaborate the information for planners and local communities. Funds from the EU UrbanSMS project were used to design a web-GIS data collection system for sustainable use of soils on the municipal level (Blumlein et al. 2012). The presented list of projects where digital soil information is widely used is not complete. In addition to the mentioned three institutions, digital soil information is regularly used by several other institutions in Slovenia due to the availability of soil information and the GIS.

7.4 The Availability of Digital Soil Information

The main sources of available soil information that have been collected and processed in the past are the following three public institutions: The Center for Soil and Environmental Science of the University of Ljubljana (CSSE; www.bf.uni-lj.si/cpvo), the Agricultural Institute of Slovenia, Department for Agroecology and Natural Resources, Centre for Soil and the Environment (AIS; www.kis.si/okenv/), and the Slovenian Forestry Institute (SFI). All three continue to collect and provide soil information. Selected databases can be provided by an institution directly or are available on the websites of government institutions. The Ministry of Agriculture, Food, and Forestry of the Republic of Slovenia (MAFF) offers access to the soil map through the website (http://rkg.gov.si/GERK/). The CSSE provides the soil data/information on request. The regularly collected, processed, and formatted data on soil contamination are sent to the Slovenian Environmental Protection Agency, where they comprise a part of the web-GIS data system. The Agricultural Institute of Slovenia has enabled access to the soil and environmental information through a web-GIS viewer that enables data querying and simple GIS operations. The English version of the system presents the DSM25 geometry with some additional information on average soil properties within soil mapping units (Fig. 7.10).

KMETIJSKI INŠTITUT SLOVENIJE
Centralni laboratorij, Center za tla in okolje

Hacquetova 17, SI1000 Ljubljana
hhtp://www.cto.kis.si; E-mail cto@kis.si

Pedološki profil **1213** *Talni tip:* **DISTRIČNA RJAVA TLA, NA KONGLOMERATU, SREDNJE GLOBOKA**

Lokacija: **GRAŠKA GORA** *Leto izkopa:* 96

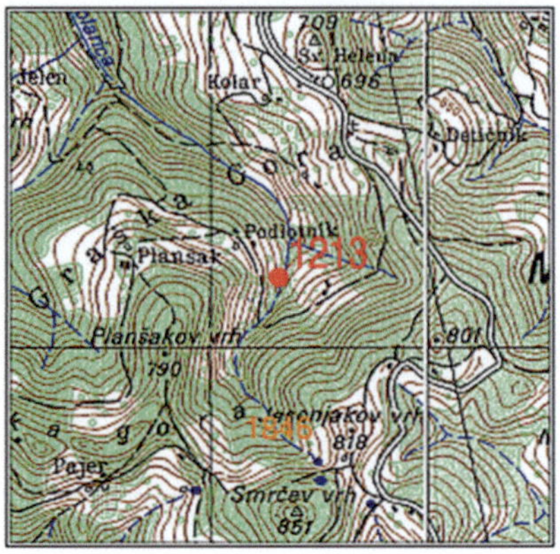

Matična podlaga:
Raba tal: TRAVNIK
Vegetacija: TRAVE

Oznaka: BV14
Opisal: Borut Vrščaj

Pregledal:
zentativnost:

TSE: *TKE:* 1846 *TKE:*

Nadm viš.: 650
Relief: HRIBOVJE,SR.PO BOČJA,KO
Neb.smer: VZHOD
Nagib: 60

GKX: 509200 *GKY:* 143200 *Z (m):* 650

Horizonti: **A-A(B)v-(B)v1-(B)v2-(B)vC-C**

Ol hor.:
Of hor.:
C/R hor.:
Opombe profila:

Horizont: **A** Globina: 0-8 (8) cm
Sturktura: GRUDIČAST DO OREŠKAST, izražena: DOBRA, obstojna: DOBRA Konzistenca: RA S DR Organska snov: HUMOZEN Barva: 10YR 3/4 Prekoreninjenost: GOSTE KORENINE Vlažnost ob opisu: SVEŽ Skelet: 3%, 2CM ZAOBLJEN

Opombe: PREHOD JASEN

Horizont: **A(B)v** Globina: 8-24 (16) cm
Sturktura: OREŠKAST, izražena: DOBRA, obstojna: DOBRA Konzistenca: L DR Organska snov: SLABO HUMOZEN Barva: 10YR 4/4 Prekoreninjenost: SREDNJE GOSTE KORENINE Vlažnost ob opisu: SVEŽ Skelet: 5%, 2CM, ZAOBLJEN

Opombe: PREHOD POSTOPEN

Horizont: **(B)v1** Globina: 24-50 (26) cm
Sturktura: OREŠKASTA DO GRUDIČAST, izražena: SREDNJA, obstojna: DOBRA Konzistenca: GO S DR Organska snov: MINERALEN Barva: 10YR 5/6 Prekoreninjenost: REDKE KORENINE Vlažnost ob opisu: SVEŽ DO VLAŽEN Skelet: 10%, 5CM, MEŠAN
Opombe: PREHOD DIFUZEN

Horizont: **(B)v2** Globina: 50-70 (20) cm
Sturktura: OREŠKASTA DO POLIEDRIČNA, izražena: SREDNJA, obstojna: SREDNJA Konzistenca: GO S DR Organska snov: MINERALEN Barva: 10YR 5/4 Prekoreninjenost: POSAMEZNE KORENINE Vlažnost ob opisu: SVEŽ DO VLAŽEN Skelet: 10%, 2CM, ZAOBLJEN
Opombe: PREHOD JASEN

Horizont: **(B)vC** Globina: 50-70 (20) cm
Sturktura: OREŠKASTA DO POLIEDRIČNA, izražena: SLABA, obstojna: SREDNJA Konzistenca: GO L DR Organska snov: MINERALEN Barva: 10YR 5/3 Prekoreninjenost: NEPREKORENINJEN Vlažnost ob opisu: SVEŽ DO VLAŽEN Skelet: 5%, 5CM, ZAOBLJEN
Opombe: PREHOD JASEN

Zap.št. hor.	Horizont / cm		pH H_2O	pH $CaCl_2$	pH KCl	Tekstura / zrnavost: Pesek %	GMelj %	FMelj %	Melj %	Glina %	Razred	P_2O_5 mg / 100g	K_2O mg / 100g	N %	C %	Org.s. %	C/N
1	A	0-8 (8)			4,4	31,1	36,8	17,2	54,0	14,9	MI	1,9	6,9	0,3	5,2	8,9	27,8
2	A(B)v	8-24 (16)			3,8	33,8	34,5	14,6	49,1	17,1	MI		2,9	0,2	2,4	4,2	21,0
3	(B)v1	24-50 (26)			3,8	33,6	32,7	12,0	44,7	21,7	I		2,4	0,1	1,4	2,4	21,8
4	(B)v2	50-70 (20)			3,9	33,7	28,5	15,1	43,6	22,7	I						
5	(B)vC	50-70 (20)															

Zap.št. hor.	Horizont / cm		Ca	Mg	K	Na	H	S	T	V %	Ca	Mg	K	Na	H
			bazični kationi v $mmol_c/100g$					$mmol_c/100g$			delež kat.na sorpt. delu tal (%)				
1	A	0-8 (8)	7,1	2,3	0,2	0,30	17,9	9,9	27,8	35,7	25,6	8,4	0,6	1,1	64,3
2	A(B)v	8-24 (16)	3,4	1,1	0,1	0,25	19,2	4,8	24,0	20,1	14,0	4,7	0,3	1,0	79,9
3	(B)v1	24-50 (26)	2,0	0,8	0,1	0,11	17,0	3,0	20,0	15,1	10,1	4,2	0,3	0,5	84,9
4	(B)v2	50-70 (20)	1,6	0,8	0,1	0,10	14,1	2,5	16,6	15,3	9,3	5,0	0,4	0,6	84,7
5	(B)vC	50-70 (20)													

Fig. 7.11 Soil profile report available from the Web GIS soil data viewer, Agricultural Institute of Slovenia 2014

Projekt: RAZISKAVE ONESNAŽENOSTI TAL S PREDLOGOM PROGRAMA UKREPOV NUJNIH SANACIJ
Naročnik: MINISTRSTVO ZA OKOLJE IN PROSTOR — Vzorčna točka: 03411
Izvajalec: BIOTEHNIŠKA FAKULTETA, ODDELEK ZA AGRONOMIJO, Center za pedologijo in varstvo okolja — Čas vzočenja: oktober 1999

Vzorčna točka: 03411

KRAJ: PRIHODI
OBČINA: JESENICE

GK: X= 427000m Y= 146000m
Nadmorska višina: 945m
Nagib: 23° (51%)
Smer pobočja: jugo-zahod

Lokacija vzorčne točke

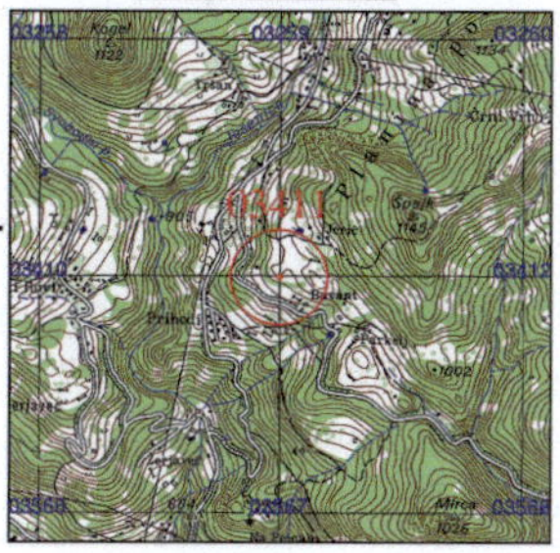

Vzorčenje: 03411/1099
Obrazec terenskega opisa: T03411_1099.tif
Čas vzorčenja: oktober 1999

Terenski opis lokacije vzorčne točke

Opisal: Marko Zupan, UNIVERZA V LJUBLJANI, BIOTEHNIŠKA FAKULTETA, Oddelek za agronomijo, Center za pedologijo in varstvo okolja

Tip tal: EVTRIČNA RJAVA TLA, na mešanih kamninah
Matična podlaga: apnenec, peščenjak, kremen
Vegetacija: trave, zeli
Raba tal: travnik/pašnik
Pedološki horizonti v profilu: A1-AB-Bv
Oddaljenost prometnic od vzorčne točke: lok. cesta:200m kolovoz:70m
Potencialni viri onesnaženja:kmetijski obrat

Makrorelief: razgiban-gričevje
Mikrorelief: sredina pobočja
Oblika mikroreliefa: nepravilna
Kamnitost: neznatna (<2%)
Skalovitost: neznatna (<2%)
Površinska org. snov: ni podatka
Dreniranost: dobra

Opombe: Vreme:oblačno,rahel dež s snegom.Točka je bila izbrana namesto točke 03731

Datum izpisa: 11.1.2002 Stran: 1 od 4
Obrazec: ROTS/LPAO/PA4

Projekt: RAZISKAVE ONESNAŽENOSTI TAL S PREDLOGOM PROGRAMA UKREPOV NUJNIH SANACIJ
Naročnik: MINISTRSTVO ZA OKOLJE IN PROSTOR — Vzorčna točka: 03411
Izvajalec: BIOTEHNIŠKA FAKULTETA, ODDELEK ZA AGRONOMIJO, Center za pedologijo in varstvo okolja — Čas vzočenja: oktober 1999

Vzorčenje: 03411/1099

Terenski opis slojev vzorčne točke

Opisal: Marko Zupan, UNIVERZA V LJUBLJANI, BIOTEHNIŠKA FAKULTETA, Oddelek za agronomijo, Center za pedologijo in varstvo okolja

Globina A (0-5cm)
Konzistenca: drobljiv
Stopnja konz.: lahko
Struktura: grudičast
Izraženost str.: srednja
Barva: 10YR 3/4
Organska snov: srednje humozen
Vlažnost ob opisu: svež do vlažen
Prekoreninjenost: goste korenine
Novotvorbe: ni novotvorb
Volumski delež skeleta: 0%

Globina B (5-20cm)
Konzistenca: drobljiv
Stopnja konz.: lahko
Struktura: oreškast
Izraženost str.: srednja
Barva: 10YR 4/6
Organska snov: slabo humozen
Vlažnost ob opisu: svež do vlažen
Prekoreninjenost: redke korenine
Novotvorbe: ni novotvorb
Volumski delež skeleta: 5%
Oblika skeleta: ostrorob
Max. velikost skeleta: 2cm

Globina C (20-30cm)
Konzistenca: drobljiv, gost
Stopnja konz.: lahko
Struktura: oreškast
Izraženost str.: slaba
Barva: 10YR 5/6
Organska snov: mineralen
Vlažnost ob opisu: svež do vlažen
Prekoreninjenost: neprekoreninjen sloj
Novotvorbe: ni novotvorb
Volumski delež skeleta: 15%
Oblika skeleta: ostrorob
Max. velikost skeleta: 10cm

Osnovni pedološki parametri

Analitski laboratorij: UNIVERZA V LJUBLJANI, BIOTEHNIŠKA FAKULTETA, Oddelek za agronomijo, Center za pedologijo in varstvo okolja

Globina	pesek	melj	glina	TRZ	TOC	org. snov	C	N	C/N	karbo nati	pH	P	K	Ca	Mg	K	Na	H	S	T	V
enota	%		%		%	%	%	%		%	CaCl2	mg / 100g		mmol / 100g							%
A (0-5cm)	40,6	45,2	14,2	I	3,5	7,2	4,2	0,37	11,4	<0,1	5,6	-1	9,8	11,4	2,39	0,22	0,07	11,85	14,1	26	54,2
B (5-20cm)	44,7	39,9	15,4	I		3,5	2	0,22	9,1	<0,1	6,2	-1	5,8	9,4	2,01	0,12	0,05	9,1	11,6	20,7	56
C (20-30cm)	44,6	37,1	18,3	I		2,1	1,2	0,15	8	2,2	6,5	-1	5	8,58	1,5	0,09	0,05	7,65	10,2	17,9	57

(izmenljivi: P, K, Ca, Mg, K, Na)

pH (v $CaCl_2$) VREDNOST (pH in $CaCl_2$)
Scale: 2 – zelo močno kisla – 3 – močno kisla – 4.5 – kisla – 5.5 – zmerno kisla – 6.5 – nevtralna – 7.2 – alkalna – 8 – zelo močno alkalna – 9

Globina	pH
A (0-5cm)	5,6
B (5-20cm)	6,2
C (20-30cm)	6,5

TEKSTURNI RAZRED (Texture Class)
lahka tla: P, IP, PI; sr. težka tla: I, MI, M, PGI; težka tla: GI, MGI, PG, MG, G

Globina	
A (0-5cm)	I
B (5-20cm)	I
C (20-30cm)	I

ORGANSKA SNOV (Organic Matter)
Scale: 0% – 10% – 20% – 30% – 35%

Globina	org. snov
A (0-5cm)	7,2%
B (5-20cm)	3,5%
C (20-30cm)	2,1%

LEGENDA: 0-1: mineralna tla; 1-2: malo humozna; 2-4: srednje hum.; 4-8: dobro hum.; 8-15: zelo dobro hum.; 15-35: ekstremno hum.; >>>> > 35: organska tla

KATIONSKA IZMENJALNA KAPACITETA - T vrednost (mmolH/100g) (Cation Exchange Capacity)
Scale: 0 – nizka – 20 – srednja – 40 – visoka – 60

Globina	T
A (0-5cm)	26
B (5-20cm)	20,7
C (20-30cm)	17,9

LEGENDA: Kalcij (Ca); Magnezij (Mg); Kalij (K); Natrij (Na); Vodik (H); >>>>> > 60mmolH/100g

ZASIČENOST Z BAZIČNIMI KATIONI - V vrednost (Base saturation)
Scale: 0% – distrièno – 50% – evtrièno – 100%

Globina	V
A (0-5cm)	54,2%
B (5-20cm)	56%
C (20-30cm)	57%

Datum izpisa: 11.1.2002 Stran: 2 od 4
Obrazec: ROTS/LPAO/PA4

Projekt: RAZISKAVE ONESNAŽENOSTI TAL S PREDLOGOM PROGRAMA UKREPOV NUJNIH SANACIJ
Naročnik: MINISTRSTVO ZA OKOLJE IN PROSTOR — Vzorčna točka: 03411
Izvajalec: BIOTEHNIŠKA FAKULTETA, ODDELEK ZA AGRONOMIJO, Center za pedologijo in varstvo okolja — Čas vzočenja: oktober 1999

Vzorčenje: 03411/1099

Anorganske nevarne snovi

Analitski laboratorij: ERICO Velenje, Inštitut za ekološke raziskave

Globina	Hg	Cd	Pb	Zn	Mo	Cu	Co	As	Ni	Cr	V	Se	Mn	Tl	Fluoridi vodotopni	Fluoridi totalni
enota	mg / kg suhe snovi															
A (0-5cm)	0,87	0,97	58,1	172	0,63	21,2	12,2	20,1	29,6	61,8	77	0,82	1210	0,77		
B (5-20cm)	0,37	0,84	48,7	140	0,56	17,4	11,4	17,1	26,2	53,6	67,4	0,83	1110	0,69		

Globina Vsebnost

Nevarna snov: As
A (0-5cm) 20,1 mg/kg
B (5-20cm) 17,1 mg/kg
pod mejo detekcije: <0,1mg/kg pod mejo določljivosti: <0,4mg/kg pod mejno vrednostjo: <20mg/kg mejna vrednost: >20mg/kg opozorilna vrednost: >30mg/kg kritična vrednost: >55mg/kg

Nevarna snov: Cd
A (0-5cm) 0,97 mg/kg
B (5-20cm) 0,84 mg/kg

Nevarna snov: Co
A (0-5cm) 12,2 mg/kg
B (5-20cm) 11,4 mg/kg

Nevarna snov: Cr
A (0-5cm) 61,8 mg/kg
B (5-20cm) 53,6 mg/kg

Nevarna snov: Cu
A (0-5cm) 21,2 mg/kg
B (5-20cm) 17,4 mg/kg

Nevarna snov: Hg
A (0-5cm) 0,87 mg/kg
B (5-20cm) 0,37 mg/kg

Nevarna snov: Mo
A (0-5cm) 0,63 mg/kg
B (5-20cm) 0,56 mg/kg

Nevarna snov: Ni
A (0-5cm) 29,6 mg/kg
B (5-20cm) 26,2 mg/kg

Nevarna snov: Pb
A (0-5cm) 58,1 mg/kg
B (5-20cm) 48,7 mg/kg
pod mejo detekcije: <0,05mg/kg pod mejo določljivosti: <0,15mg/kg pod mejno vrednostjo: <85mg/kg mejna vrednost: >85mg/kg opozorilna vrednost: >100mg/kg kritična vrednost: >530mg/kg

Nevarna snov: Zn
A (0-5cm) 172 mg/kg
B (5-20cm) 140 mg/kg
pod mejo detekcije: <0,1mg/kg pod mejo določljivosti: <0,3mg/kg pod mejno vrednostjo: <200mg/kg mejna vrednost: >200mg/kg opozorilna vrednost: >300mg/kg kritična vrednost: >720mg/kg

LEGENDA:
Dejanska vsebnost elementa.
Vsebnost snovi je zelo majhna in pod mejo detekcije instrumenta.
Vsebnost snovi je zelo majhna in pod mejo dolčljivosti.
Analitski postopek in inštrumenti so zaznali manjšo količino snovi, katere skupna vsebnost je pod MEJNO vrednostjo.
Vsebnost snovi v tleh je povečana in nad MEJNO vrednostjo, a še pod OPOZORILNO vrednost.
Vsebnost snovi v tleh je presežena glede na OPOZORILNO vrednost in je manjša od KRITIČNE vrednosti.
>>>>> Vsebnost snovi presega KRITIČNO vrednost.

Datum izpisa: 11.1.2002 Stran: 3 od 4
Obrazec: ROTS/LPAO/PA4

Projekt: RAZISKAVE ONESNAŽENOSTI TAL S PREDLOGOM PROGRAMA UKREPOV NUJNIH SANACIJ
Naročnik: MINISTRSTVO ZA OKOLJE IN PROSTOR — Vzorčna točka: 03411
Izvajalec: BIOTEHNIŠKA FAKULTETA, ODDELEK ZA AGRONOMIJO, Center za pedologijo in varstvo okolja — Čas vzočenja: oktober 1999

Vzorčenje: 03411/1099

Organske nevarne snovi

Analitski laboratorij: Zavod za zdravstveno varstvo Maribor, Inštitut za varstvo okolja

Globina	PCB	Aceto klor	Ala klor	Aldrin	Diel drin	Endr in	Drini	alfa-HCH	beta-HCH	delta-HCH	gama-HCH	HCH spojine	Atra zin	Cia nazin	Hepta klor	Klordan -cis	Klordan -trans
enota	mg / kg suhe snovi																
A (0-5cm)	[0,0005]	[0,003]	[0,003]	[0,001]	[0,001]	[0,001]	[0,001]	[0,001]	[0,001]	[0,001]	[0,001]	[0,001]	[0,003]	[0,003]	[0,001]	[0,001]	[0,001]

Globina	DDD (o,p)	DDD (p,p)	DDE (o,p)	DDE (p,p)	DDT (o,p)	DDT (p,p)	Vsota DDT, DDD in DDE	Metola klor	Heksa klor benzen	Desetil-atrazin	Desizo propil-atrazin	Prome trin	Propa zin	Sebuti lazin	Simaz in	Terbu tilazin	Terbu trin
enota	mg / kg suhe snovi																
A (0-5cm)	[0,001]	[0,001]	[0,001]	0,005	[0,001]	[0,001]	0,005	[0,003]	[0,001]	[0,003]	[0,003]	[0,003]	[0,003]	[0,003]	[0,003]	[0,003]	[0,003]

Globina	Antra cen*	Benzo (a)antr acen*	Benz o(a)pi ren*	Benzo (ghi) perilen*	Benzo (k)fluor anten*	Fenan tren*	Fluora nten*	Indeno (1,2,3-cd) piren*	Krize n*	Naftal en*	Acena ften	Acena ftilen	Benzo (b)fluor anten	Dibenzo (a,h)antra cen	Fluo ren	Piren	PAH	PAH*
enota	mg / kg suhe snovi																	
A (0-5cm)	[0,005]	[0,005]	[0,005]	[0,005]	[0,005]	<0,01	[0,005]	[0,005]	[0,005]	0,026	[0,005]	[0,005]	[0,005]	[0,005]	[0,005]	[0,005]	0,03	0,026

Opombe: Analiza opravljena; vsebnost pod mejo določanja (LOQ):<x,xxx
Analiza opravljena; vsebnost pod mejo detekcije (LOD):[x,xxx]

Globina Vsebnost

Nevarna snov: Atrazin
A (0-5cm) <0,003 mg/kg POD MEJO DETEKCIJE
pod mejo detekcije: <0,003mg/kg pod mejo določljivosti: <0,005mg/kg pod mejno vrednostjo: <0,01mg/kg mejna vrednost: >0,01mg/kg opozorilna vrednost: >3mg/kg kritična vrednost: >6mg/kg

Nevarna snov: DDT/DDD/DDE
A (0-5cm) 0,005mg/kg

Nevarna snov: Drini
A (0-5cm) <0,001 mg/kg POD MEJO DETEKCIJE

Nevarna snov: HCH spojine
A (0-5cm) <0,001 mg/kg POD MEJO DETEKCIJE

Nevarna snov: PAH*
A (0-5cm) 0,026mg/kg

Nevarna snov: PCB
A (0-5cm) <0,0005 mg/kg POD MEJO DETEKCIJE

Nevarna snov: Simazin
A (0-5cm) <0,003 mg/kg POD MEJO DETEKCIJE

LEGENDA:
Dejanska vsebnost elementa.
Vsebnost snovi je zelo majhna in pod mejo detekcije instrumenta.
Vsebnost snovi je zelo majhna in pod mejo dolčljivosti.
Analitski postopek in inštrumenti so zaznali manjšo količino snovi, katere skupna vsebnost je pod MEJNO vrednostjo.
Vsebnost snovi v tleh je povečana in nad MEJNO vrednostjo, a še pod OPOZORILNO vrednost.
Vsebnost snovi v tleh je presežena glede na OPOZORILNO vrednost in je manjša od KRITIČNE vrednosti.
>>>>> Vsebnost snovi presega KRITIČNO vrednost.

Datum izpisa: 11.1.2002 Stran: 4 od 4
Obrazec: ROTS/LPAO/PA4

◀ **Fig. 7.12** Standardized soil contamination report from late nineties: *Page 1* sampling location, *Page 2* soil description and basic analytical data, *Page 3* heavy metal content, and *Page 4* the concentration of selected organic compounds. The standard four-page soil contamination report presents concentrations of heavy metals and organic contaminants measured in soil samples from depths of 0–5 cm and 5–20 cm. In the tables, the analytical values are assessed according to the three contamination classes of the Decree on Dangerous Substances in Soil. *Grey bars* show the concentration of the contaminant, while the *green*, *yellow*, *red*, and *purple bars* illustrate the contamination threshold values. The threshold values under the detectable limits of the method used are indicated with *blue bars*

The system allows soil profile reports to be downloaded (Fig. 7.11). The Slovenian version of the web viewer offers access to much more complete information on soils and environmental information related to agricultural production. The content is dynamic and often changed. In addition to the general public soil data viewer, an educational portal and several project-oriented soil data portals are available (Fig. 7.12).

8 Soil Degradation

Borut Vrščaj, Blaž Repe, and Primož Simončič

Abstract

The extent and degree of the main soil degradation processes that have been well identified and assessed in Slovenia are presented within this chapter: soil sealing, soil contamination, soil organic matter (SOM) content and decline, erosion, and soil compaction in forests. Soil sealing, the most important form of degradation, was especially intensive between 2002 and 2007. Intensive housing, construction of infrastructure, and the expansion of new commercial areas in the suburbs of cities raised concern among soil experts already at that time. The total increase in the urbanized surface area in this period was 19,790 ha, or a 22.5% increase over the initial urban surface area. Primarily agricultural land was urbanized. Soil contamination research intensified in the late 1980s when soil contamination monitoring was extended from localized areas to a regular programme in order to systematically cover the entire country in 1999. The soils of Slovenia are predominantly contaminated with Cd, Pb, Ni, and Zn only in the hinterlands of limited industrial areas, ironworks, smelters, and coal-fired power plants. Contamination with other heavy metals occurs rarely and in a hot spot pattern. The lack of systematic monitoring of SOM content data in Slovenian soils led to empirical expert knowledge-based spatial modelling to assess areas with a lower/higher potential risk of SOM content decline in agricultural soils. On the contrary, the carbon stocks in soils under forest are presented with data collected during the systematic soil monitoring campaign. Water erosion and wind erosion in Slovenia are neither frequent nor widespread but have been severe under occasional conditions in selected areas. Soil compaction has been studied in soils under forest. Several experiments on the impact of heavy harvesters in undisturbed Cambisols have demonstrated the negative effects of heavy machinery as regards the resulting wheel ruts, even a few years after use has ended.

B. Vrščaj (✉)
Department of Agricultural Ecology and Natural Resources, Centre for Soil and Environment Research, Agricultural Institute of Slovenia, Ljubljana, Slovenia
e-mail: Borut.Vrscaj@kis.si

B. Repe
Geography Department, University of Ljubljana, Faculty of Arts, Ljubljana, Slovenia
e-mail: blaz.repe@ff.uni-lj.si

P. Simončič
Department of Forest Ecology, Slovenian Forestry Institute, Ljubljana, Slovenia
e-mail: primoz.simoncic@gozdis.si

B. Vrščaj et al., *The Soils of Slovenia*, World Soils Book Series,
DOI 10.1007/978-94-017-8585-3_8

8.1 Soil Degradation

Many types of soil degradation and threats to soil are widely recognized, with the following considered the main threats: soil erosion, organic carbon depletion, soil compaction, soil sealing, soil contamination, soil acidification, and landsliding. Four of these are considered to be prominent and have an important impact in Slovenia: soil sealing, soil contamination, landsliding, and soil erosion. The extent and the degree of some of these threats have been well identified (soil contamination) and assessed (landsliding, soil sealing, and erosion) in Slovenia. The others have not been assessed in a comprehensive manner or have not been measured using a long-term systematic approach. The latter is especially important as regards assessment of organic matter decline risks and trends. This chapter briefly presents the data on and the extent of the main soil degradation processes in Slovenia.

8.2 Soil Sealing

Soil urbanization is a process of land use or land cover changing from agricultural, seminatural, and natural land uses to urban, traffic, industrial, and related land uses. A large portion of urban land use is sealed, in other words, covered with an impervious material such as concrete and/or asphalt. Sealed soil loses the capacity to perform virtually all soil functions and services except for providing support for housing and infrastructure. Important soil ecosystem functions and capacities such as filtering, buffering, carbon sequestration, and biomass production are to a great extent impaired or even prevented. Soil loss due to urbanization is largely considered to be finite; the soil is destroyed and gone for the era of human civilization. In addition to the direct impact on soils, the surrounding agricultural and natural soils may also be influenced by urbanization. The contamination, fragmentation, and change in the water flow are frequent consequences that are difficult to avoid. Soil sealing is practically an irreversible process. A large extent of soils are still sealed worldwide daily. In the European Union, the total sealed area in 2006 was 2.3% of the EU territory or 200 m^2 per inhabitant (Global Soil Forum 2013). A large extent of soils was and still is being sealed due to the construction of motorways and other infrastructure (Jiménez-Beltrán and Töpfer 2000).

After becoming an independent state, soil urbanization and soil sealing accelerated in Slovenia. These processes were especially intensive after the year 2000, mainly between 2002 and 2007 before the first signs of the global economic crisis appeared in 2007. The intensive housing construction boom, the construction of motorways and other infrastructure, and the expansion of new commercial areas in the suburbs of cities raised concern among the experts already at that time. To evaluate the extent of soil sealing, an analysis was carried out by the Agricultural Institute of Slovenia using the 2002, 2005, and 2007 digital land use data (Vrščaj 2008). The study employed a series of digital land use databases covering the entire territory of Slovenia (MKGP 2002a, MKGP 2007; MKO 2011a, 2012; MKGP/MAFF 2014). The main input was polygon land use databases on a scale of 1:5000 (MKGP 2007, 2005, 2002a). They were elaborated by the interpretation of high-resolution aerial ortho-photo imagery (0.5-m cell size) using a well-documented land use/land cover interpretation manual. The manual is regularly updated and improved with new examples of land use classes (MKGP 2009, 2002b, 2011, 2013). Despite improvements in the accuracy of the land use polygon delineation in the databases of different years, as well as minor changes in the classification keys, the temporal comparison of land use changes provides important information on the dimension of soil sealing in Slovenia at that time.

According to Vrščaj (2008), the extent of urban land, including traffic areas, in Slovenia in the period 2002–2007 increased from 88,070 ha in 2002 to 107,919 ha in 2007. The total increase was 19,790 ha or 22.5% of the initial urban surface. Primarily, agricultural land (65.2%) and forests (24.4%) were urbanized. In the opinion of many, the extent of urbanization in that period was very large. Taking into account the duration between the creation of digital ortho-photo imagery and the creation of the database, a total soil loss of 11 ha per day was estimated. For Slovenia, a country with a modest extent of high-quality agricultural soils, such growth in urbanization was disconcerting. In the next step of the study, the agricultural quality of the soils of the urbanized areas was investigated. The findings were not surprising. The soils of better quality in the lowlands and on favourable terrain were urbanized to a significantly larger extent. Figure 8.1 shows a comparison of the extent of existing soils and sealed soils, both having the same agricultural quality. Agricultural soil quality is expressed in terms of soil quality points ranging from 7 (i.e. marginal agricultural quality) to 100 (i.e. prime agricultural land). The positive differences indicate that soil of the listed quality was sealed to a larger degree when compared to the total extent of soils of the same quality. And vice versa, negative values indicate that soils of a certain quality were sealed to a lower degree compared to the total extent of soils of the same quality. The positive values on the right side of the chart marked from 54 to 96 points show that agricultural land of better quality was sealed.

The land take in Slovenia was compared to selected EU countries. Compared to Germany, where the degree of urbanization was app. 90 ha a day, the extent of urbanization in Slovenia was 17.6-fold larger, in relative terms. In addition to that, a significant proportion of Slovenia is mountainous and since high-quality agricultural land is limited, the findings were even more serious.

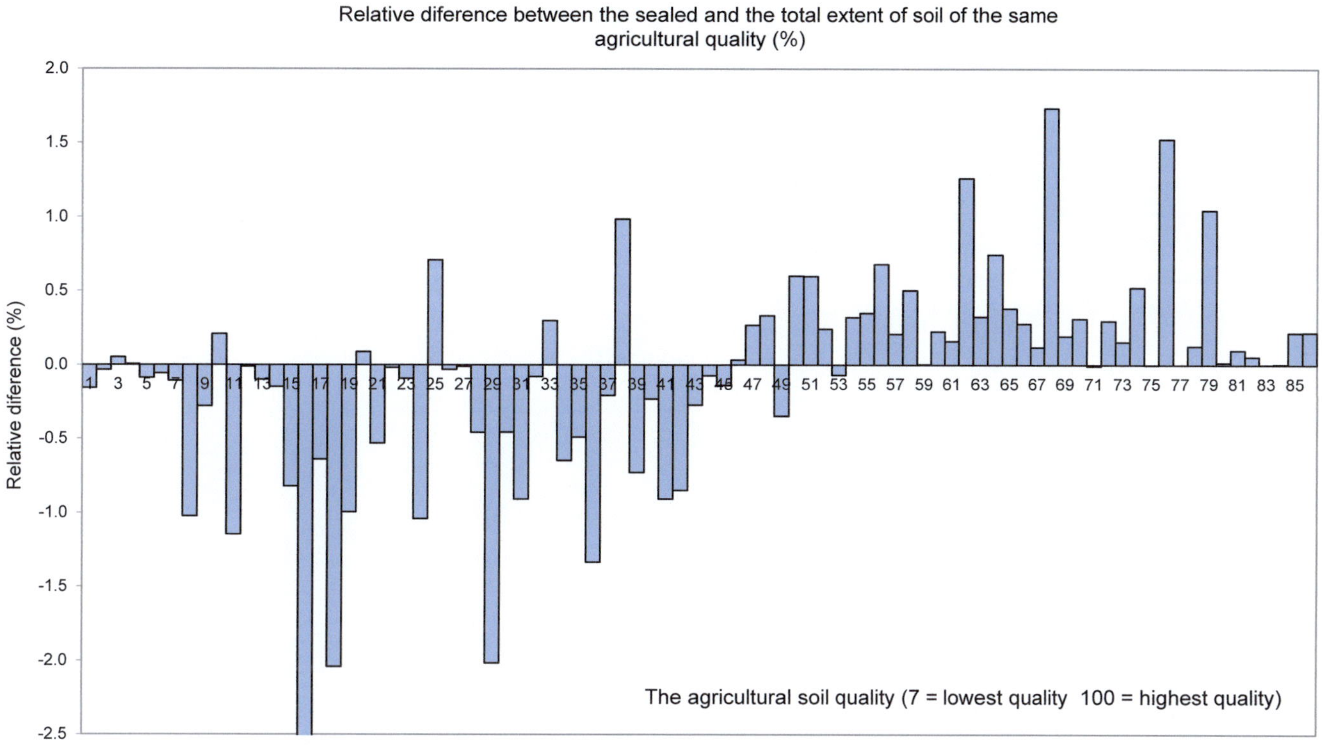

Fig. 8.1 Comparison of the degree of urbanized soils in relation to the agricultural soil quality

Urbanization in Slovenia has a very dispersed pattern as a result of not very rational spatial planning, which is carried out on a municipal level. In practice, housing and development is frequently spontaneous especially as regards individual housing and accompanying traffic infrastructure. This is one of the important reasons for the high and still increasing fragmentation of agricultural land.

The land use change studies carried out on the international level does not detect the land take and fragmentation adequately. The methodologies used to assess the land use changes at the continental European level using the CORINE Land Cover data at a scale of 1:100,000 (Prokop et al. 2011) are not suitable for detecting the total extent of individually small but numerous land use changes. This and other 1:100,000 scale European-wide evaluations (Gardi et al. 2014) have identified much more modest soil sealing in comparison with a study where a 1:5000 scale land use/land cover map was utilized.

The changes and amendments to the Slovenian legislation before 2007 and especially in 2010 were frequently in favour of development or, in other words, the accelerated sealing of soils. After the identification of the problem and in combination with the inadequate food security of the country, the legislation was improved in favour of limiting sealing and lowering the fragmentation of the quality agricultural soils (Republika Slovenija 2011) (Fig. 8.2).

The picture above shows one of the studies for planning a large motorway junction in an area of irrigated Eutric Cambisols. Being aware of the value of these fertile soils, which in this area are traditionally used for hops production, the local inhabitants, farmers, and experts organized a local initiative to protect the best soils and to direct the motorway planning to lower quality soils.

Figures 8.3 and 8.4 show fields between or close to existing industry facilities that were sealed between 2002 and 2007. Such soil sealing was expected among lay persons as well as soil and agricultural experts and is, in spite of the considerable total extent, still considered an acceptable sacrifice necessary for the faster economic development of the country.

The aerial picture above (Fig. 8.5) combined with GIS-derived sealed areas presents quite a frequent example of the sealing of high-quality agricultural soils in the suburbs of smaller cities and rural centres. Deep and loamy Eutric Cambisols previously used for hop fields and arable land were used to build a typical shopping centre consisting of a large parking lot and a large building. In such cases, the discussion, if it existed at all, was in favour of big stores and comfortable shopping.

Figure 8.6 explains the urbanization and land fragmentation pattern that occurs due to the urbanization of terraced Calcaric Cambisols in the suburbs and backgrounds of the coastal city of Koper. In the past, this intensive agricultural area supplied the large cities (e.g. Trst/Triest) with high-

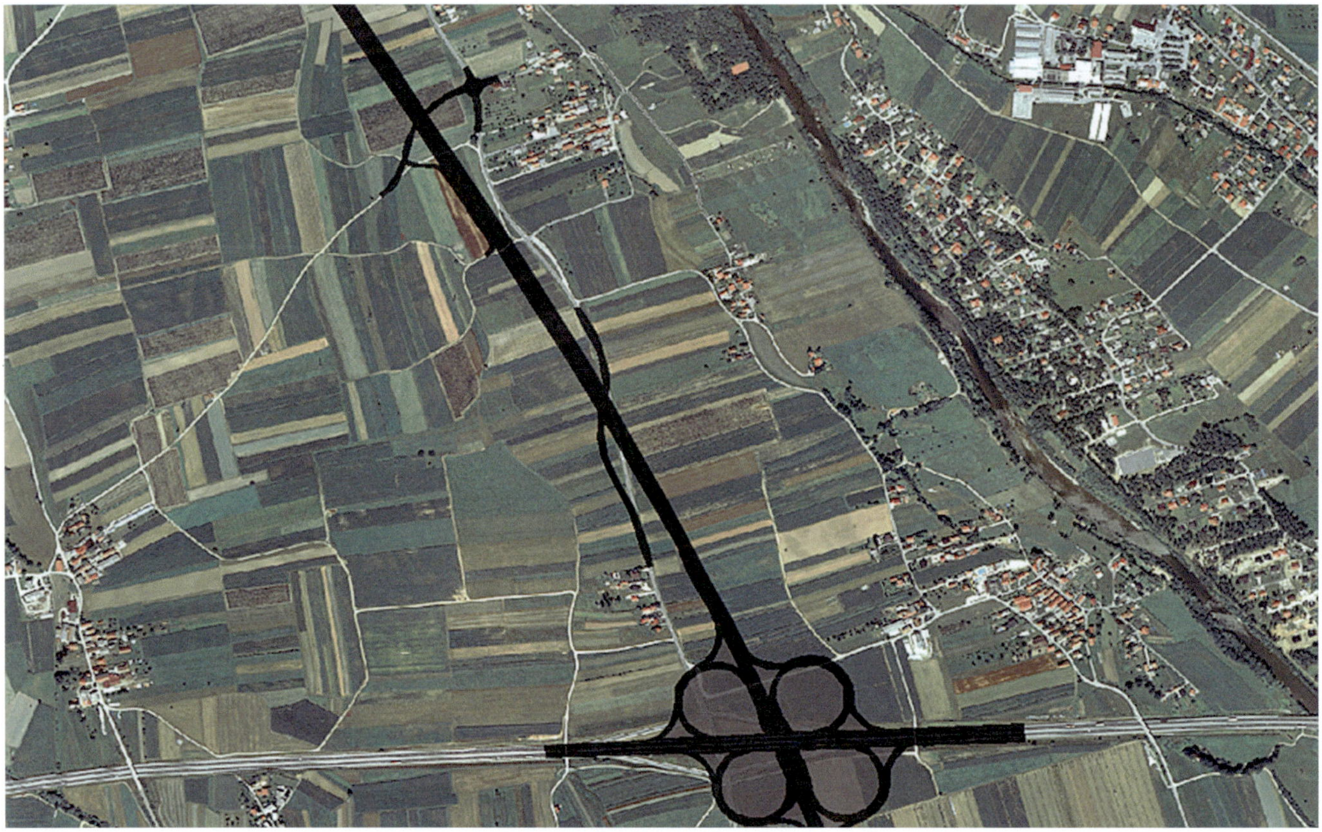

Fig. 8.2 Planning a new motorway crossing good quality agricultural soils (Vrščaj 2011)

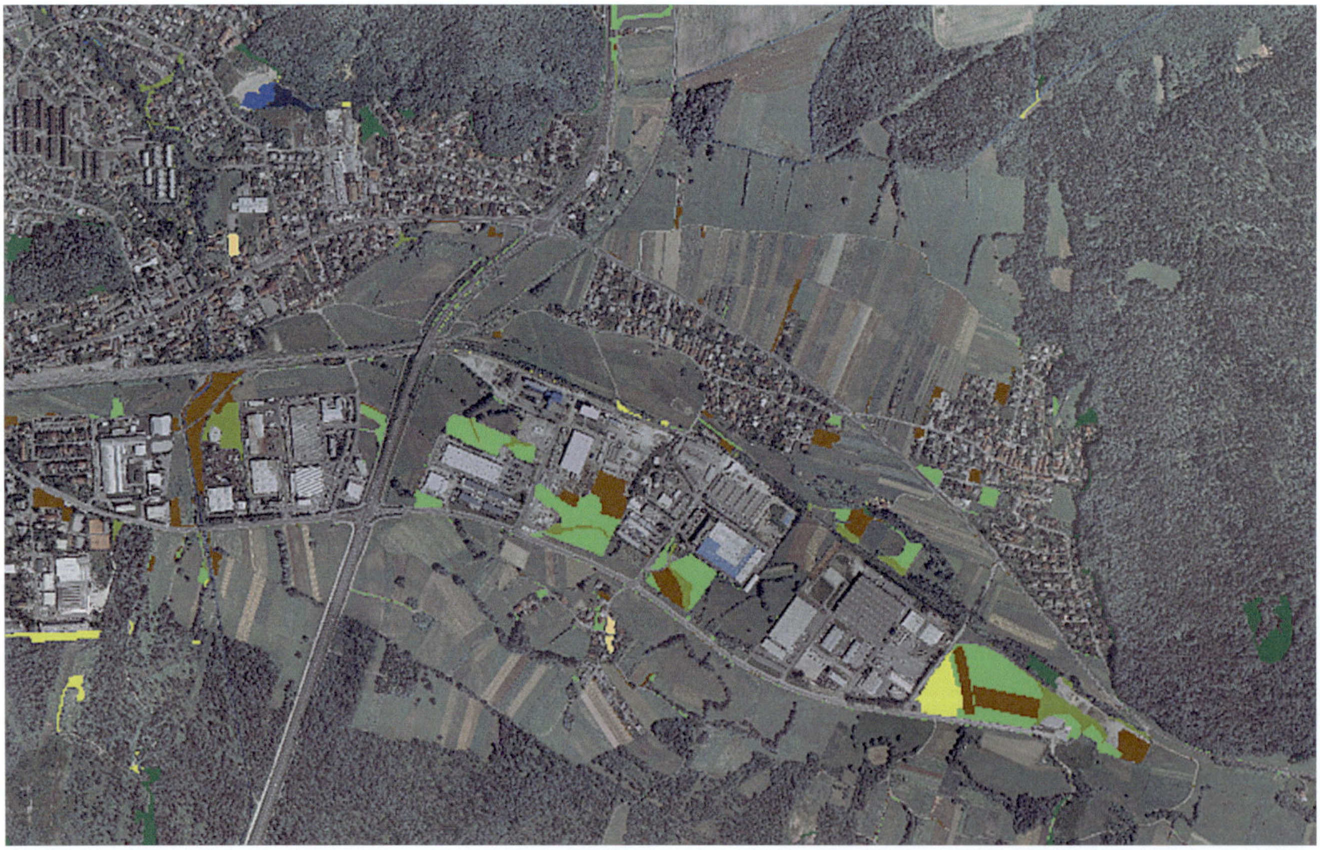

Fig. 8.3 Enlargement of the existing industrial areas north of Ljubljana (GURS 2011, n. processed in GIS)

Fig. 8.4 New development in the suburbs. Industrial facilities have extended to prime agricultural land of Eutric Cambisols (GURS 2011, n. processed in GIS)

quality Mediterranean vegetables, fruits, wine, and olive oil. For centuries, the people built and maintained terrace-supporting walls made of flysch sandstone by hand. In doing so, they created a characteristic cultural landscape. These soils are affected by two processes: sealing and the erosion. The land is urbanized by the individual housing units that are predominantly located on the ancient terraces and especially on hilltops and ridges in order to enjoy the views. The crumbly Bv horizon of the Calcaric Cambisols of this area is very erodible. The area is being afforested as well. Tree roots frequently destroy ancient walls and trigger erosion. If not stopped on time, erosion rapidly degrades a Calcaric Cambisol into a Regosol (Figs. 8.7, 8.8 and 8.9).

The nature of the soil sealing combined with the total extent of sealed soils makes sealing the most important form of soil degradation in Slovenia.

8.3 Soil Contamination

Soil contamination measurements and assessments have a longer tradition in Slovenia in comparison with the assessment of other soil-related threats. Industrial areas with old environmental burdens gave rise to concerns about the effects of contaminated soils on human health and the environment among experts, citizens, environmentalists, and NGOs quite early. Soil contamination research intensified in the late eighties, when the lead in research activities was assumed by the Centre for Soil and Environmental Science of the University of Ljubljana (CSES). A sampling methodology was developed, standard soil analytical methods were harmonized, and the first soil contamination data collection and data processing procedures were designed. Dense soil sampling was carried out in the industrial city of Celje and its hinterlands, where the soils are heavily contaminated by zinc, cadmium, and lead due to emissions from a zinc smelter. The collected data were digitized and stored as a separate soil information layer within the emerging soil information system of that time (Vrščaj et al. 1989). The contamination was evaluated in the light of the legislation, and the soil pollution map was elaborated using different interpolation techniques (Lobnik et al. 1989b). The research was extended to the areas of the industrial cities of Jesenice, Ljubljana, Maribor, and the port of Koper. In the early nineties, the soil contamination assessment projects focused on agricultural land (Lobnik et al. 1992a, b, 1993). The

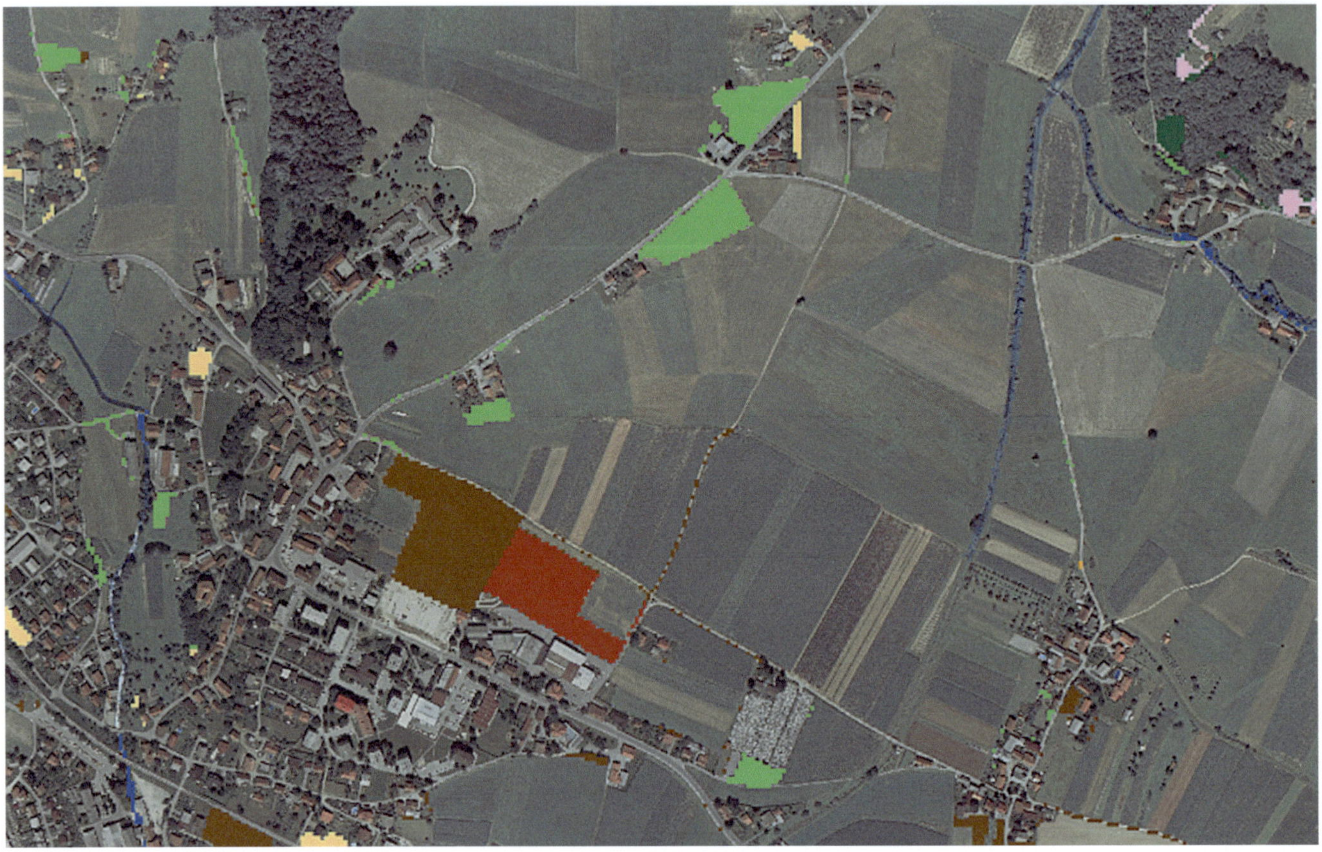

Fig. 8.5 Areas of sealed soil due to the construction of a new shopping centre in the Savinja River valley, central Slovenia (GURS 2011, n. processed in GIS)

general picture of the state of soil contamination in Slovenia was misleading due to the sampling of primarily the known contaminated areas. The imbalanced results and interest in assessing the contamination in other areas led to the decision to extend the soil contamination monitoring to a regular programme that was to systematically cover the entire country. The complete, contaminated, and uncontaminated territory of Slovenia had to be sampled in a uniform and regular grid in order to yield representative information on the soil contamination in the entire country. In 1999, a soil contamination monitoring programme was initiated within the first National Environmental Protection Programme. A regular sampling grid with 2689 pre-defined locations was designed. The grid density was 2 × 2 km in the agricultural areas below 600 m.a.s.l. and 4 × 4 km in forests and above 600 m.a.s.l. The grid was dense, and the high number of locations initially planned to be investigated annually was lowered in the following years due to budgetary reasons. In spite of that, the total number of locations systematically sampled between the 1989 and 2007 reached 368.

Below we present the state of soil contamination as summarized in a comprehensive and well-illustrated report prepared by Zupan et al. (2008). As of 2007, the soils have been sampled at 368 locations in three depths 0–5 cm, 5–20 cm, and 0–20 cm (Fig. 8.10).

The selection of contaminants that are analysed in each soil sample is representative and encompasses the most frequent pollutants of industrial, traffic, and agricultural origins. The standard analytical set is composed of ten heavy metals (arsenic, cadmium, cobalt, chromium, copper, mercury, molybdenum, nickel, lead, and zinc), fluorides, persistent organic insecticides such as the total of "drins" (i.e. aldrin, dieldrin, and endrin); hexachlorocyclohexane (HCH-α, β,γ, and δ), the active substances of selected persistent pesticides used in the past (DDT, DDD, DDE, simazine, atrazine), the sum of seven different polychlorinated biphenyls (PCBs), and ten different polycyclic aromatic hydrocarbons (PAH, e.g. anthracene, phenanthrene, naphthalene, and benzo[a]pyrene).

The measured values of contaminants in soils were evaluated against the threshold values defined in the Decree on the Limit, Warning, and Critical Concentrations of Dangerous Substances in Soil (hereinafter: the Decree on Dangerous Substances in Soil) (Republika Slovenija 1996) (Table 8.1).

Fig. 8.6 New shopping centres in an area of Eutric Cambisols, Savinja River valley, central Slovenia (*photo* B. Vrščaj)

Heavy metals are the main soil contaminants in Slovenia. The distribution of elevated cobalt (Co) concentrations in the southern suburbs of the city of Maribor indicates the industrial emission of Co, while other randomly distributed locations show hot spot local contamination. The same contamination pattern prevails in the case of chromium, copper, mercury, and molybdenum. The concentrations of molybdenum, mercury, and copper rarely exceed the warning value. Although increased in the topsoil, their concentrations are below the warning value on 95, 97, and 97% of the locations, respectively. The concentrations of nickel (Ni) are elevated in the coastal area due to the naturally Ni-enriched parent material. At the locations around the city of Maribor and the lignite mining area of Trbovlje, Ni enrichment in the topsoil was detected. Lead and zinc concentrations are raised in the topsoils in the hinterlands of the city of Ljubljana and the city of Celje, and in the southern and eastern suburbs of Maribor. In the area of Jesenice, the soils are strongly contaminated mainly by the ironworks emissions.

Tables 8.2, 8.3, 8.4, 8.5, 8.6, 8.7, and 8.8 present the state of heavy metal soil contamination per soil depth and per contamination class, as defined by the Decree. The main heavy metal contaminant is cadmium. The concentration of 22% of the samples from a depth of 0 to 5 cm exceeded the warning value and an additional 14% exceeded the critical value. The concentrations of cadmium at a depth of 5–20 and 0–20 cm are significantly lower. The pattern and the concentrations at different soil depths as well as land use indicate that aerial emissions from industry were the main sources of Cd in soils. Several locations indicate hot spot contamination caused by largely unidentified activities. The spatial pattern of arsenic (As) contamination is present at several locations in the Trbovlje coal mining area where a power plant is situated. The limit and warning threshold values for arsenic are exceeded there. The irregular distribution of other locations with elevated As concentrations indicate spatially limited hot spot contamination (Fig. 8.11).

The distribution of elevated cobalt (Co) concentrations in the southern suburbs of the city of Maribor indicates the industrial emission of Co, while other randomly distributed locations show hot spot local contamination. The same contamination pattern prevails in the case of chromium, copper, mercury, and molybdenum. The concentrations of molybdenum, mercury, and copper rarely exceed the warning value. Although increased in the topsoil, their concentrations are below the warning value on 95, 97, and 97% of the locations, respectively. The concentrations of nickel (Ni) are elevated in

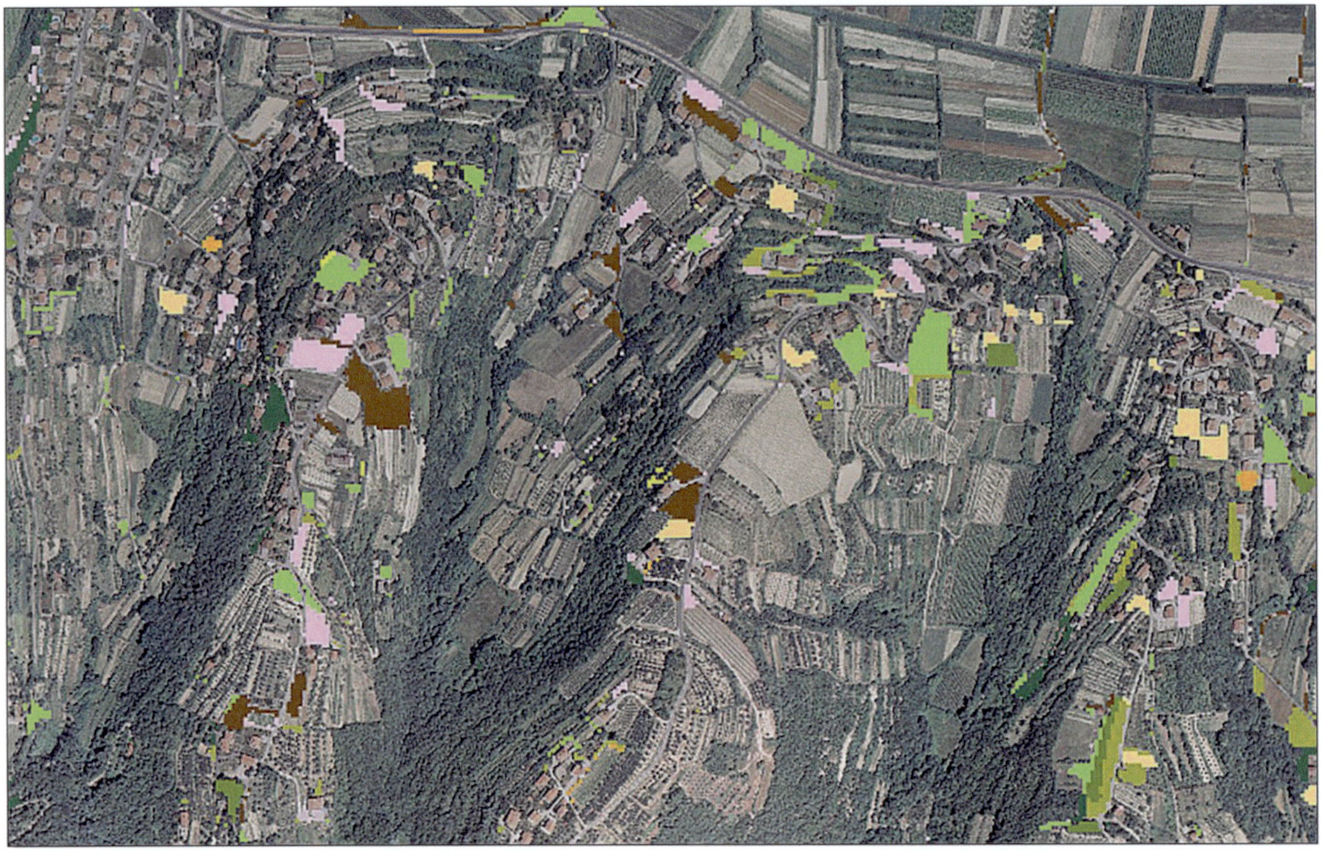

Fig. 8.7 Areas of land sealed by housing in the hinterland of the city of Koper. The terraced Eutric Cambisols in the coastal region of Slovenia are among the oldest cultural landscapes of the country (GURS 2011, n. processed in GIS)

Fig. 8.8 Urbanization—land fragmentation due to housing (*photo* B. Vrščaj)

Fig. 8.9 Soil sealing due to infrastructure development (*photo* B. Vrščaj)

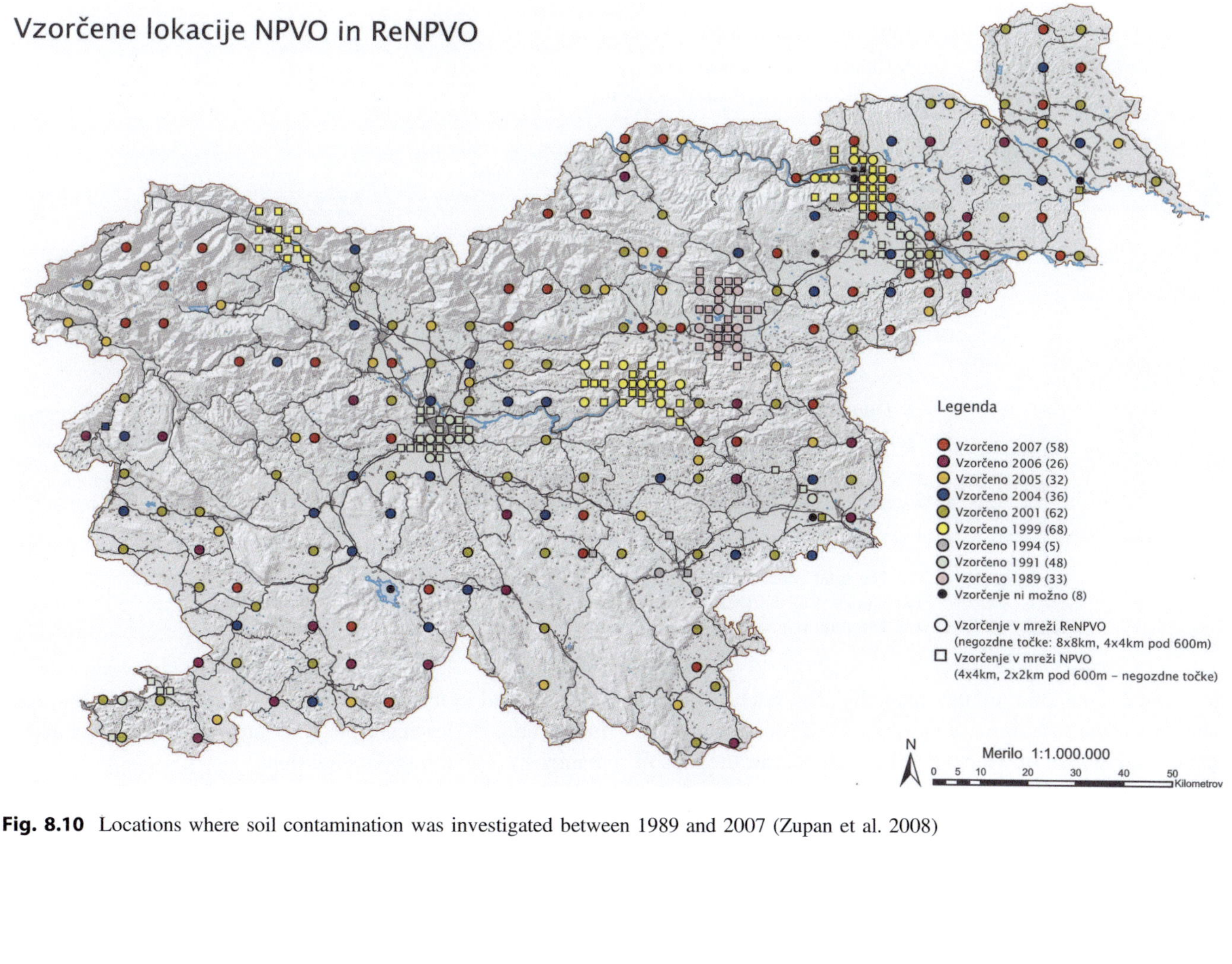

Fig. 8.10 Locations where soil contamination was investigated between 1989 and 2007 (Zupan et al. 2008)

Table 8.1 Threshold values as set by the decree on dangerous substances in soil (Republika Slovenija 1996)

Contaminant	Limit value	Warning value	Critical value
	mg/kg dry soil		
1. Heavy metals-aqua regia extraction			
Cadmium and its compounds. Expressed as Cd	1	2	12
Copper and its compounds. Expressed as Cu	60	100	300
Nickel and its compounds. Expressed as Ni	50	70	210
Lead and its compounds. Expressed as Pb	85	100	530
Zinc and its compounds. Expressed as Zn	200	300	720
Chromium and its compounds. Expressed as Cr	100	150	380
Hexavalent Cr. Expressed as Cr^{6+}	–	–	25
Mercury and its compounds. Expressed as Hg	0.8	2	10
Cobalt and its compounds. Expressed as Co	20	50	240
Molybdenum and its compounds. Expressed as Mo	10	40	200
Arsenic and its compounds. Expressed as As	20	30	55
2. Other inorganic compounds			
Fluorides. Total	450	825	1200
3. Aromatic compounds			
Volatile phenols	0.1	20	40
Benzene	0.05	0.5	1
Ethyl benzene	0.05	25	50
Toluene	0.05	65	130
Xylene	0.05	12.5	25
4. Polycyclic aromatic hydrocarbons (PAH)			
Total PAH concentration	1	20	40
5. Chlorinated hydrocarbons			
5a. Polychlorinated biphenyls (PCBs)			
The total concentration of the PCB (2)	0.2	0.6	1
5b. Insecticides based on chlorinated hydrocarbons			
DDT/DDD/DDE (3)	0.1	2	4
Drin's (4)	0.1	2	4
HCH compounds (5)	0.1	2	4
5c. Other plant protection products			
Atrazine	0.01	3	6
Simazine	0.01	3	6
6. Other compounds			
Hydrocarbons derived from petroleum (mineral oils)	50	2500	5000

(1) The total concentration of PAHs is the sum of naphthalene, anthracene, phenanthrene, fluoranthene, benzo (a) anthracene, chrysene, benzo (a) pyrene, benzo (ghi) perylene, benzo (k) fluoranthene and indeno (1,2,3) pyrene
(2) The total concentration of PCBs is the sum of PCB 28, 52, 101, 118, 138, 153, and 180
(3) The total concentration of the sum of DDT, DDD, and DDE
(4) The sum of the total concentration of aldrin, dieldrin, and endrin
(5) The total concentration of the sum of a-HCH, b-HCH, g-HCH, and d-HCH

the coastal area due to the naturally Ni-enriched parent material. At the locations around the city of Maribor and the lignite mining area of Trbovlje, Ni enrichment in the topsoil was detected. Lead and zinc concentrations are raised in the topsoils in the hinterlands of the city of Ljubljana and the city of Celje, and in the southern and eastern suburbs of Maribor. In the area of Jesenice, the soils are strongly contaminated mainly by the ironworks emissions.

The results of the monitoring of soil contamination with organic compounds in Slovenia indicate that the

Table 8.2 Cadmium concentrations at three soil depths with regard to the threshold values

Cadmium (n = 368)	Threshold value	Soil depth (cm)		
	mg/kg dry soil	0–5 (%)	5–20 (%)	0–20 (%)
Values below the level detectable by the method	<0.1	0	0	0
Concentrations < limit value	<1	64	69	86
Concentrations > limit and < warning value	≥ 1	22	21	13
Concentrations > warning and < critical value	≥ 2	14	9	1
Concentrations ≥ critical value	≥ 12	0	1	0

Table 8.3 Arsenic concentrations at three soil depths with regard to the threshold values

Arsenic (n = 368)	Threshold value	Soil depth (cm)		
	mg/kg dry soil	0–5 (%)	5–20 (%)	0–20 (%)
Values below the level detectable by the method	<2	2	0	1
Concentrations < limit value	<20	85	79	98
Concentrations > limit and < warning value	≥ 20	11	17	1
Concentrations > warning and < critical value	≥ 30	1	3	0
Concentrations ≥ critical value	≥ 55	1	1	0

Table 8.4 Cobalt concentrations at three soil depths with regard to the threshold values

Cobalt (n = 368)	Threshold value	Soil depth (cm)		
	mg/kg dry soil	0–5 (%)	5–20 (%)	0–20 (%)
Values below the level detectable by the method	<2	0	0	0
Concentrations < limit value	<20	78	76	83
Concentrations > limit and < warning value	≥ 20	21	24	17
Concentrations > warning and < critical value	≥ 50	1	0	0
Concentrations ≥ critical value	≥ 240	0	0	0

Table 8.5 Chromium concentrations at three soil depths with regard to the threshold values

Chromium (n = 368)	Threshold value	Soil depth (cm)		
	mg/kg dry soil	0–5 (%)	5–20 (%)	0–20 (%)
Values below the level detectable by the method	<5	0	0	0
Concentrations < limit value	<100	94	88	95
Concentrations > limit and < warning value	≥ 100	4	10	4
Concentrations > warning and < critical value	≥ 150	2	2	1
Concentrations ≥ critical value	≥ 380	0	0	0

contamination is low, and very low to none in the case of some organic contaminants. The spatial distribution of the concentrations of persistent organic insecticides, HCH compounds, and persistent derivatives of DDT indicate that Slovenian soils are to a large extent not polluted with these compounds. The total sum of "drins" was detected by a method used in 34 samples (10%). In the rest of the samples, the concentrations were below the detectable limit of the method. In the case of DDT, only 17 samples out of the 368 exceeded the limit value, and none of them exceeded the warning value. Hexachlorocyclohexane (α, β, γ, and δ HCH) was found in 22% of the samples. In general, the concentrations of organic contaminants are close to the detectable limit of the method or slightly above. The same holds true for the PCBs, with the exception of one well-known location in south-eastern Slovenia in the vicinity of a former electric capacitors factory where heavy leakage occurred. The limit value for PCBs (0.2 mg/kg) was exceeded there. The measured concentrations of PAH are under the detectable limit of the method or well below the limit value, with the exception of 3 locations out of a total of 368 samples. At 44 locations, the concentrations of atrazine (13%) and at 15 locations (5%) the concentrations of simazine exceeded the limit value. In the large majority of other samples, the presence of simazine and atrazine was not detected by the method chosen.

Table 8.6 Nickel concentrations at three soil depths with regard to the threshold values

Nickel (n = 368)	Threshold value	Soil depth (cm)		
	mg/kg dry soil	0–5 (%)	5–20 (%)	0–20 (%)
Values below the level detectable by the method	<5	0	0	0
Concentrations < limit value	<50	78	73	85
Concentrations > limit and < warning value	≥ 50	11	14	8
Concentrations > warning and < critical value	≥ 70	10	12	7
Concentrations ≥ critical value	≥ 210	1	1	0

Table 8.7 Lead concentrations at three soil depths with regard to the threshold values

Lead (n = 368)	Threshold value	Soil depth (cm)		
	mg/kg dry soil	0–5 (%)	5–20 (%)	0–20 (%)
Values below the level detectable by the method	<5	0	0	0
Concentrations < limit value	<85	86	95	96
Concentrations > limit and < warning	≥ 85	3	3	0
Concentrations > warning and < critical	≥ 100	10	2	4
Concentrations ≥ critical value	≥ 530	1	0	0

Table 8.8 Zinc concentrations at three soil depths with regard to the threshold values

Zinc (n = 368)	Concentrations	Soil depth (cm)		
	mg/kg dry soil	0–5 (%)	5–20 (%)	0–20 (%)
Values below the level detectable by the method	<10	0	0	0
Concentrations < limit values of	<200	93	98	93
Concentrations > limit and < warning value	≥ 200	4	2	1
Concentrations > warning and < critical value	≥ 300	2	0	6
Concentrations ≥ critical value	≥ 720	1	0	0

To conclude, following the results of the systematic soil contamination investigations in a regularly spaced grid, the soils of Slovenia can be considered to be contaminated only in the hinterlands of limited industrial areas. They are primarily contaminated with heavy metals in the hinterlands of ironworks and smelters and coal-fired power plants. The predominant heavy metal contaminants are Cd, Pb, Ni, and Zn. Contamination with other heavy metals occurs rarely and in a hot spot pattern. The soils are not contaminated with organic compounds; the concentrations rarely exceed the limit value, if at all, in the case of some contaminants.

8.4 Soil Organic Matter Content

8.4.1 The Soil Organic Matter Content in Agricultural Soils

Soil organic matter (SOM) is one of the most important soil quality indicators. The SOM content and form guide the processes in soils and to the largest extent define the soil properties and suitability. Many people simply say that SOM is the basis of soil fertility. The information on SOM content and trends (whether the SOM content is increasing or decreasing) is important for guiding agricultural production, for adapting soil tillage practices and other agro-technical measures, for designing climate change mitigation measures, as well as for assessing the national CO_2 balance. Unfortunately, Slovenia has no countrywide programme to systematically collect and process SOM information. In order to assess trends as regards SOM content in the agricultural soils and to identify the areas where the soil SOM depletion risk is high, such a programme needs to be introduced in the coming years. Being aware of these needs and due to the lack of initial information, the Ministry of the Environment and Spatial Planning financed a preliminary study to bring together the available/accessible data and to assess the state of SOM content in the agricultural soils of Slovenia.

In the study (Sušin et al. 2009), the data on 11,173 soil samples were obtained from seven different regional soil laboratories operated by the agricultural extension service and the laboratory of the Agricultural Institute of Slovenia. One of the main activities of the laboratories is to perform the soil analytics needed to elaborate the soil fertility recommendations for individual farmers. The bulk of the soil samples are analysed for pH, plant available phosphorus (P),

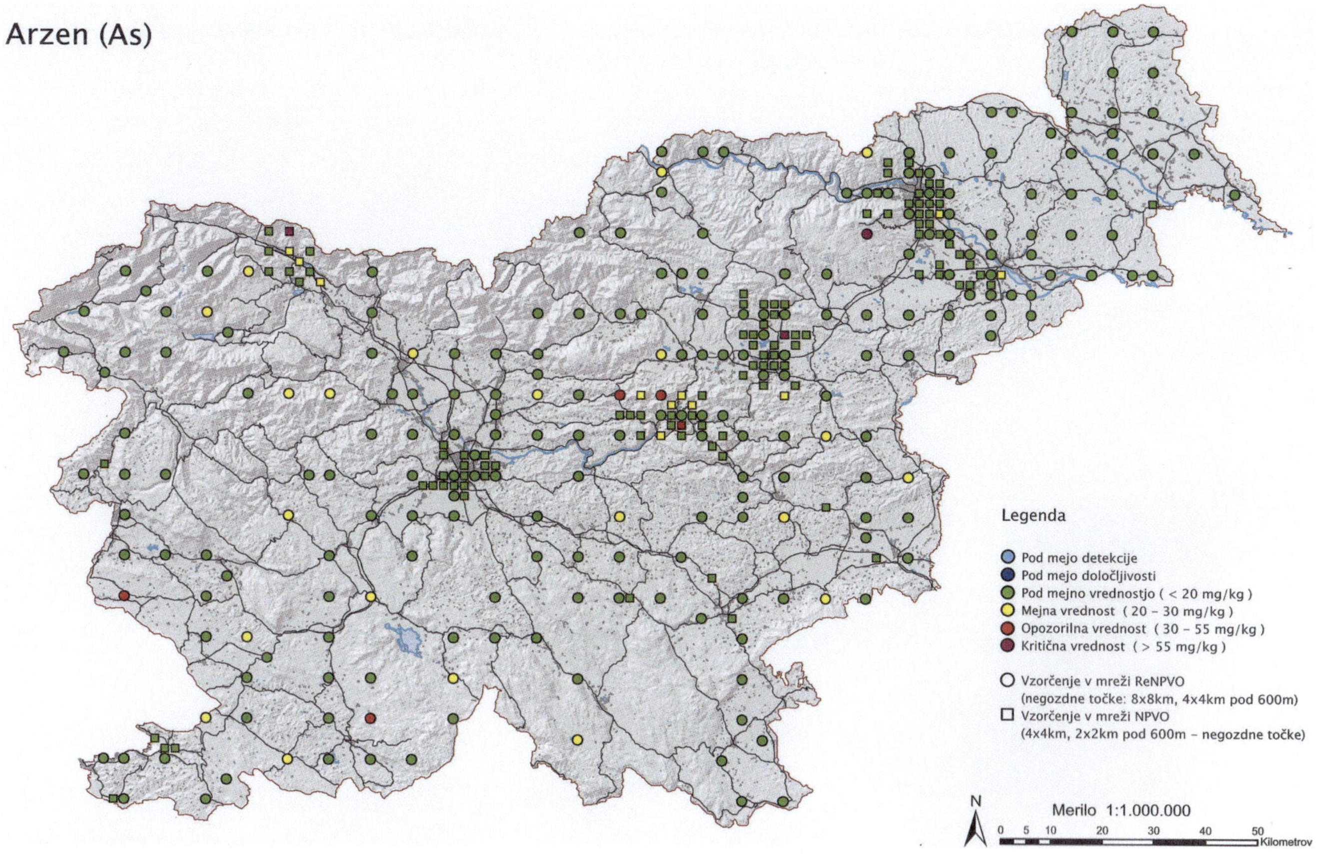

Fig. 8.11 Distribution of Arsenic in soil (Zupan et al. 2008). See Table 8.3 to interpret the colour of the dots

Table 8.9 Soil organic matter (%) in agricultural soils (Sušin et al. 2009)

SOM content (%)	Olive groves	Fields, hop fields, gardens	Orchards	Vineyards	Meadows and pastures	Total
n	73	5790	1000	1175	3135	11,173
Min.	0.6	0.4	0.4	0.2	0.4	0.2
Max.	4.5	66.0	11.5	8.3	57.6	66.0
Average	2.4	3.9	3.3	2.6	6.0	4.3
Std. dev.	0.9	2.7	1.4	1.1	3.6	3.0
CV%	36.7	68.0	42.2	41.8	59	69.2

potassium (K), and selected micronutrients such as boron (B) and magnesium (Mg). The number of samples where the SOM is analysed is significantly lower in comparison with those where the pH, P, and K are analysed. The soils for which the data were obtained were sampled between the years 2004 and 2009. The data were geocoded, spatially analysed in the GIS, and processed to derive basic statistics. The average SOM content in the topsoil (0–25 cm) of all agricultural soils in Slovenia is 4.3% (std. dev. = 3.0). As expected, the SOM content significantly varies from 0.3%, e.g. sandy vineyard soils, to 66.0% in bog meadows and fields. Table 8.9 shows detailed statistics of SOM content in relation to the different groups of agricultural land use.

These results do not allow firm conclusions on SOM content to be made. Namely the sampling methodologies were not entirely uniform nor were the SOM analytical procedures entirely comparable. Nevertheless, the results of the study roughly indicate the approximate SOM content in the agricultural soils of Slovenia.

8.4.2 An Attempt to Identify Areas with Agricultural Soils at Risk of SOM Content Decline

Climate change has triggered several adaptations to the hitherto agricultural production in Slovenia. The new

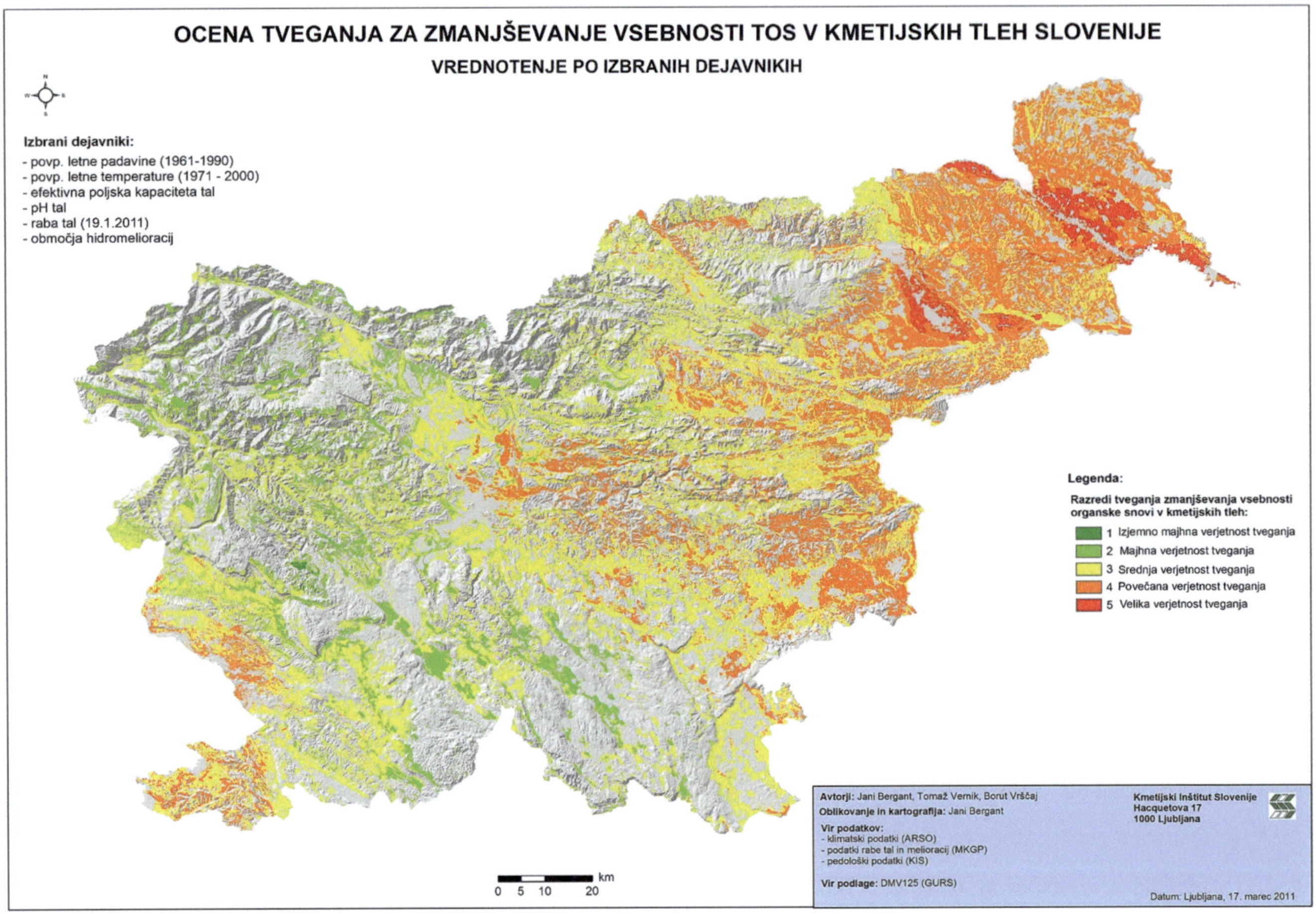

Fig. 8.12 Visualization of the SOM decline risk in agricultural areas

programmes that are currently designed need more accurate and detailed information on soils. Due to more frequent and severe droughts between the years 2000 and 2013, information on water holding capacity as well as SOM content became crucial to determining the adaptation measures and programmes and to steer climate change mitigation activities on the country level. The important information that has to be elaborated in the near future is spatial information on the dynamics and trends of SOM content in agricultural soils. Within the preparatory activities for acquiring such information, a preliminary study was conducted to spatially estimate the SOM depletion risk on agricultural land. Due to the unavailability of long-term SOM measurements, an expert knowledge-based modelling approach to estimating the risk of SOM decline had to be utilized (Verbič et al. 2011).

Among the main factors that influence SOM content in Slovenia are precipitation and temperatures, land use, soil tillage and fertilization, relief, soil type/soil properties, as well as exposure in relation to solar radiation. The input information was rasterized land use data (1:5000) and the digital soil map at a scale of 1:25,000, raster average annual precipitation, mean annual temperatures, relief data (elevation model and exposure), solar radiation, soil water holding capacity, soil pH, and the extent of hydro-ameliorated and irrigated areas. All datasets were resampled to 12.5-m cell size and standardized regarding the values. For each dataset, a factor/weight was defined regarding its contribution to the SOM mineralization potential. An empirical expert knowledge-based spatial model was designed and implemented using raster-based GIS programming. The visualized result of the model was a qualitative map showing areas with a lower/higher potential risk of SOM content decline. The map is considered to be a preliminary assessment that will be used for more appropriate design of an SOM content monitoring programme as well as to guide climate change adaptation measures in agriculture.

Fig. 8.12 presents a qualitative estimation of the SOM decline risk within agricultural land use areas in Slovenia. The central part of the country, having a continental climate and soils with suitable water holding capacity, shows a low-to-very low SOM decline risk. The risk is estimated to be higher in the area with a Mediterranean climate in the coastal region, south-western Slovenia. The highest risk is in east and north-eastern Slovenia, predominantly due to the

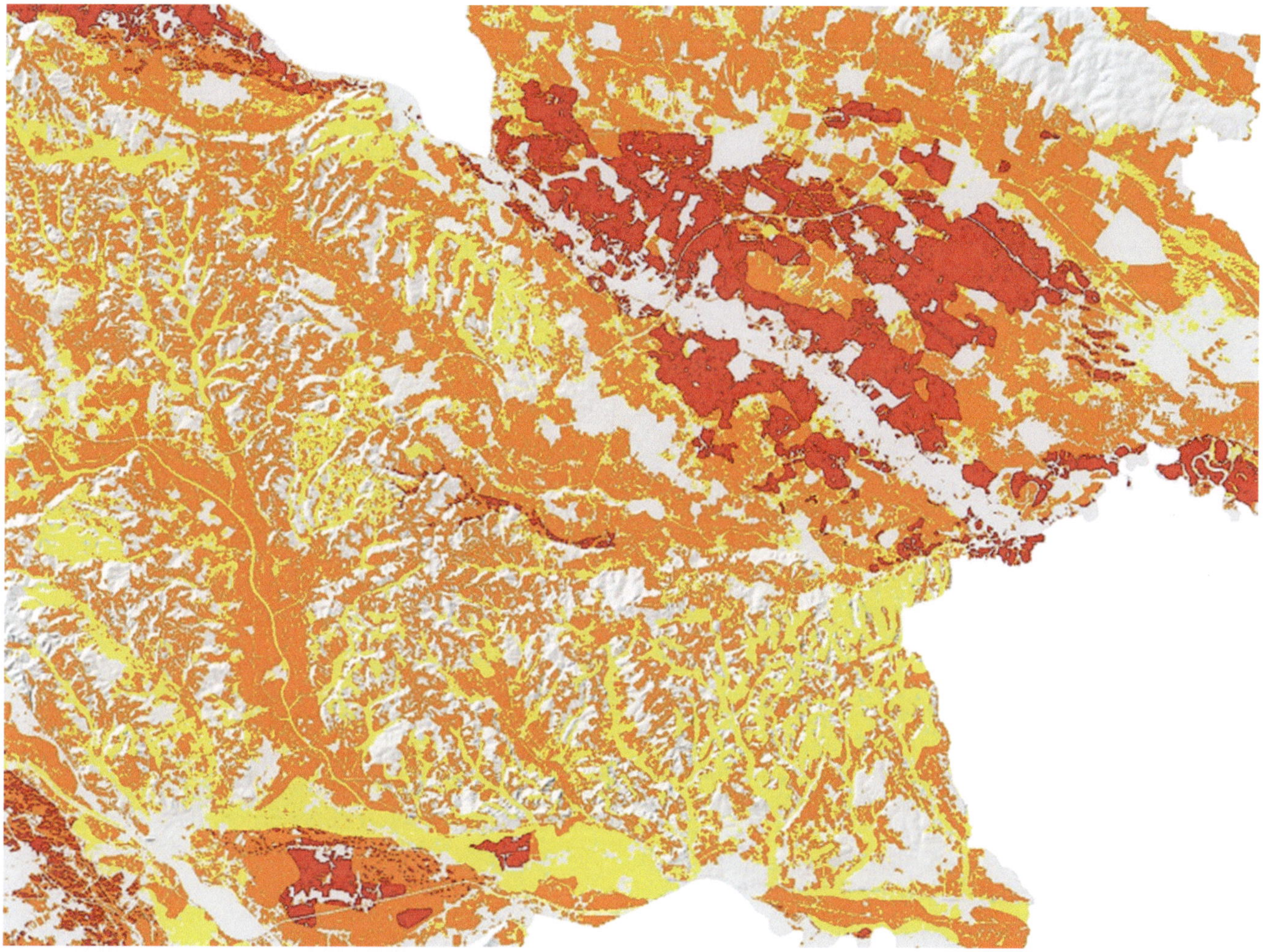

Fig. 8.13 Potential risk of SOM content decline in the agricultural land of north-eastern Slovenia

low mean annual precipitation, higher mean annual temperature, and especially due to the significant extent of coarse-textured Cambisols.

According to the model, the area presented by Fig. 8.13 is the area at greatest risk of SOM content decline. The risk is estimated to be very high in the areas of sandy and gravelly Dystric Cambisols (dark red), increased in the areas of Eutric Cambisols on marl (yellow), and moderate on Eutric and Dystric Gleysols in river valleys (light brown).

8.4.3 Carbon Stocks in Litter and Soils

For the national assessment of carbon stocks in litter and mineral soils for the UNFCCC and Kyoto Protocol (KP) reporting according to the GHG GGP (2003), a total of 909 soil profiles (mineral soil) descriptions with analytical data were used. The soil database includes data on physical and chemical soil properties, but no data on soil bulk density, so pedotransfer functions were used.

The litter was calculated only for soils under forests as a part of naturally preserved soil layers. A separate estimate of the carbon stock in soils under forests in O_l, O_f, and Oh sub-horizons was provided due to the fact that each organic sub-horizon was sampled within a 25 × 25 cm frame within the regularly spaced locations of the 8 × 8 km national grid in forests. In addition to this network, more intensive forest soil monitoring was carried out two times in 1995/96 and 2005/06 within an EU project (BioSoil Demonstration Project, Forest Focus), and within the 8 × 8 km grid of the national KP survey in 2007. The volume of the roots and coarse fragments (soil skeleton >2 mm) were subtracted from the volume of the soil sample. The carbon stock in the litter was calculated according to the GPG and national approach described in the National Inventory Report (NIR), in the LULUCF chapter, Mekinda et al. (2014) (Table 8.10).

Soil—Non-forest Land Use Areas

The carbon stock in the mineral part of the soil (SOM) was calculated according to GPG 2003 and as a part of NIR, UNFCCC, and KP national reports (2014). Because no

Table 8.10 Average carbon stock in the litter (from the 8 × 8 km grid survey)

Layer	Average carbon stock [t ha^{-1}]	*n*
O_l horizon	1.44 ± 0.15	143
O_{fh} horizon	8.85 ± 1.42	145
Litter (O_l + O_{fh})	10.41 ± 1.50	143

Source NIR (2014)

Table 8.11 Average carbon stock (SOC) at a depth of 40 cm in the mineral part of the soil

Land use	Average carbon stock [t ha^{-1}]	Number of sampling locations
Cropland annual	80.66	15
Cropland perennial	78.89	15
Grassland annual	89.64	10
Grassland perennial	69.24	9
Wetlands	113.68	11
Settlements	44.82	GL_a × 0.5

Source Mekinda et al. (2014)

measurements of bulk density were available in older soil datasets, the soil bulk density [g/cm3] was estimated from the following transfer function.

In 2012, the average carbon stock was estimated by additional field sampling on non-forest land, and previous estimates were corrected. The carbon stock in the mineral part of the soil (SOM) was calculated for a soil depth of 0–40 cm. For soil horizons with lower boundaries below 40 cm, a constant value of the carbon stock within the horizon was assumed (Table 8.11).

On the basis of data on green areas in the urban land of the municipality of Ljubljana, the expert judgment provided was that the soil carbon stock in settlements equals half the value of the soil carbon stock in the category *grassland annual*.

Carbon Stocks in Soils—Forests

For the determination of carbon stocks in soils under forests, we used more detailed information from the 8 × 8 km grid survey due to the fact that Slovenia includes forest management in KP commitments, which is the reason for a more structured approach to data collection, the so-called higher level tier (Figs. 8.14 and 8.15).

Data sources in forestry at the national level include the 4 × 4, 8 × 8, and 16 × 16 km grids in order to carry out forest monitoring and national reporting activities.

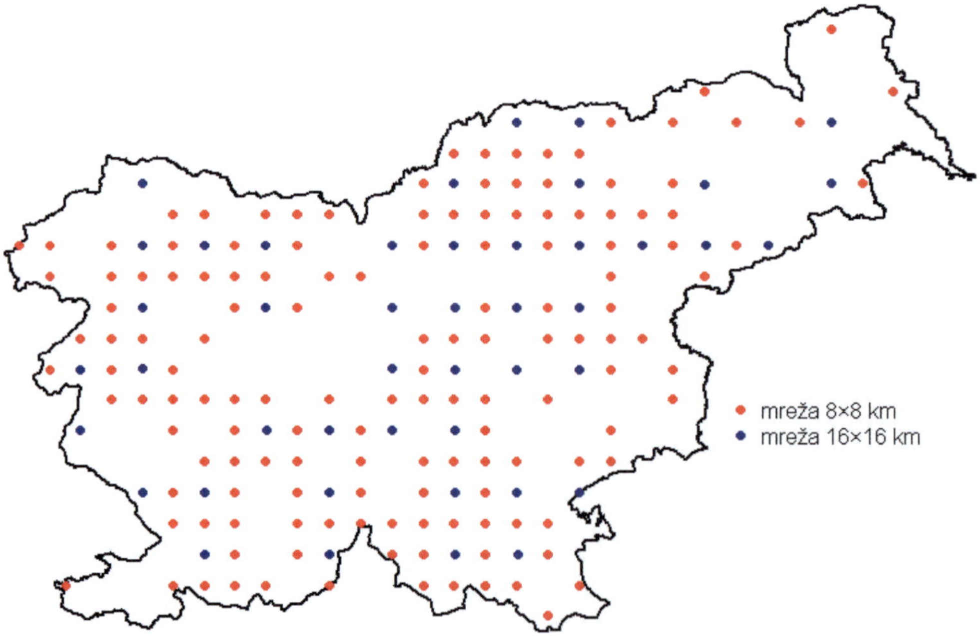

Fig. 8.14 8 × 8 km national grid, forest area

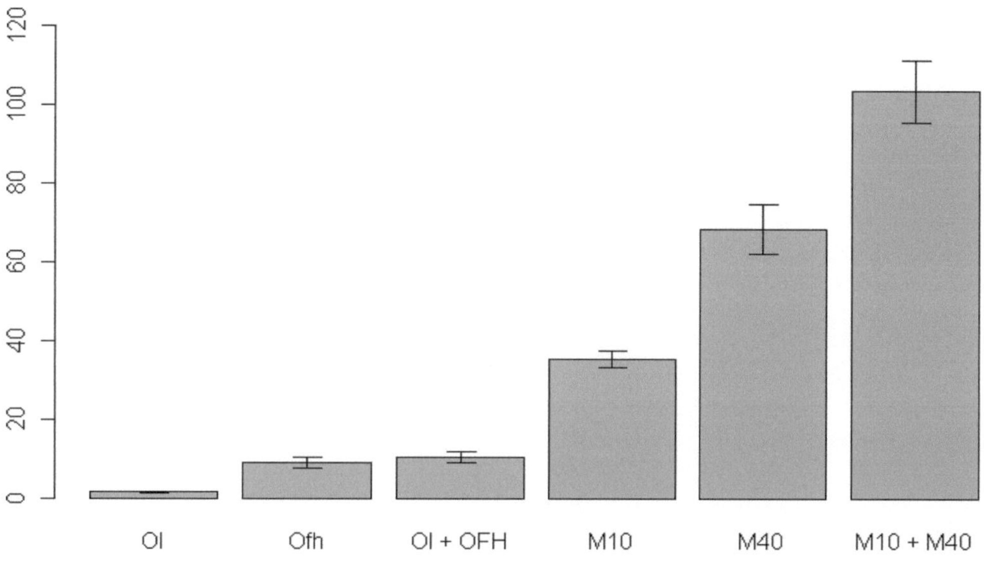

Fig. 8.15 Histogram of two carbon storage pools: litter and mineral soil (t/ha)

Table 8.12 Average carbon stock in mineral soils for forest land (from the 8 × 8 km grid survey)

Layer	Average carbon stock [t ha^{-1}]	The number of sampling locations
M_{10} horizon	35.25 ± 2.06	141
M_{40} horizon	68.32 ± 6.22	136
Mineral soils	103.31 ± 7.90	136

Source Žižek et al. (2013)

The Carbon Stock Changes in Forest and Agricultural Soils

In accordance with the GPG 2003, regarding carbon stock changes in soils, Slovenia has applied the approach in accordance with which "a pool is not a source", under which it is assumed that when forest remains forest, the carbon stock in the soil organic matter does not change, regardless of any change in forest management, types, and disturbance regimes; in other words, the carbon stock in mineral soil remains constant as long as the land remains forest. The results of our preliminary research for the period 1996–2006 (Kobal et al. 2013) show relatively stable carbon stocks in soils under forests. In the 20-year period until 2013, no major fluctuation in the forest management regime and no major disturbance event that could affect the soil carbon stock occurred. In addition, soil also depends on the climate, bedrock factors and on the source of carbon on the climate, bedrock factors and on the source of carbon coming from dead wood and litter. Since those two pools seem to be stable, the soil pool is therefore assumed to be stable (Table 8.12).

Changes in the Carbon Stock in Soils Under Forests—A Comparison of Two Forest Soil Monitoring Surveys (1996, 2006)

For 1996, three soil sub-samples were taken for the organic (litter) and from each plot (Large Scale Soil Survey of forest soil condition in Europe, 1997). For the mineral part of the soil, samples were taken with a soil auger (∅ = 7 cm) at a fixed depth (0–5, 5–10, and 10–20 cm) and for the organic layer within a 25 × 25 cm square.

Sub-samples were taken 5 m from the plot centre, 120° clockwise, and were combined for laboratory analysis (composite samples). The volume of roots and coarse fragments (soil skeleton >2 mm) was subtracted from the volume of the soil sample. In 2006, a soil survey was carried out according to the methodology of the BioSoil Demonstration Project. Soil sub-samples for the organic and mineral layers were taken as in 1996, with 5 replicate samples in each plot (the centre of the plot and the four directions of the compass). The volume of the roots and coarse fragments was subtracted from the volume of the soil sample. A paired t-test was performed for the evaluation of temporal changes (Fig. 8.16).

Carbon stock changes in the litter slightly increased from 1996 to 2006, when national forest soil surveys were carried out as a part of the ICP Forest and BioSoil Soil Demonstration Projects. The differences with regard to litter stock changes are not statistically significant ($p = 0.205$). Carbon stock changes in the mineral soils for the same period decreased, but the changes were not significant ($p = 0.052$).

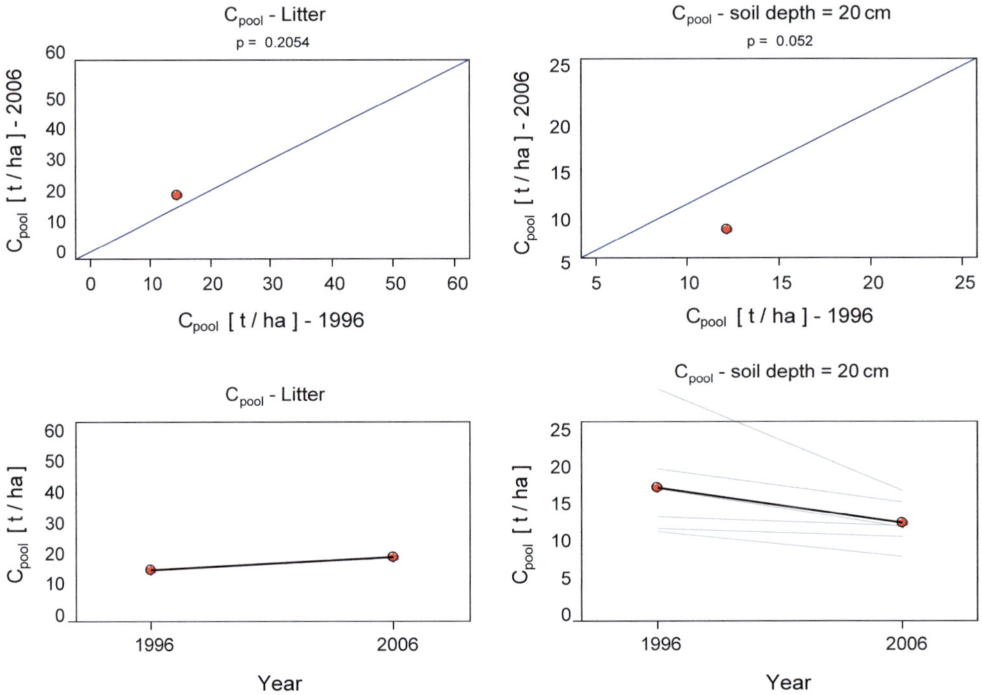

Fig. 8.16 Carbon stock in soils under forests and litter (1996–2006)

The existing results on a limited number of sampling plots show relatively stable carbon stocks in soils under forests for the observed period, with insignificant differences between the organic—litter layers and also regarding the mineral layer. Taking into account the currently valid values found in the literature for central European forests, we can conclude that there are no significant changes in soils under forests within a period of 5–10 years, provided the land use has not changed. Within the UNFCCC and KP processes for forest management (LULUCF), Slovenia has applied the approach in accordance with which "a pool is not a source" when reporting on net carbon stock changes in the litter and mineral soil carbon pools under forest management. Analysis based on data from demonstration projects and forest soil surveys showing no statistically significant difference in carbon pools for the period 1996–2006 was provided.

Case Study at a Plot in Brdo: Assessment of Organic Matter Changes in the Soil of the Plot Under Different Climate Change Scenarios Through the Application of The Yasso07 Model—A Case Study (Kobal et al. 2014)
To assess the changes in the total amount of organic carbon (C_{org}) as a part of organic matter storage in soils under forests over time, the Yasso07 model was used at the Brdo site. The site is within Brdo Park as a part of the Brdo Estate with Hunting Grounds. Mature stands of Scotch pine are dominant, and the prevailing soil type is *Dystric Cambisol.* Litter ranking was normal farming practice, especially in the vicinity of the surrounding settlements and ended in the previous century.

The Yasso07 model describes the decomposition of organic matter in the forest soil by dividing litter inputs into different components with varying decomposition rates. The temporal change in the soil C_{org} in various scenarios of future climate change was predicted. The difference between the measured amount and the model-predicted amount of C_{org} in the soil for the current climate on the Brdo plot is 6.4 tC/ha (88.6 tC/ha measured vs. 95.0 tC/ha predicted). Taking into consideration the climate change scenarios for Slovenia, i.e. an increase in air temperature and change in precipitation, the C_{org} stock is expected to decrease in the future according to Yasso07 projections in all climate change scenarios. The estimate of the 100-year decrease in C_{org} is the largest for the scenario in which an increase in temperature and precipitation is expected and the smallest when a small temperature increase and precipitation decrease are predicted. Assuming stable litter input due to sustainable forest management practises, a larger influence on C_{org} decrease was predicted for the temperature change compared to the precipitation change. However, many uncertainties are included in the model estimates, ranging from litter input estimates, climate change uncertainties, climate–litter production feedbacks, starting value estimates, etc. (Fig. 8.17).

Change in the stock of C_{org} *in the soil during the 1986–2012 period, calculated using the Yasso07 model, taking into account the estimated input of litterfall on the Brdo plot, and climate data from the main meteorological station located at Brnik.*

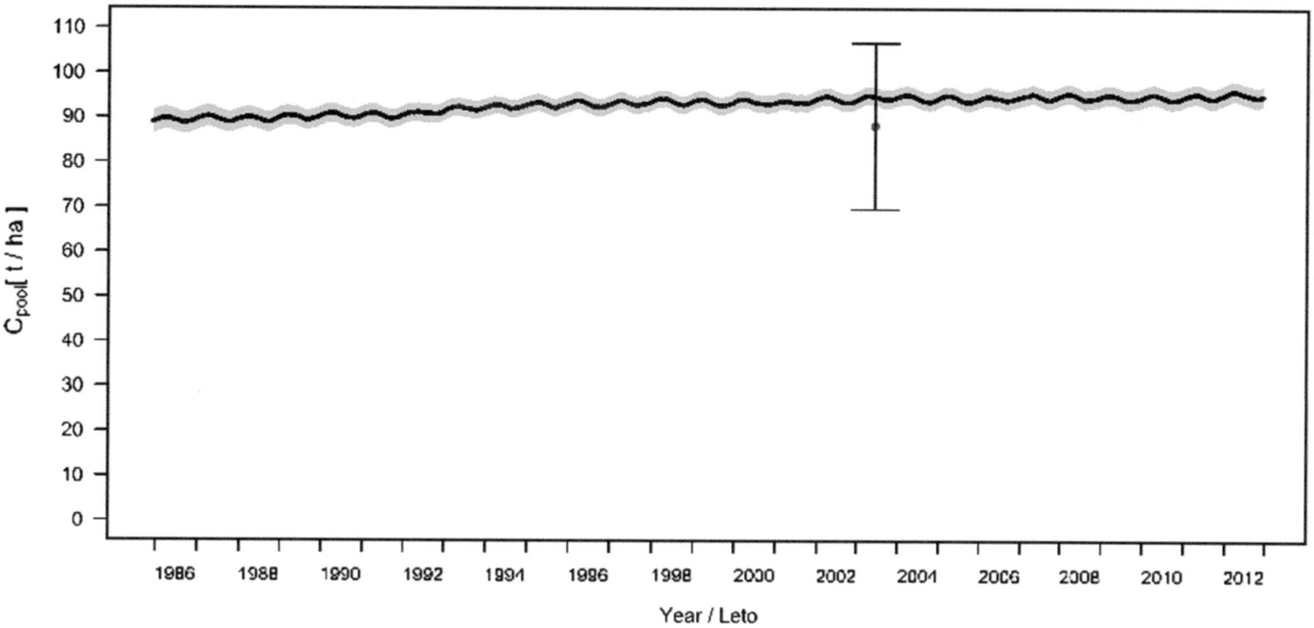

Fig. 8.17 Change in the stock of C_{org} in the soil during the 1986–2012 period (M. Kobal 2014)

8.5 Landslides

Different factors can cause imbalances in the weakly or unconsolidated material on the surface, which is a part of the pedosphere or weathered parent material. This results in a shift, movement, slip, creep, or slide of a larger bulk of soil, weathered material, and/or parent material, which is called a landslide. This slide occurs in the contact between the pedosphere and lithosphere.

Slovene terminology also distinguishes *usadi* (i.e. landslips) from *zemeljski plaz* (landslides) (Ribičič 2002, p. 2; Zorn and Komac 2008). The Dictionary of the Slovenian Literary Language (Bajec 1998) states that a landslip is "*a tearing or sliding of soil on a steep slope*" or "*a bulk of soil that tears and is separated from its origin, and slips down the slope*". A landslip is a shallow landslide that materializes during or immediately after a strong rainfall, with a sudden tearing of the soil and a downslope slide of an entire bulk. Landslips most commonly form on the steepest part of the slope; the slide starts rapidly and the initial movement ends in a few hours. The term landslip also refers to the depression in the slope that remains visible after the slide.

Landslides in Slovenia are by far the most common in the higher and lower hills of the central, pre-Alpine, sub-Pannonian, and sub-Mediterranean parts. They appear on a variety of softer clastic rocks and parent materials (sandstone, mudstone, flysch, shale, slates, schists) that disintegrate completely, deeply, rather rapidly, and result in a large amount of weathered and poorly consolidated material. The process of weathering forms an even and uniform contact between the formed soils and hard rock. Since most of these rocks are impermeable, following a rainfall, this contact is moistened and saturated, which enables the initial slide conditions. Landslides are more frequent where geological strata inclination coincides with slope inclination. Because of the extreme, even sheer inclinations, landslides generally do not form in the mountainous areas of Slovenia. More commonly, individual pieces of mechanically weathered parent material fall or larger rockfalls occur. Apart from the plains and basins, hillslope processes do not occur in the karstic areas (Mikoš et al. 2004).

The inclination and appropriate geological structure are always the main reasons for the occurrence of landslides. On the other hand, the triggers initiating landslides are very different in Slovenia (Zorn and Komac 2002). Among these trigger events are weather conditions and rainfall (extreme quantities, duration), quick changes in temperature (rapid snow melting), earthquakes, and inappropriate human interventions in the environment (e.g. the undermining of slopes, construction on slopes, forest clear-cutting, mining, etc.).

Slovenia's largest landslides have denominations: Koseč (in the mountains of Krn in the Julian Alps), Slano blato (in the Vipava River valley), Stovžje (around the settlement of Bovec in the Julian Alps), and Macesnik (in the upper part of the Savinja River valley). They are of very dissimilar dimensions. The areas range from 10 to 25 ha; they are around 10-m thick, 100–400 m wide, and 600–2200 m long. Estimated volumes range from 700,000 to 2,000,000 m^3 of labile material (Zorn and Komac 2007). Comparison to the list of the world's largest landslides ("List of landslides—Wikipedia, the free encyclopedia", 2014) once again confirms the fact of the smallness of Slovenia and nearly everything in it. Slovenia's largest landslides, although very problematic, causing relatively considerable material damage and even casualties, are of small absolute proportions. Slovenia's slopes are indeed very steep, but are short and there is in fact very little total soil cover or weathered material. However, the number of such can be extreme, as numerous small landslides can occur in a short period. Evidence of this fact can be found in the Haloze hills in July and August 1989. Between 3rd and 4th July, up to 200 mm of rain fell in only 24 h. This quantity is extremely above average for this part of the country. On the entire area of 106 km^2, there was a myriad of 5000 small landslides and landslips. Their average length was 12.4 m and average width 12.0 m. The average density was 47 landslides per km^2. Since all of them were triggered only on agricultural land, there were in fact 120 small landslides per km^2. This event is evidence of the violent acceleration of geomorphic processes. In just a few hours, an enormous total amount of material moved that would otherwise have taken centuries under normal, everyday circumstances. Even more interesting is the fact that after 20 years, there is absolutely no evidence in Haloze of this event. Small landslides, typically less frequent than the above-mentioned event, are just a usual part of moulding the surface in the destructive fluviodenudational system (Natek 1990, 1996) (Fig. 8.18).

Fig. 8.18 Landslip threatens. A Eutric Cambisol area close to the city of Velenje (*photo* B. Vrščaj)

Studies have shown that predominantly agricultural areas with larger patches of natural vegetation, mixed forests, pastures, and deciduous forests significantly affect landslide appearance. Natural grasslands, coniferous forests, agricultural areas with fragmented ownership as well as a fragmented land use structure, irrigated fields, and non-contiguous urban areas do not show a significant correlation withlandslides. Most of Slovenia's landslides occur in agricultural areas and on pastures. Meadows and pastures are prone to landslides due to their slope position. They can both quite effectively function on steeper slopes where all other agricultural activity is seriously hindered. However, grassland without the help of trees cannot effectively protect the surface against earth movements and landslides. Grass roots penetrate to only a shallow extent. Naturally, landslides appear most frequently on barren and at the same time inclined areas, but this combination is relatively uncommon in Slovenia (Komac and Ribičič 2005).

According to Zorn and Komac (2004), 6% or 1214.2 km^2 of Slovenia is under direct threat of landslides. They occur everywhere independently of altitude, on the entire variety of inclinations and on the already-mentioned geological types. The risk of landslides is maximal in the Alpine regions, where they spread over 1/5 (21.0%) of the area. Within this region, the largest threat is in the western Karavanke Mountains (34.3%), followed by the eastern Karavanke Mountains (24.6%), and the Kamniško-Savinjske Alps (21.9%). In the pre-Alpine region, 48.3% of the area in the Cerkljansko, Škofjeloško, and Polhograjsko hills is under such threat, and 39.4% of the area in the Posavsko hills. High-risk areas are also located in the Ložniško and Hudinjsko hills (29.3%) and in the Velenjsko and Konjiško hills (23.8%). In the sub-Pannonian region, low hills are at the highest risk. About 54.0% of the Mount Boč and Macelj areas are under threat of landslide, 50.3% of the areas in the Haloze hills, and 43.4% of the Voglajnsko and Zgornjesotelsko lower hills. Some areas around the Ljubljana basin and in the Idrijsko hills are also under threat. Landslides threaten the sub-Mediterranean part of Slovenia less (7.2% of the area). Even less threatened are karstic areas (Mikoš et al. 2004; Zorn and Komac 2008).

People perceive landslides as exceptional and catastrophic but as a non-repeatable event. Not in terms of dimensions, but in terms of recurrence, landslides are a regular and permanent phenomenon in Slovenia. Their influence is more or less local. The damage is material,

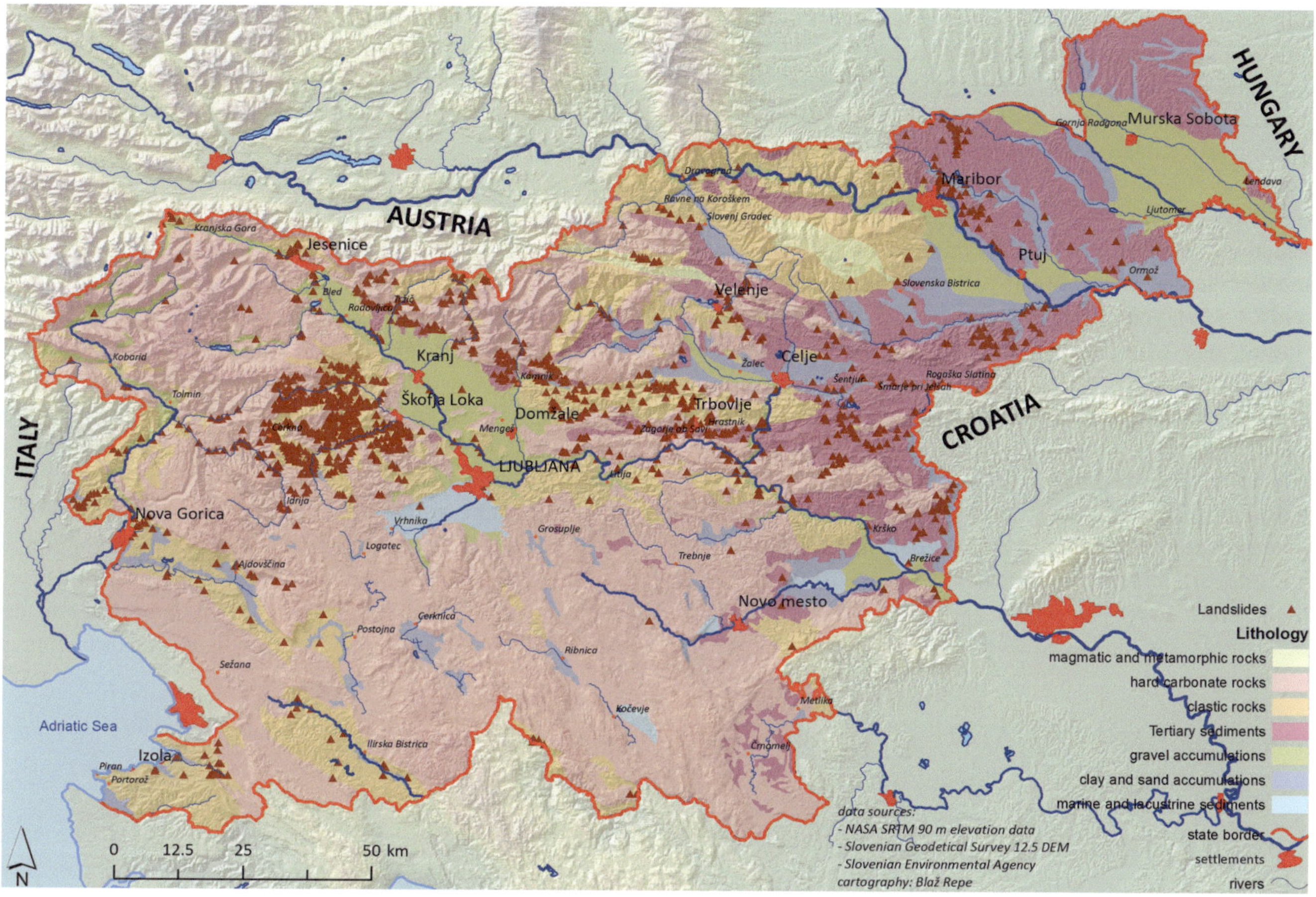

Fig. 8.19 Threat of landslide occurrence (ARSO, n.d.)

mostly to residential and husbandry buildings and agricultural land; they disturb traffic and the energy and water supply. Casualties are rare. Komac and Jemec (2007) estimate that 18% of the inhabitants of Slovenia live in areas are threatened by landslides. The most tragic case of a landslide in Slovenia occurred in the Log pod Mangartom settlement, between 15 November 2000 and 17 November 2000. A landslide occurred on 15 November 2000 below Stovžje, and started to creep down the valley of the Mangart Brook as a debris flow. The landmass was 1.5 km long, with at least 300,000 m^3 of material. The next three days were marked by heavy rainfall. The landslide saturated with water and liquefied has moved for a second time. This time it rushed down the valley and in a few minutes reached the settlement. Seven people were killed, 18 residential and 8 husbandry buildings were destroyed. The total sum of damaged infrastructure, roads, and agricultural land cost EUR 10.8 million to repair (Zorn and Komac 2002).

In Slovenia, a considerable amount of the financial burden is accounted for by the cost of reconstruction and clearing activities following landslides; much too little is invested in conservation, preventive activities, and the regulation of human intervention in nature (Fig. 8.19).

8.6 Erosion

Much the same as is described in the previous section on landslides is factually very similar with regard to water soil erosion. Areas threatened by soil water erosion processes are substantial. However, in the past the assessments varied greatly and they are considerably different compared to today's figures. Researchers quote different portions of Slovene territory affected by water soil erosion: 42–44% (Zemljič 1972; Kolbezen 1979; Horvat 2002), 66% (Mikoš et al. 2004), 80% (Zorn and Komac 2005a), and the exceptional 95% (Lazarević 1981). The reason for this large range of assessments can be attributed to different definitions of the terms erosion, water erosion, and water soil erosion; also different were the factors taken into the account in these assessments. The main reason lies in not distinguishing between the soil water erosion threat and the actual occurrence of erosion processes. If we compare the modelled amounts of eroded material, they vary to a lesser extent. These calculations for the county as a whole (collected by Zorn and Komac 2007) are 4.18 t/ha/year (Zemljič 1972), 3.13 t/ha/year (Lazarević 1981), and 3.70–4.52 t/ha/year (Zorn and Komac 2005b). There are very few direct measurements of soil water erosion, since every author cites the fact that there is insufficient research on erosion in Slovenia. Ravbar (1975) carried out one of the oldest measurements on a karstic surface near the city of Novo Mesto, where water eroded 22 t/ha/year. Horvat and Zemljič (1998) also made measurements. On a 29° inclined slope in a mixed-type forest the erosion was 6.3×10^{-3} t/ha/year; on a meadow, it was 39×10^{-3} t/ha/year; on a field with planted potatoes, it was 3.5 t/ha/year; and on a ploughed and barren field 22.4 t/ha/year. Erosion was measured in a vineyard near Maribor at a 14.9° inclination. On the permanently grassed surface, the erosion was 0.156 t/ha/year and on the periodically grassed surface 10.76 t/ha/year (Vršič et al. 2000). On a ploughed field near Kranj Komac and Zorn (Zorn and Komac 2005b) measured the level of erosion to be 36.0 t/ha/year. As a part of his dissertation thesis, Zorn (2007) systematically carried out the measurements for the Slovenian part of Istria (Istra). Between the years 2005 and 2007, in the hills above Koper in an olive grove with a 5.5° inclination, the erosion was 90–118 t/ha/year; on an overgrown meadow with a 9.4° inclination, the erosion was 0.9–1.68 t/ha/year; in a forest with a 7.8° inclination, the erosion was 3.91–4.21 t/ha/year and with a 21.4° inclination it amounted to 1.15–4.15 t/ha/year (Zorn and Petan 2007) (Fig. 8.20).

All the modelled and measured values indicate high erosion rates in Slovenia. On the other hand, estimations on the European level (MESALES, Modèle d'Evaluation Spatiale de l'ALéa Erosion des Sols-Regional Modelling of Soil Erosion Risk) show very unimportant erosion sensitivity (the lowest class) and a slight risk of erosion (the second lowest class) (Le Bissonnais et al. 2002). At the same time, PESERA (PESERA-Pan-European Soil Erosion Risk Assessment) designates most of Slovenia as the lowest classes of actual water soil erosion, with losses from 0 to 1 t/ha/year (Kirkby et al. 2004).

The question can then be asked: Which assumption is correct? Is soil erosion in Slovenia high and problematic or are the rates low and insignificant? To come closer to the answer and the actual state of erosion in Slovenia, factors of

Fig. 8.20 Fighting the soil erosion—a terraced area in Slovenske Gorice (*photo* T. Kralj)

water soil erosion have to be examined. One can gain a partial impression of the erosion factors summarized according to the Universal Soil Loss Equation (USLE) or the Revised USLE (RUSLE) from the factors of soil formation. However, this needs to be addressed in more detail. Using the USLE or RUSLE, the calculation of average annual losses (*A*) due to water soil erosion can be obtained by:

$$A = R \times K \times LS \times C \times P$$

(Renard 1997; Wischmeier and Smith 1978)

where *R* is the rainfall and run-off factor; *K* is the soil erodibilty factor; *LS* is the topographic factor of slope inclination and length; *C* is the cover and management factor; and *P* is the support and practice factor. Furthermore, the factors can be divided into two parts (Repe 2002). The first three factors (*R, K, LS*) are natural factors, upon which humans have practically no or very little direct influence. They are natural, physical facts that considerably change over a longer time span. The last two factors (*C, P*) are subject to the significant or complete influence of humans over the environment. The most important is the land use type and with it the associated removal of the natural vegetal cover. Slovenia‘s natural vegetation is forest. If humans had not intervened in the territory of Slovenia, more than 90% of the country would be forested (Repe 2007b). The soil erosion would equal the natural rates of erosion or geological erosion. The factors of water soil erosion can therefore be divided into potential and actual (Repe 2002, 2007c).

Slovenia‘s precipitation and rainfall are above average on the European and global scale. Over the last 30 years, 1580 mm of precipitation fell annually. Some parts, especially mountainous areas with extremely steep slopes, receive over 3600 mm of precipitation annually. With the rise in continental influences towards the east and north-east, the annual amounts are declining, but on the other hand more and more of this precipitation falls in the form of short and intensive (mostly) summer showers and thunderstorms (ARSO 2006). This implies the very strong and high erosive potential of precipitation. The potential is high not only in terms of amounts, but also as regards distribution. Also according to rocks and their erodibility, threats are present and serious. Highly erodible rocks (partially consolidated weathered material and soft rocks, mostly flysch, marl, and unconsolidated sediments) cover 47% of the country. An additional 11% is covered by medium erodible rocks (medium-hard rocks, with shallow soil or weathered material cover, mostly clastic rocks). From the geological and lithological point of view, Slovenia is highly threatened as far as water soil erosion and soil degradation are concerned (Čarman et al. 2007). Factors that directly influence soil erodibility (Wischmeier et al. 1971) are soil texture, organic matter content, soil structure, and permeability. In terms of the most common soil texture classes, most of our soils can be regarded as medium (heavy) soils. The moderate weathering conditions result in silty and fine sandy material that lacks coherence and plasticity. They are all very susceptible to water soil erosion. Slovenia's soils are also not very abundant in organic matter. Most of them, 56.2%, can be classified as soils with low-to-very low (less than 2%) organic matter content. The decomposition rate of organic matter is relatively high, which is appropriate to the relatively warm and humid climate. Although forest and grassland land use is prevalent, the organic carbon content is quite low. The aforementioned texture and organic matter conditions do not create favourable solid and erosion-resistant structure types. Most common are blocky and sub-angular blocky structural aggregates (ARSO 2009). In particular in agricultural land, they are less resistant due to the organic matter deficiency. Of the available factors combatting water erosion, soil permeability is the most promising. Most of Slovenia's soils are very well permeable. Proof of this statement lies in the extremely high proportion (84.9%) of automorphic soil, i.e. soil with normal water drainage, where water is not retained in the profile for very long. Nevertheless, the permeability is problematic with some Cambisols and Luvisols, although they are automorphic, due to the naturally compact and dense clayey horizons. The most problematic are of course Gleysols and Stagnosols (Stagnic Cambisols or Luvisols). In terms of their properties, Gleysols are extremely prone to erosion, but they appear only levelled, even on the bottom of concave ground. The worst-case scenario is with Stagnosols or soils with stagnic properties (3.7%). In a wet state, they are smeary. When dry, they become brittle and impermeable. Since they can also form on slopes, they are by far the most at risk to water soil erosion. The soil types most sensitive to erosion can be found generally on level ground (15.8%) or at very low inclinations. Nevertheless, there is a reoccurring trend of droughts, especially in north-eastern Slovenia. Dried-out soils harden on the surface or are pulverized, which strongly increases the chances of water soil erosion. Apart from the Stagnosols, the problems with erosion are most commonly noticed with regard to Regosols, some types of Leptosols, and some more developed Cambisols (especially eutric and dystric) (Fig. 8.21).

The last of the potential factors is topography. Inclination and slope length drastically work in favour of water soil erosion. A consequence of the hilly (low and high) and mountainous topography are high to extreme inclinations (see Table 3.3). Slope lengths are relatively short since the surface of Slovenia is geologically very young. According to the digital elevation model (25 × 25 m resolution), the average range of lengths is 440 m. Average slope length is 90.5 m. In addition, LS values are small, and the maximum is 13.8, with an average of 4.47.

Fig. 8.21 Rill erosion in the hinterland of the city of Maribor, north-western Slovenia (*photo* B. Vrščaj)

From all these, we can legitimately conclude that the threat of water soil erosion in Slovenia is extremely high, very widespread, and agrees with the cited figures in the literature. There is a very high potential threat of water soil erosion in Slovenia. However, there are also two additional factors from the equation, the cover/management factor and the support/practice factor, which have to be taken into account. Thereby, a partial conclusion can be made:

- Extreme and sheer inclinations in the mountainous regions are void of vegetation, completely rocky, and soil-less.
- High and low hills are mainly forested or at least covered with some types of grassland.
- Most of the agricultural land, especially fields, is concentrated on the plains.

Slovenia had the lowest level of forest cover in the middle of the nineteenth century, when it declined to (a still very high) 36% (Repe 2007b). Ever since then, the forested areas have been constantly increasing. Hand in hand with the industrial revolution, people abandoned lands and agricultural production on less suitable areas (steeper slopes, higher altitudes, karstic surfaces, more distant to homes or market), any areas demanding high, mostly labour inputs with low harvest or income. These processes have not yet concluded even today. The most frequent and important land use changes are converting agricultural areas into grasslands, and grasslands into forest. Apart from that, there is also very efficient grassing of different groves and orchards (vine, fruit, olive), which are traditionally found on slopes (Repe 2002, 2009). There is also another negative aspect, the extremely fragmented division of parcels into a true myriad of tiny fragments of ownership. To illustrate this, in the year 2010, the average size of an agricultural production unit comprised 6.4 ha of land on which 5.6 livestock units were reared (Kutin Slatnar et al. 2012).

On the basis of this paragraph, we can again conclude that erosion on agricultural land must not be neglected, and it can

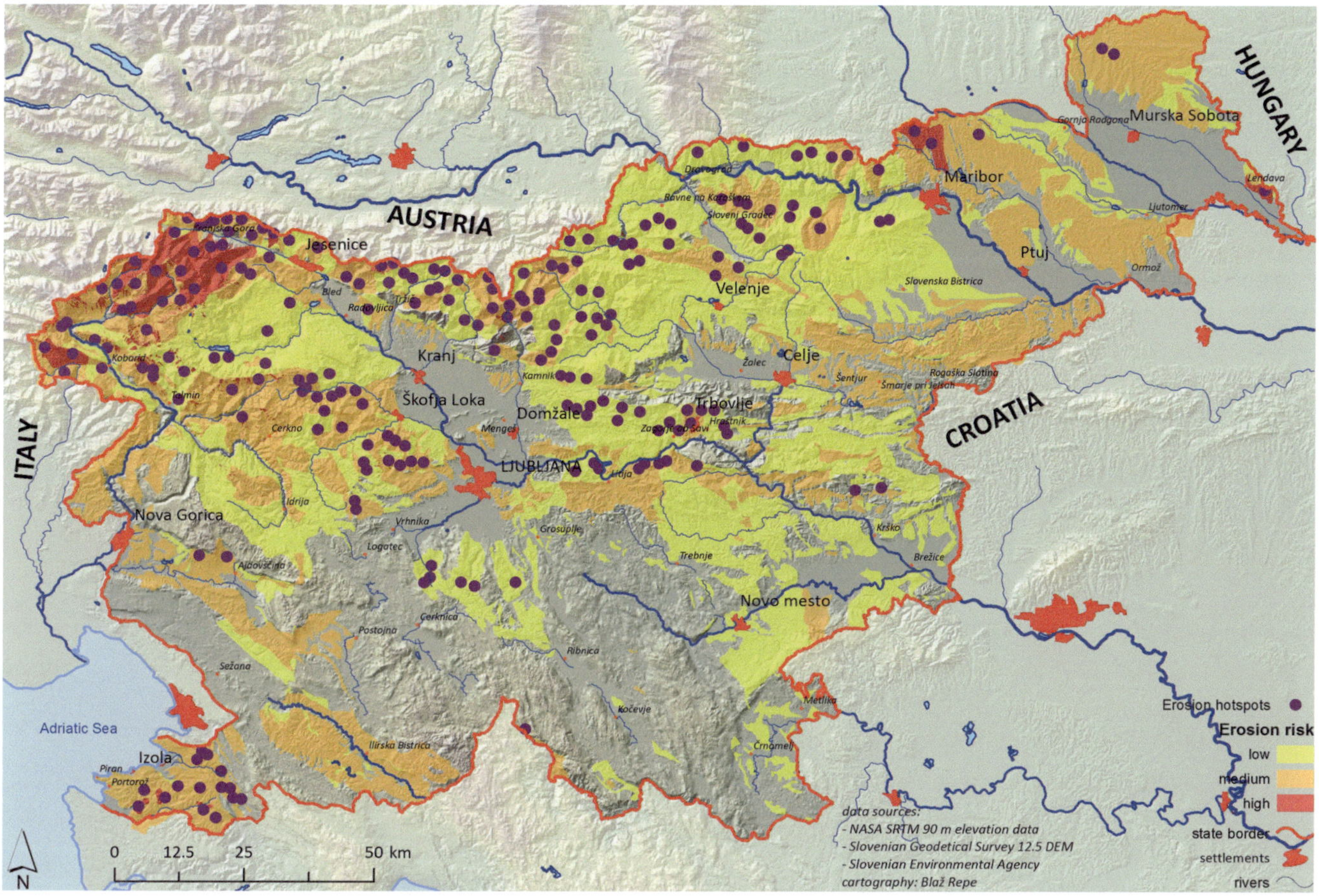

Fig. 8.22 Soil water erosion occurrence (ARSO, n.d.)

be the most intensive directly on the fields. Locally, it can cause substantial damage, but is incomparable to the damage caused by a torrential flow (Zorn and Komac 2005b). Of course, there is no better protection against water erosion than dense forest cover. The root system of the vegetation retains soils at their origin even on very steep slopes, and this vegetal cover is extremely abundant in Slovenia.

There is (again) a certain erosion duality in Slovenia. On the one hand, there is this enormous threat of water soil erosion. In other words, there is extremely high potential water soil erosion, due to natural factors. To prove this, there are many forests indicated as protective. Their area continues to grow. In the year 2009, there were 110,112 ha of protective forests, which is 10% more than in the year 2008 and 77% more than in 2000 (Gale et al. 2011). On the other hand, taking Slovenia as a whole, the appearance of erosion processes or anthropogenic accelerated or actual water soil erosion is not a significant type of soil degradation. This brings two additional things into perspective. Firstly, this does not mean that erosion does not occur. There are many erosion hot spots or even badlands, but they are all of small extent and local "importance". Secondly, the potential clearance of the vegetal cover or forest clear-cutting would trigger irreversible erosion processes of immense proportions. All of which confirms the duality of the potential (threat) and actual (soil loss) erosion hypothesis (Repe 2002) (Fig. 8.22).

8.6.1 Wind Erosion

Wind (soil) erosion in Slovenia is neither very frequent nor widespread, and its consequences do not seriously affect Slovenian soil cover. The reasons are numerous: the surface is rough and very diverse; the vegetal (forest) cover is dense and above average; the agricultural areas are not prevalent, just to name a few. In general, winds are of lesser strength and not very frequent. Slovenia lies in the relative calmness of the Alps. Nevertheless, that does not mean that locally erosion cannot periodically occur where extremely strong winds do form at exposed locations in the mountains and on agricultural land in the sub-Mediterranean regions, namely the very strong Bora wind (Repe 2002). The Vipava River valley stands out in this respect.

There is very little data about wind soil erosion. It is measured only occasionally and not systematically. Malovrh (1955) discussed the conditions of the Koper hinterland in

Table 8.13 Wind soil erosion in an olive grove, over a period between 18 November 2005 and 24 November 2005

	Material removed from the erosion test field			Material removed from the entire olive grove		
Unit	Erosion test field 1	Erosion test field 2	Average of both test fields	According to erosion test field 1 (kg)	According to erosion test field 2 (kg)	According to the average of both erosion test fields (kg)
g/m^2	75.31	53.23	64.28	222.54	158.75	191.64
mm	0.07	0.05	0.06			

Source Zorn (2008, p. 273)

Fig. 8.23 Road construction permanently alters natural soils under forests (*photo* R. Robek)

1954. The velocity of the Bora wind at 85 km/h removed 10 cm of soils, in some places down to the exposed vine roots. Zorn (2008) measured wind soil erosion in the olive groves of Slovenian Istria (Slovenian: *Istra*). In one week, from 1 m^2 of barren ground 64.28 g of material was removed, which equals 0.06-mm lowering of the ground (Table 8.13).

Special mention has to be made of the wind erosion conditions in the Vipava River valley in February 2012. Due to the very arid conditions due to the lack of precipitation in January 2012, the upper soil on the tilled and barren fields dried out and resulted in poor structure (ARSO 2012). An extremely strong Bora wind caused serious damage to the agricultural areas, especially fields and vineyards (*Extreme gusts of wind at Ajdovščina 04/02/2012 = Ekstremni sunki burje v Ajdovščini 04/02/2012*, 2012). Mostly the humus part of the topsoil with a cloddy structure was removed, transported, and deposited in the irrigation ditches and river beds. In January and February 2012, a Bora wind of

Fig. 8.24 Off-road timber transport by a forwarder on soils with good bearing capacity after a longer dry period, Trnovo plateau (*photo* P. Simončič)

hurricane proportions removed around 20,000 m^3 of fertile soil (Črv and Turk 2012). About 1200 ha of fields were seriously damaged, and 3–10 cm of topsoil were blown away, which amounts to approximately 530 tons/ha, collectively 600,000 tons. Wind soil erosion damage was estimated to be EUR 600,000 (up to EUR 3000/ha). This is an amount approximately equal to the cost of 56 km of wind brakes, which are one the most effective conservation practices against wind soil erosion (Škvarč 2012; Zorn 2012).

8.7 Impacts of Timber Transport on Soils Under Forests

Infrastructure construction and off-road operations of logging machinery are the two major sources of soil damage due to forestry activities in Slovenia. Although soil alterations due to road construction activities are permanent, nowadays new disturbances are well controlled and rather limited in forests (Fig. 8.23). Uncontrolled operation of heavy machinery on natural preserved soils under forests can severely affect their physical properties and the biological activities therein and alter the aesthetic value of the entire site. In poorly managed commercial forests, over 30% of the soils may be disturbed due to wood transport (Wästerlund and Hassan 1995).

In Slovenian forests, several surveys of stand and forest soil damage due to the use of heavy machinery in forest operations have been conducted (Košir and Robek 2000; Mali and Košir 2006; Kraigher Report; Simončič 2010; Mihelič 2014). The purpose of those studies was to establish and test different methods of assessment of forest soil damage, to analyse forest soil damage regarding the technologies used (tractors, harvesters, and forwarders), and to estimate the adequacy of the choice of different machines for selected site and stand conditions, soil types, weather preconditions, and harvesting intensity.

As a result of the research activities in this field and increased public sensitivity to environmental issues, forestry decision-makers have recently raised awareness of the suitability of forest operation activities: which soil types allow a certain technical approach, what kind of technology should be used, and whether such activities are economically justified (Fig. 8.24).

Ruts are the most evident impact of off-road vehicle locomotion. It has been determined that soil damage is smaller using cut-to-length technology than when skidder and chainsaw approaches are applied (Mihelič 2014). With

Fig. 8.25 Skidding timber in winter conditions cannot prevent long-lasting soil damage (*photo* R. Robek)

regard to soil damage when using cut-to-length technology in dry weather, soil compaction and rutting are not problematic and do not cause irreversible changes to soils under forests (Fig. 8.25). Significantly higher soil damage has been observed when forwarders and harvesters operate in inappropriate weather periods, especially after rain or snowmelt periods (Fig. 8.25) in deeper soils (wet Cambisols on acid bedrock). In a case study on the Osankarica research plot in the Pohorje area, the impacts of forest machinery on soils under forests and fine root growth with the use of minirhizotrons and ingrowth soil cores were studied (Železnik 2011). The experiment was carried out in undisturbed soils (Cambisols), nearby a vehicle rut, within the vehicle rut of a harvester, and within the vehicle rut of a harvester and forwarder. The results show the negative effect of heavy machinery in the wheel rut, even a few years after its use ended (Kraigher 2011).

The research results contribute to the preparation of background directives for the use of heavy forest machinery on differently vulnerable forest sites. Together with forestry practitioners, a best management practice for harvesting operations has recently been adopted in Slovenia (Krč et al. 2014).

About the Book

The book opens with an introduction of the basic data regarding Slovenia, its position within Europe, and division into the distinctive units that encompass all the country's natural features. It continues with the factors impacting soil formation. That chapter reveals the general and soil-specific factors: the parent material, topography, climate, waters, vegetation, time, and the land use. The fourth chapter on the classification of Slovenian soils briefly describes the diagnostic horizons, qualifiers, soil types, and relationship of soil types to the WRB reference soil groups. The fifth chapter introduces the most important soil types of Slovenia. It contains soil characterizations, soil profile descriptions, and representative soil profile data sets. Thereafter, some specific or soil types unexpected in Slovenia are presented in the next chapter. The seventh chapter spatially represents the extent of Slovenia's reference soil groups according to the WRB. Illustrated with maps and graphs, it provides an insight into the geographical distribution of the WRB soil reference groups in Slovenia. Special consideration is devoted to the human actions that threaten soils. Ranked according to importance, a somewhat unfortunate position is assigned to soil sealing, contamination, the decline in organic matter, landslides, erosion, and compaction. Chapters on soil research in the past, soil surveying and mapping, our research institutes and enterprises, soils in our school system, and the most important recent projects, have been added. The last chapter provides a historical review of activities related to the collection, digitalization, and formation of soil information. The information on the availability of the soil information on Slovenia should prove to be a valuable conclusion to the book.

This book on Slovenian soils is a review of what kind, where, and how much soil is available in Slovenia; what the major threats to soils are, what the main research activities are, as well as what the status of the available soil information is.

The purpose of the book is to provide an overview of the state of Slovenian soils and a comparison in the international context. Additionally, it should serve as an important summary that can be used to direct the soil survey, soil research, data collection, and data processing towards designing sustainable soil management and better soil protection in Slovenia in the future.

It should be borne in mind that is an important strategic and ethical question how much and what kind of soil will be left to future generations.

B. Vrščaj et al., *The Soils of Slovenia*, World Soils Book Series,
DOI 10.1007/978-94-017-8585-3

The Future

Good, fertile soils are scarce in Slovenia. In this mountainous and hilly country with shallow and sandy soils in the lowlands, we should be particularly sensitive to the need for adequate protection and sustainable soil management. Public and institutional awareness of how important a natural resource soil is and especially why it needs better protection and more sustainable management is growing rapidly in Slovenia.

The Ministry of the Environment and Spatial Planning (MESP) actively celebrated the International Year of Soils in 2015. It has financed several soil awareness-raising events in 2015 and 2016 and has scheduled a permanent programme of activities related to soil protection. The MESP has introduced a national "Soil in the Environment" logotype (Fig. A.1) and requested that soil experts summarize the environmental aspects of soils in a series of soil posters (Fig. A.2).

The Ministry of Agriculture, Forestry and Food (MAFF) has recently been devoting more attention to sustainable agricultural production. It has launched several research projects related to soil quality monitoring and carbon sequestration/carbon balance in agriculture and forestry. The 2014–2020 Rural Development Programme now includes several measures that directly or indirectly integrate soil protection and sustainable agricultural production.

The sensitivity of the general Slovene public and the interest of the media in soils have slowly but steadily been growing in recent years. Soil is gradually gaining attention and recognition—which within ten years, according to some of the authors, will be equal to the attention devoted to air and water quality. A good example is the recent case of planning a new motorway through the fertile Savinja Valley. Not only the local people but also national-level environmentalists, NGOs, agriculturalists, and lay people have expressed strong opposition to the sealing of good agricultural soil. The topic was presented in prime headlines in the national media.

The numerically modest soil science community in Slovenia is witnessing these important changes. Soil experts

Fig. A.1 The "Soil in the Environment" logo (MESP 2015)

B. Vrščaj et al., *The Soils of Slovenia*, World Soils Book Series,
DOI 10.1007/978-94-017-8585-3

Fig. A.2 The "Earth without soil?", "Is soil a living space?", and "How do we treat soils?" are three posters from the MESP 2015 soil poster series (MESP 2015)

are encouraged by the promising development of soil awareness on the part of the state: from the "Soil? Not a topic!" of previous decades to a situation where many activities and ongoing soil projects have been carried out recently.

The increasing attention of governmental institutions, the public, and media sends an important message: soil, a common good, is back on the agenda.

References

ARSO (n.d.) Spatial data distribution for INSPIRE directive. Spatial data distribution for INSPIRE directive

Antić M, Jović N, Avdalović V (1980) Pedology. Naučna knjiga, Beograd

ARSO (2005) Climatology of Slovenia, average precipitation, period: 1961–1990. Slovenian Environment Agency, Ljubljana, Slovenia

ARSO (2006) Climatic conditions of Slovenia (period: 1970–2000). Slovenian Environment Agency, Ministry of the Environment and Spatial Planning, Ljubljana, Slovenia

ARSO (2009) Soils. Environ Rep 2009, Environmental report 2009, 17

ARSO (2012) Bora wind and frost from January 28th till February 14th with high waves and low sea temperatures in the first half of February 2012 [WWW Document]. Natl Meteorol Serv Slov. URL http://meteo.arso.gov.si/uploads/probase/www/climate/text/sl/weather_events/burja-mraz_feb12.pdf. Accessed 1.4.15

Bajec A (ed) (1998) Dictionary of standard Slovenian, DZS dictionaries. Research Center of the Slovenian Academy of Sciences and Arts, Fran Ramovš Institute of the Slovenian Language, Ljubljana

Bat M, Beltram G, Cegnar T, Dobnikar-Tehovnik M, Grbović J, Krajnc M, Mihorko P, Rejec Brancelj I, Remec-Rekar S, Uhan J, Bat M, Frantar P, Komavec T, Marolt D, Ulaga F, Robič M, Brancelj A, Makovec T (2003) Richness of Slovenia's waters. Ministry of the Environment, Spatial Planning and Energy: Slovenian Environment Agency, Ljubljana

Beltram G (1996) The conservation and management of wetlands in Slovenia. In the context of European policy related to wetlands. Ph. D. thesis, Universiteit Brussel

Beltram G (2003) Wetlands. In: Uhan J, Bat M (eds) Richness of Slovenia's waters. Ministry of the Environment, Spatial Planning and Energy: Slovenian Environment Agency, Ljubljana

Bergant J, Vrščaj B (2014) Karta potencialnih naravnih rastišč poletne gomoljike. Geogr Inf Sist V Slov 2012–2014(2014):95–103

Bertalanič R (2007) Catastrophic winds in the world in 2006. Ujma 106–115

Blumlein P, Schug B, Wieshofer I, Huber S, Parolin M, Biasioli M, Medved P, Vernik T, Vrščaj B, Sabielec G, Kozak J, Fulajtar E, Jaensch S (2012) Tla v mestu. Startegija upravljanja z urbanimi tlemi [Soils in the city. Urban soil management strategy]. Mestna občina Celje/Kmetijski inštitut Slovenije [Municipality of celje, Agricultural Institute of Slovenia], Celje/Ljubljana, Slovenija [Slovenia]

Čarman M, Mikoš M, Pintar M (2007) Different aspects of soil erosion in Slovenia. In: Knapič M (ed.) Thematic strategy for soil protection: conference at the World soil day, 5 Dec 2007. Soil Science Society of Slovenia, Ljubljana, pp 39–51

Čergan Z, Jejčič V, Knapič M, Modic Š, Moljk B, Poje T, Simončič A, Sušin J, Urek G, Vrščaj B, Žerjav M (2008) Koruza [Maize]. Kmecki glas, Ljubljana, Slovenia [Slovenia]

Ćirić M (1984) Pedologija [Pedology]. Svijetlost, Sarajevo

Ćirić M (1984b) Uputstvo za izradu namjenskih pedoloških karata krupnog mjerila I.dio [Instructions for elaboration of large scale soil maps, Part I.]. Sarajevo

CKFF (2010a) Map of habitat type - Banjščice. Slovenia's habitat type maps

CKFF (2010b) Map of habitat type - Murska Sobota, Lendava. Slovenia's habitat type maps

CKFF (2010c) Map of habitat type - Haloze West. Slovenia's habitat type maps

CKFF (2010d) Map of habitat type - Ljubljansko barje 2010. Slovenia's habitat type maps

Črv R, Turk M (2012) Meliorations and the importance of anti-erosion barriers in the Vipava River Valley [WWW Document]. URL http://www.zrc-sazu.si/sites/default/files/crv_turk_melioracije_vipavska_dolina.pdf. Accessed 1.4.15

Čuden I (2009) Klasifikacija gozdnih tal na apnencu v skladu s slovensko in WRB klasifikacijo [Classification of forest soils on limestone according to the Slovenian and WRB classification]. Univerza v Ljubljani, Biotehniška fakulteta, Oddelek za gozdarstvo in obnovljive gozdne vire [University of Ljubljana, Bitechnical Faculty, Department of Foresty and Renewable Forest Resources], Ljubljana, Slovenia [Slovenia]

Čustović H (2009) Discussion on soil data and soil information systems

Dolinar M (2006) Spatial interpolation of sunshine duration. In: Perko D, Nared J, Čeh M, Hladnik D, Krevs M, Podobnikar T, Šumrada R (eds) Geographical information systems in Slovenia 2005–2006. Založba ZRC, pp 61–71

Dolinar M, Frantar P, Kurnik B (2008) Water balance in Slovenia 1971–2000. Presented at the water management strategy from the aspect of climatic changes (Mišič's Hydrology Days 2008), p 7

EUROSTAT (2010) Results on EU land cover and use published for the first time (Land Use/Cover Area frame Survey No. 145/210), Land Use/Cover Area frame Survey. Eurostat Press Office

EWN-Si (2001) EUROWATERNET - SLOVENIJA. Digital information (waters)

Extreme gusts of wind at Ajdovščina 04/02/2012 = Ekstremni sunki burje v Ajdovščini 04/02/2012 (2012)

FAO (2006a) World reference base for soil resources, 6th edn. Food and Agriculture Organization of the United Nations, Rome

FAO (2006b) World reference base for soil resources 2006, 2nd edn. International soil classification system for naming soils and creating legends for soil maps. FAO, Rome

FAO (2007) Land evaluation. Towards a revised framework, Land and water discussion paper. FAO, Rome

FAO, IUSS (2014) World reference base for soil resources 2014. World soil resources report. FAO, Rome

B. Vrščaj et al., *The Soils of Slovenia*, World Soils Book Series,
DOI 10.1007/978-94-017-8585-3

FAO, UNESCO (1997) Soil map of the world. Revised legend, 2nd edn. International soil classification system for naming soils and creating legends for soil maps. ISRIC, Waageningen
Filipovski Ǵorǵi (2000) Soils of the Republic of Macedonia. Svijetlost, Skopje
Frantar P (2007) Geographical overview of water balance of Slovenia 1971–2000 by main river basins. Acta Geogr Slov 47:25–45
Frantar P, Hrvatin M (2005) Discharge regimes in Slovenia from 1971 to 2000. Geogr Bull 77:115–127
Furlan D (1960) Klimatska razmejitev Slovenije [Climatic delimitation of Slovenia]. Geogr Vestn 45–57
Gabrovec M (1996) Solar radiation and the diverse relief of Slovenia. Geogr Zb 47–68
Gabrovec M, Hrvatin (1998) Surface. In: Fridl J (ed) Geographical atlas of Slovenia: the county in space and time. DZS, Ljubljana, pp 74–77
Gale Š (2011) Week of Slovene forests 2011 [WWW Document]. Week Slov For 2011. URL http://www.stat.si/novica_prikazi.aspx?id=3928. Accessed 1.4.15
Gale Š, Lešić M, Kutin Slatnar B (2011) Tree, forest, wood. Statistical Office of the Republic of Slovenia, Ljubljana
Gams I (1972a) Prispevek h klimatogeografski delitvi Slovenije [Contribution to climate-ogeographic division of Slovenia]. Geogr Obz 1–9
Gams I (1972b) A contribution to the climatic division of Slovenia. Geogr Horiz 1–9
Gams I (1974) Karst: a historical, natural and geographical outline. Slovenska matica
Gams I (1996) Geografske značilnosti Slovenije [Geographic features of Slovenia]. Mladinska knjiga, Ljubljana, Slovenia
Gams I (1998) Relief. In: Gams I, Vrišer I (eds) Geography of Slovenia. Slovenska matica, Ljubljana, Slovenia, pp 24–54
Gams I (2003) Karst in Slovenia in space and time. Založba ZRC, ZRC SAZU, Ljubljana
Gardi C, Panagos P, Van Liedekerke M, Bosco C, De Brogniez D (2014) Land take and food security: assessment of land take on the agricultural production in Europe. J Environ Plan Manage 1–15. doi:10.1080/09640568.2014.899490
Gartner D (1999) Average yearly amount of precipitation in the period 1961–1990
Geodetska uprava RS [Slovenian Geodetical Survey] (1984) Pravilnik za ocenjevanje tal pri ugotavljanju proizvodne sposobnosti vzorčnih parcel. Obvezno navodilo za izvajanje pravilnika za ocenjevanje tal pri ugotavljanju proizvodne sposobnosti vzorčnih parcel [Procedure for soil evaluation for identification of the production capability of agricultural lands]
Gerrard J, Gerrard JG, Gerrard AJ (1992) Soil geomorphology: an integration of pedology and geomorphology, 1993 edn. Springer, London; New York
Global Soil Forum (2013) Soil sealing
Gračanin M (1951) Pedologija. III dio - Sistematika Tala [Pedology, part III—soil systematics]. Sveučilište u Zagrebu, Zagreb, Hrvaška [Croatia]
GURS (2011) Barvni digitalni orto foto 0.5 m [Color digital ortho—photo 0.5 m] (DOF5)
Habič P (1992) Kras and karst in Slovenia. In: Pak M, Orožen Adamič M (eds) Slovenia, geographic aspects of a new independent European Nation. The Association of the Geographical Societies of Slovenie, pp 31–39
Hočevar A, Petkovšek Z (1988) Meteorology: the fundaments and some applications. Partizanska knjiga, Ljubljana, Slovenia
Horvat A (2002) Erosion. In: Natural disasters and protection against them, Ljubljana, Administration of the Republic of Slovenia for civil protection and disaster relief. Ljubljana, Slovenia, pp 411–424
Horvat A, Zemljič M (1998) Anti-erosion role of mountain forest. In: XIX. For. stud. days, conference proceedings "Mountain Forest". University of Ljubljana, Biotechnical Faculty, Department of Forestry and Renewable Resources, pp 267–274
Husnjak S (2014) Sistematika tala Hrvatske [Soil taxonomy of Croatia], Udžbenici sveučilišta u Zagrebu. Hrvatska sveučilišna naklada, Zagreb, Hrvaška [Croatia]
Ilešič S (1970) Klimatska območja Jugoslavije [Climatic zones of Yugoslavia]. Geogr Obz 4–10
Jaecks Vidic N (1994) Pedogenesis and soil-age relationships of soils on glacial outwash terraces in the Ljubljana Basin. Doktorska dizertacija [Doctoral thesis]. University of Colorado, Department of Geological Sciences, Boulder, Colorado, USA
Jaecks Vidic N, Lobnik F (1996) Rates of soil development of the chronosequence in the Ljubljana Basin, Slovenia. Geoderma
Jamnik B, Smrekar A, Vrščaj B (2009) Plot gardening in Ljubljana, geography of Slovenia. Založba ZRC, Ljubljana
Jenny H (1941) Factors of soil formation, a system of quantitative pedology. McGraw Hill Book Company, New York; London
Jiménez-Beltrán D, Töpfer K (2000) "Down to earth". Soil degradation and sustainable development in Europe. A challenge for the 21st century. The storyline. Presented at the United Nation convention to combat desertification fourth conference of the parties, Bonn
Kajfež Bogataj L (2007) Chapters from climate (agrometeorology) [WWW Document]. URL http://web.bf.uni-lj.si/agromet/Poglavja%20iz%20klime.pdf
Kastelec D, Rakovec J, Zakšek K (2007) Solar energy in Slovenia. ZRC, ZRC SAZU, Ljubljana
Kikec T (2014) Agricultural drought in Pomurje Region and possibilities of adaptations. In: Zorn M, Komac B, Ciglič R, Pavšek M (eds) (Ne)prilagojeni. Založba ZRC
Kirkby MJ (1978) Hillslope hydrology, 1 edn. Wiley, Chichester; New York
Kirkby M, Jones R, Irvine B, Gobin A, Govers G, Cerdan O, Van Rompaey A, le Bissonnais Y, Daroussin J, King D, Montanarella L, Grimm M, Vieillefont V, Puigdefabregas J, Boer M, Kosmas C, Yassoglou N, Tsara M, Mantel S, van Lynden G, Hunting J (2004) Pan-European soil erosion risk assessment: the PESERA map, Version 1. Oct 2003
KIS (1962) Pedološka karta Ljubljanskega barja 1962 [Soil map of the Ljubljana Marshes 1962]
Kmetijski inštitut Slovenije, Ministrstvo za kmetijstvo in okolje (2014) Poročilo o stanju kmetijstva, ribištva, gozdarstva in živilstva v letu 2013 [The report on the state of agriculture, fisheries, forestry and food industry in 2013]. Kmetijski inštitut Slovenije, Ljubljana, Slovenia
Kobal M, Eler K, Simončič P, Kraigher H (2014) Assessment of organic matter changes in the soil of the Brdo plot under different climate change scenarios through the Yasso07 model application. Acta silvae et ligni, ISSN 2335-3112. [Št.] 103, p. 21–34, doi: 10.20315/ASetL.103.2
Kočevar H, Vidic NJ (1998) Fundaments of geology (selected chapters). Biotechnical Faculty, Department of Agronomy, Ljubljana
Kočevar H, Vidic N (2003) Izbrana poglavja iz osnov geologije [Selected chapters from geology]. Univerza v Ljubljani, Biotehniška fakulteta, Oddelek za Agronomijo [University of Ljubljana, Biotechnical Faculty, Department of Agronomy], Ljubljana, Slovenia [Slovenia]
Kodrič M (1958) Crvenica na slovenskom Krasu [Terra Rossa in Slovenian Kast]. Zemlj Biljka VIII:41–45
Kokalj Ž, Oštir K (2006) Land cover map of Slovenia from landsat satellite imagery. Geogr Bull 78:85–95
Kokalj Ž, Špes M, Oštir K (2004) Evaluation of landscape-ecological types of Slovenia with regard to land cover produced by landsat satellite imagery classification. Diploma thesis, Ž. Ramšak

Kolbezen M (1979) Rock mass movement by the streams of the Eastern and South Eastern Pohorje Mountains as a consequence of soil erosion. Geogr Bull 51:73–83

Kolbezen M (1998) Hydrography. In: Gams I, Vrišer I (eds) Geografija Slovenije. Slovenska matica, Ljubljana, Slovenia, pp 139–172

Kolbezen M, Pristov J (1998) Surface streams and water balance of Slovenia. Ministry of Agriculture and the Environment, Ljubljana

Komac M, Jemec M (2007) Landslides and landmass movement on a sloped surface and their prevention in Slovenia. In: Knapič M (ed) Thematic strategy for soil protection: conference at the world soil day, 5th Dec 2007. Pedološko društvo Slovenije, Ljubljana, pp 25–39

Komac M, Ribičič M (2005) Landslide occurrence prediction with analysis of satellite images and other spatial data. Geological Survey of Slovenia

Košir Ž (1979) Ecological, phyto-sociological and forest management characteristics of the Gorjanci region in Slovenia. Research Reports. Forest Wood Sci Tech 17(1):1–242

Košir Ž (1994) Ecological and phyto-sociological conditions of the mountainous and hilly SW outlaying regions of Pannonia. Zveza gozdarskih društev Slovenije, Ljubljana, Ministry of Agriculture, Forestry and Food, 149

Košir B, Robek R (2000) Značilnosti poškodb drevja in tal pri redčenju sestojev s tehnologijo strojne sečnje na primeru delovišča Žakenc. ZGiL, 62:87–115

Kotar M (2009) Pogovoru o odnosu do kmetijske zemlje v Uradu predsednika Republike Slovenije. Prečiščeno in avtorizirano besedilo magnetograma [Discussion on the attitude to agricultural land. The Office of the President of the Republic of Slovenia. Consolidated and authorized text transcript.]

Kraigher H (2013) Klimatske spremembe ter vplivi antropogenih motenj na primarno produkcijo v gozdnih tleh. (05.2009/04.2012) URN:NBN:SI:doc-UB58R6J2 from http://www.dlib.si

Kralj T (2008) Primerjava sistemov za razvrščanje tal na izbranih tleh v Sloveniji [Comparison of soil classification systems on selected soils in Slovenia]. Doktorska disertacija [Doctoral Thesis]. Univerza v Ljubljani, Biotehniška fakulteta, Oddelek za agronomijo [University of Ljubljana, Biotechnical faculty, Department of Agronomy], Ljubljana

Kralj T, Grčman H (2009a) Harmonizacija klasifikacije z WRB [Harmonising the soil classification with the WRB] (Raziskovalno poročilo No. 3.2.1.2), Poročila CRP projekta V2 051 DKT-TIS. Kmetijski inštitut Slovenije [Agricultural Institute of Slovenia], Ljubljana

Kralj T, Grčman H (2009b) Predlog obveznega navodilo za klasifikacijo tal [Proposal for a mandatory instruction for soil classification]. Raziskovalno poročilo [Research report] No. 3.2.1.2, Poročila CRP projekta V2 051 DKT-TIS [Reports of the CRP V2 051 DKT-TIS project]. Kmetijski inštitut Slovenije [Agricultural Institute of Slovenia], Ljubljana

Kralj T, Grčman H (2010) Harmonizacija klasifikacije z WRB. Delovni sklop 3. Algoritmi digitalne kartografije in pretvorbene funkcije za potrebe nacionalnega talnega informacijskega sistema [Harmonising the soil classification with the WRB. WP3. Algorithms of digital soil mapping and conversion tools for the national soil information system]. Raziskovalno poročilo [Research report], Raziskovalna poročila Kmetijskega inštituta Slovenije [Agricultural Institute of Slovenia Research Reports]. Kmetijski inštitut Slovenije, CL, Center za tla in okolje [Agricultural Institute of Slovenia, CL, Centre for Soil and Environment Research], Ljubljana, Slovenija = Slovenia

Kranjc-Kušlan S (1995) Underground water balance in Republic of Slovenia: annual report (No. of Contract: R7-6414-215/95). Institute for geology, geotechnics and geophysics

Krč J, Beguš J, Primožič J, Levstek J, Papler-Lampe V, Klun J, Mihelič, M (2014) Vodila dobrega ravnanja pri strojni sečnji. Ljubljana: BF, Odd. za gozdarstvo in obnovljive gozdne vire: 38 p.

Kunaver J (1989) Determination of soil erosion at the local scale. Geogr Horiz 36:15–19

Kutin Slatnar B, Krajnc A, Lojović E, Stele A (2012) An inventory of agriculture 2010—every farm counts!. Statistical Office of the Republic of Slovenia, Ljubljana

Kutnar L, Zupančič M, Robič D, Zupančič N, Žitnik S, Kralj T, Tavčar I, Dolinar M, ZRNEC C, Kraigher H (2002) The delimitation of the regions of provenance of forest tree species in Slovenia based on ecological regions, Research Reports. Forest Wood Sci Tech, 67:73–117

Lampič B (2002) Soil and vegetation vulnerability. Dela 18:619–632

Lazarević R (1981) Erosion on agricultural land in Yugoslavia. Geogr Iugosl 7–17

Le Bissonnais Y, Montier C, Jamagne M, Daroussin J, King D (2002) Mapping erosion risk for cultivated soil in France. CATENA 46:207–220. doi:10.1016/S0341-8162(01)00167-9

Leskošek M (1993) Gnojenje: za velik in kakovosten pridelek, za zboljšanje rodovitnosti tal, za varovanje narave [Fertilization for high and quality yields, soil fertility improvement and nature protection]. ČZP Kmečki glas, Ljubljana, Slovenia [Slovenia]

Leskošek M, Mihelič R (1998) Smernice za strokovno utemeljeno gnojenje [Guidelines for fertilization], 1. del-Poljedeljstvo in travništvo. Ministrstvo za kmetijstvo, gozdarstvo in prehrano Republike Slovenije [Ministry of Agriculture Food and Forestry, Ljubljana, Slovenia [Slovenia]

Linacre E, Geerts B (1999) Water balance of the continents [WWW Document]. Water Balance Cont. URL http://www-das.uwyo.edu/~geerts/cwx/notes/chap10/continents.html. Accessed 1.2.15

List of All Countries in Europe|Bridgat.com [WWW Document] (2011) URL http://countries.bridgat.com/Europe.html. Accessed 1.2.15

List of countries and dependencies by area (2014) Wikipedia free encycl

List of landslides—Wikipedia, the free encyclopedia [WWW Document] (2014) URL http://en.wikipedia.org/wiki/List_of_landslides. Accessed 1.4.15

Lobnik F, Stepančič D, Rupreht J, Prus T, Šporar M, Vidic N, Vrščaj B, Zupan M, Hafner V (1987) Pedološko kartiranje SRS II.del: list Postojna 4 [Soil mapping in Slovenia—section Postojna 4]. Raziskovalno poročilo [Research report], Raziskovalna poročila CPVO. Univerza Edvarda Kardelja v Ljubljani, Biotehniška fakulteta, VTOZD za agronomijo [University of Ljubljana, Biotechnical Faculty, Agronomy Department], Ljubljana, Slovenia [Slovenia]

Lobnik F, Fajt N, Zupan M, Vrščaj B, Vidic N, Šporar M, Hrustel-Majcen M (1989a) Kontrola rodovitnosti tal in gnojenja v zahodni Vipavaski dolini/Soil fertility control in West Vipava Valley. Raziskovalno poročilo/Research report, Raziskovalna poročila CPVO. Univerza Edvarda Kardelja v Ljubljani, Biotehniška fakulteta, VTOZD za agronomijo/University of Ljubljana, Biotechnical Faculty, Agronomy Department, Ljubljana, Slovenija

Lobnik F, Hrustel-Majcen M, Zupan M, Vrščaj B, Rupreht J, Šporar M, Hodnik A, Trobiš-Lednik M, Prus T, Kastelec D, Kočevar H, Hudnik V, Vučko H (1989) Soil pollution map of Celje County (Poročilo za MOP za leto 1992). Univerza v Ljubljani, Biotehniška fakulteta, Oddelek za Agronomijo, Ljubljana, Slovenia

Lobnik F, Prus T, Turk I, Vrščaj B, Zupan M, Hudnik V (1992a) Raba in varstvo tal v Sloveniji [Land use and soil protection in Slovenia]. Presented at the 3. posvetovanje onesnaževanje in varstvo okolja: geologija in tehnika za okolje, Zavod za tehnično izobraževanje, Ljubljana, Slovenia [Slovenia], pp 129–139

Lobnik F, Zupan M, Turk I, Hodnik A, Rupreht J, Šporar M, Prus T, Vidic NJ, Trobiš-Lednik M, Vrščaj B, Leskošek M, Hudnik V, Šiftar A (1992b) Ugotavljanje onesnaženosti tal in monitoring v R Sloveniji: Model na izbranih lokacijah [Determination and monitoring of soil contamination in Slovenia—a model implemented on selected locations] (Poročilo za MOP za leto 1992). Univerza v Ljubljani, Biotehniška fakulteta, Oddelek za Agronomijo [University of Ljubljana, Biotechnical Faculty, Department of Agronomy], Ljubljana, Slovenia [Slovenia]

Lobnik F, Vrščaj B, Zupan M (1993) Podatki tal in talni informacijski sistem (TIS): onesnaženost tal v Sloveniji [Soil data and soil information system (SIS): soil contamination in Slovenia]. Presented at the Kakovost vode z vidika uporabnosti za namakanje kmetijskih zemljišč, Portorož, 27. - 28. septembra 1993, Zavod za tehnično izobraževanje Ljubljana: Ministrstvo za kmetijstvo in gozdarstvo, Portorož, Slovenija [Slovenia], pp 1–15

Lobnik F, Prus T, Rupreht J, Šporar M, Vrščaj B, Tič I, Knapič M (1994) Nacionalni program namakanja Republike Slovenije [National irrigation programme]. Okolje V Slov Environ Slov 1994:41–51

Lobnik F, Rupreht J, Šporar M, Prus T, Vrščaj B (1995a) Presoja vplivov na tla kot sestavni del okolja z valorizacijo pridelovalnega potenciala kmetijskih zemljišč: detajlna študija za avtocesto Divača Koper, odsek Kastelec Srmin [An impact assessment on soil as an integral part of the environment and the valorisation of the productive potential of agricultural land: detailed study of the highway Koper Divača, Section Kastelec Srmin]. Raziskovalna poročila CPVO (CSES research reports). Center za pedologijo in varstvo okolja, Biotehiška fakulteta, Univerza v Ljubljani [Center for Soil and Environment Protection, Biotechnical Faculty, University of Ljubljana], Ljubljana, Slovenia [Slovenia]

Lobnik F, Rupreht J, Šporar M, Prus T, Vrščaj B, Knapič M, Tič I (1995b) Presoja vplivov na tla kot sestavni del okolja z valorizacijo pridelovalnega potenciala kmetijskih zemljišč: detajlna študija za avtocesto Celje Ljubljana, odsek: Zadobrova Blagovica [An impact assessment on soil as an integral part of the environment and the valorisation of the productive potential of agricultural land: detailed study of the highway Celje Ljubljana, section Zadobrova Blagovica]. Raziskovalna poročila CPVO (CSES research reports). Center za pedologijo in varstvo okolja, Biotehiška fakulteta, Univerza v Ljubljani [Center for Soil and Environment Protection, Biotechnical Faculty, University of Ljubljana], Ljubljana, Slovenia [Slovenia]

Lobnik F, Rupreht J, Šporar M, Prus T, Vrščaj B, Tič I (1995c) Presoja vplivov na tla kot sestavni del okolja z valorizacijo pridelovalnega potenciala kmetijskih zemljišč: detajlna študija za avtocesto Celje Ljubljana, odsek: Šentjakob Domžale [An impact assessment on soil as an integral part of the environmenta and the valorisation of the productive potential of agricultural land: detailed study of the highway Celje Ljubljana, Section Šentjakob Domžale]. Raziskovalna poročila CPVO (CSES research reports). Center za pedologijo in varstvo okolja, Biotehiška fakulteta, Univerza v Ljubljani [Center for Soil and Environment Protection, Biotechnical Faculty, University of Ljubljana], Ljubljana, Slovenia [Slovenia]

Lobnik F, Rupreht J, Šporar M, Prus T, Vrščaj B, Tič I (1995d) Presoja vplivov na tla kot sestavni del okolja z valorizacijo pridelovalnega potenciala kmetijskih zemljišč: detajlna študija za avtocesto Celje Ljubljana, odsek: Šentjakob Domžale: dodatna študija [An impact assessment on soil as an integral part of the environmenta and the valorisation of the productive potential of agricultural land: detailed study of the highway Celje Ljubljana, section Šentjakob Domžale—detailed study]. Raziskovalna poročila CPVO (CSES research reports). Center za pedologijo in varstvo okolja, Biotehiška fakulteta, Univerza v Ljubljani [Center for Soil and Environment Protection, Biotechnical Faculty, University of Ljubljana], Ljubljana, Slovenia [Slovenia]

Lobnik F, Prus T, Rupreht J, Šporar M, Vidic N, Vrščaj B, Tič I (1996a) Primerjalna študija poteka tras AC na osnovi presoje vplivov na tla kot sestavnega dela okolja in valorizacije pridelovalnega potenciala kmetijskih zemljišč: za odsek: Cogetinci Radmožanci: študija za izbor variante [A comparative study of potential motorway routes and an impact assessment on the soil and productive potential of agricultural land: for the section: Cogetinci Radmožanci: a study for the selection of variants]. Raziskovalna poročila CPVO [CSES research reports]. Center za pedologijo in varstvo okolja, Biotehiška fakulteta, Univerza v Ljubljani [Center for Soil and Environment Protection, Biotechnical Faculty, University of Ljubljana], Ljubljana

Lobnik F, Prus T, Zupan M, Vrščaj B, Simončič P, Grčman H, Ivačič M, Tič I, Rupreht J, Šporar M (1996b) Poročilo o vplivih na okolje za lokacijski načrt za pododsek AC Razdrto Vipava: vpliv na tla kot sestavni del okolja in na pridelovalni potencial kmetijskih zemljišč: posnetek stanja tal pred izgradnjo avtoceste [Environmental impact assessment for subsection of motorway Razdrto-Vipava: the impact on the solid as an integral part of the environment and production potential of agricultural land: a soil quality assessment prior to construction of the highway]. Raziskovalna poročila CPVO (CSES research reports). Center za pedologijo in varstvo okolja, Biotehiška fakulteta, Univerza v Ljubljani [Center for Soil and Environment Protection, Biotechnical Faculty, University of Ljubljana], Ljubljana

Lovrenčak F (1970) Soils and vegetation of Raka. Geogr Bull 42:91–95

Lovrenčak F (1976) New soil classification (some news from the field of soil geography). Geogr Bull 48:181–190

Lovrenčak F (1994a) Pedogeografija [Pedogeography]. Univerza v Ljubljani, Filozofska fakulteta, Oddelek za geografijo [University of Ljubljana, Faculty of Arts, Department of Geography], Ljubljana, Slovenia [Slovenia]

Lovrenčak F (1994) Soil geography. Geography Department, Faculty of Arts, University of Ljubljana, Ljubljana, Slovenia

Lovrenčak F (2007) Upper forrest line of the Slovenian Alps, high Karstic Plateaus and Prokletije, Razprave Filozofske fakultete. Znanstvenoraziskovalni inštitut Filozofske fakultete, Ljubljana

Malovrh V (1955) Meteorological observations of winds at Črni Kal, annual report of the meteorological service (1955). Hydrometeorological institute of Republic of Slovenia, Ljubljana, Slovenia

Mali B, Košir B (2007) Soil damage by harvesters and forwarders when cutting and skidding wood (CTL-technology). GV, 65(3):131–142

Marinček L, Čarni A (2002) Commentary to the vegetation map of forest communities of Slovenia in a scale 1:400.000. Založba ZRC, ZRC SAZU, Biološki inštitut Jovana Hadžija, pp. 79

Melik A (1935) Slovenija - Geografski opis [Slovenia—a geographical description]. Slovenska matica, Ljubljana, Slovenia

Mekinda M T, Logar M, Verbič J, Simončič P, Mali B, Kobal M, Hladnik D, Žižek L, Skudnik M, Malešič I, Kranjc A, Stare R (2014) Slovenia's national inventory report 2014 : emission inventories 1986–2012 : submitted under the United Nations framework convention on climate change and the Kyoto protocol. Ljubljana: Ministry for Agriculture and the Environment, SEA, 2014. Ilustr. http://unfccc.int/national_reports/annex_i_ghg_inventories/national_inventories_submissions/items/8108.php.

Mihelič M (2014) Economy and environmental aspects of forest harvesting for energy biomass : Doctoral dissertation. Ljubljana: [M. Mihelič], XVI, p. 285

Mikkelsen J, Cools N, Langohr R, Kobal M, Urbančič M, Kralj T, Simončič P (2006) Navodila za opis talnega profila za projekt BIOSOIL [BIOSOIL Project Instructions for Soil Description], Gozdarski inštitut Slovenije [Slovenian Forestry Institute], Ljubljana, Slovenija [Slovenia]

Mikoš M, Brilly M, Ribičič M (2004) Floods and landslides in Slovenia. Acta Hydrotech 22:113–133

Ministry of Agriculture and the Environment (1999) Ministry of agriculture and the environment [WWW Document]. Digit Soil Map Slov. URL http://rkg.gov.si/GERK/documents/Pedoloska Karta.zip. Accessed 8.14.13

MKGP (2002a) Raba kmetijskih zemljišč Republike Slovenije 1:5.000 [Land use digital database (1:5.000)] 2002.10.25. Raba

MKGP (2002) Interpretacijski ključ; podroben opis metodologije zajema dejanske rabe kmetijskih in gozdnih zemljišč. MKGP, Ljubljana, Slovenia

MKGP (2005) Raba kmetijskih zemljišč Republike Slovenije 1:5.000 [Land use digital database (1:5.000)] v.2005.08.04. Raba

MKGP (2007) Raba kmetijskih zemljišč Republike Slovenije 1:5.000 [Land use digital database (1:5.000)] v. 2007.09.08. Raba

MKGP (2009) Interpretacijski ključ; podroben opis metodologije zajema dejanske rabe kmetijskih in gozdnih zemljišč. MKO, Ljubljana, Slovenia

MKGP (2014) Raba kmetijskih zemljišč Republike Slovenije 1:5.000 [Land use digital database (1:5.000)]. Raba

MKGP, CPVO (2001) Digitalna pedološka karta Slovenije 1:25.000 (PK25) [Digital soil map of Slovenia 1:25,000 (DSM25)]

MKGP/MAFF (2014) Raba kmetijskih zemljišč Republike Slovenije 1:5.000 [Land use digital database (1:5,000)] v.13.01.2014. Raba

MKO (2011) Interpretacijski ključ; podroben opis metodologije zajema dejanske rabe kmetijskih in gozdnih zemljišč. MKO, Ljubljana, Slovenija [Slovenia]

MKO (2012) Grafične enote kmetijske rabe 1:5.000

MKO (2013) Interpretacijski ključ; podroben opis metodologije zajema dejanske rabe kmetijskih in gozdnih zemljišč. MKO, Ljubljana, Slovenija [Slovenia]

Moore ID, Gessler PE, Nielsen GA, Peterson GA (1993) Soil attribute prediction using terrain analysis. Soil Sci Soc Am J 57, NP. doi:10.2136/sssaj1993.572NPb

Mršić N, Novak T, Devetak D (1997) The animals of our soils: an introduction to soil zoology—general overview of soil favna with systematics and ecology. Tehniska zalozba Slovenije, Ljubljana

Mršić N, Novak T, Devetak D (1997b) Živali naših tal: Uvod v pedozoologijo - sistematika in ekologija s splošnim pregledom talnih živali [Animals in our soils: an introduction to pedozoology—systematics and ecology with general review of soil anilmals]. Tehniska zalozba Slovenije, Ljubljana, Slovenia [Slovenia]

Natek K (1990) Geomorphological properties of landslips in the Haloze Hills. Ujma 4:11–15

Natek K (1996) Threat to the surface of the North-Eastern Slovenia and its regeneration capabilities according to the regeneration of the surface in the Haloze hills after the catastrophic thunderstorm in July 1989. In: The lower drava region with Prlekija: the possibilities of the regional and spatial development. Presented at the 17th assembly of the Slovene geographers, Ptuj 23rd–26th Oct 1996, Association of the Geographical Societies of Slovenia, Ptuj

Natek K (2002) Landscape-ecological division of Slovenia. In: Špes M (ed) Environmental vulnerability study (methodology and aplication), acta geographica Slovenica. Založba ZRC, Ljubljana, Slovenia, pp 23–37

Natek K (2007) Periglacial landforms in the Pohorje mountains. Dela 247–263

Natek K, Natek M (2008) Slovenia: a portret of a country. Natek in ostali, Ljubljana

Neugebauer V, Filipovski Ǵ, Ćirić M, Živković M (1963) Klasifikacija zemljišta Jugoslavije [Yugoslav soil classification]. VIII kongres JDPZ, Ohrid

Ogorelec B, Mišić M, Cimerman F, Faganeli J (1983) Sédiments quaternaire du forage effectue dans le Baie de Koper, Adriatique du nord, Yougoslavie. Rapp Comm Int Explor Sci Mer Medit 28:251–253

Ogrin D (1996) The climate types in Slovenia. Geogr Bull. Bulletin of the Association of the Geographical Societes of Slovenia 68:39–56

Ogrin D (2012) Drought—a hundred year constant, that comes always as a surprise. Delo 4

Ogrin D, Plut D (2009) Aplicative physical geography of Slovenia, 1st edn. Znanstvena založba Filozofske fakultet, Ljubljana

Oset F (1987) Kategorije, talni tipi, bonitetni razredi, raba tal in velikost parcel na posameznih pedosekvencah v Savinjski dolini. Diplomska naloga/Diploma thesis. Univerza Edvarda Kardelja v Ljubljani, Biotehniška fakulteta, VTOZD za agronomijo/University of Ljubljana, Biotechnical Faculty, Agronomy Department, Ljubljana, Slovenija

Pagon P (2006) Morning thermic wind in barcola bay near trieste: the applicability analysis within the sight of sailing. Geography Department, Faculty of Arts, University of Ljubljana, Ljubljana, Slovenia

Perko D (2001) Analysis of the surface of Slovenia using the 100-meter digital elevation model. Založba ZRC

Petauer M (2005) Study of soils as landscape factor in the Celje basin: application for schools. Master thesis, Geography Department, Faculty of Arts, University of Ljubljana, Ljubljana, Slovenia

Petauer M (2011) The role of soils in raising public awareness on the example of the Savinjska Statistical Region. Ph.D. dissertation, Geography Department, Faculty of Arts, University of Ljubljana, Ljubljana, Slovenia

Peucker TK, Douglas DH (1975) Detection of surface-specific points by local parallel processing of discrete terrain elevation data. Comput Graph Image Process 4:375–387. doi:10.1016/0146-664X(75)90005-2

Plenčar M, Vrabec M, Novak M (2009) Geology of Slovenia. Geological Survey of Slovenia, Ljubljana

Podobnikar T (2010) Numerical definition, composition and verification of morphometric factors of soil formation in cartographic soil units, algorithms of digital cartography and transforming functions for the need of the national soil information system (project CRP V2 0511. DS 1, Digital cartography algorithm development). Ljubljana, Slovenia

Povše M, Repe B (2013) Proučevanje zaslanjenih prsti slovenske obale (diplomsko delo) [Investigation of salt affected soils of the Slovenian coast (B.Sc. thesis)]. Diplomsko delo, Oddelek za geografijo, Filozofska fakulteta, Univerza v Ljubljani, Ljubljana, Slovenia [Slovenia]

Prestor J, Rikanović R, Janža M (2002) Underground waters. Presented at the natural disasters and protection against them, Administration for Civil Protection and Disaster Relief, Ministry of Defence, Ljubljana, Slovenia, pp 200–205

Prokop G, Jobstmann H, Schönbauer A (2011) Report on best practices for limiting soil sealing and mitigation its effects. European Comission, DG EnvironemntUmveltBundesamt, Austria

Prus T (1983) Priročnik za pedološka raziskovanja za znanstveno-raziskovalni tabor v Lučah 1983/Soil survey handbook for scientific research camp in Luče 1983. Univerza Edvarda Kardelja v Ljubljani, Biotehniška fakulteta, VTOZD za agronomijo/University of Ljubljana, Biotechnical Faculty, Agronomy Department, Ljubljana, Slovenija

Prus T (2000) Klasifikacija tal Slovenije - študijsko gradivo [Slovenian soil classification—lecture notes]. Center za pedologijo in varstvo okolja, Ljubljana, Slovenija [Slovenia], p 22

Prus T, Vrščaj B (1997) Use of soil information system in the environmental impact assessment of highway construction in Slovenia. In: Hodgson S (ed) Presented at the EGIS/MARI '94, Geographical information '97: from research to application through cooperation, Amsterdam, pp 385–393

Prus T, Lobnik F, Vrščaj B (1985a) Pedološke raziskave agromelioracijskega območja Povšnik - Seruč pri Litiji [Soil investigations in Povšnik—Seruč near Litija amelioration area]. Raziskovalno poročilo/Research report, Raziskovalna poročila CPVO. Univerza Edvarda Kardelja v Ljubljani, Biotehniška fakulteta, VTOZD za agronomijo/University of Ljubljana, Biotechnical Faculty, Agronomy Department, Ljubljana, Slovenija

Prus T, Vidic N, Vrščaj B, Lobnik F (1985b) Tla sekcije Cerkno [Soils of Cerkno section]. Raziskovalno poročilo/Research report. Raziskovalna poročila CPVO. Univerza Edvarda Kardelja v Ljubljani, Biotehniška fakulteta, VTOZD za agronomijo [University of Ljubljana, Biotechnical Faculty, Agronomy Department], Ljubljana, Slovenija

Prus T, Rupreht J, Šporar M, Vidic N, Vrščaj B, Lobnik F (1986a) Pedološke raziskave hidromelioracijskega kompleksa Radomlja Rovšca II [Soil investigations in Radomlje Rovšca Hidromelioration Area, vol II]. Raziskovalno poročilo [Research report], Raziskovalna poročila CPVO. Univerza Edvarda Kardelja v Ljubljani, Biotehniška fakulteta, VTOZD za agronomijo [University of Ljubljana, Biotechnical Faculty, Agronomy Department], Ljubljana, Slovenia [Slovenia]

Prus T, Šporar M, Vidic N, Vrščaj B, Lobnik F (1986b) Pedološke raziskave hidromelioracijskega kompleksa Slivnica pri Grosuplju [Soil investigations in Hidromelioration Area Slivnica pri Grosuplju]. Raziskovalno poročilo [Research report]. Raziskovalna poročila CPVO. Univerza Edvarda Kardelja v Ljubljani, Biotehniška fakulteta, VTOZD za agronomijo [University of Ljubljana, Biotechnical Faculty, Agronomy Department], Ljubljana, Slovenia [Slovenia]

Radinja D (1990) Geographical characteristic of the Trieste bay and the surrounding inland: maritime and sub maritime outline of both regions according to the present state of social development. Presented at the Primorje: zbornik 15. zborovanja slovenskih geografov, Portorož, 24–27 oktobra 1990, Portorož, pp 7–12

Radišek J, Muršec M, Vrščaj B (2014a) Izdelava pedološke karte in karte bonitetnih vrednosti ter interpretacija pedoloških podatkov za potrebe pridobivanja bonitetnih ocen kmetijskih, gozdnih in drugih zemljišč v občini Braslovče [Elaboration of soil map, soil data interpretation and land evaluation in the Municipality Braslovče], Priprava kmetijskih strokovnih podlag Občine Braslovče. ProTellus d.o.o, Kmetijski inštitut Slovenije [ProTellus d.o.o. and Agricultural Institute of Slovenia], Braslovče, Slovenija [Slovenia]

Radišek J, Muršec M, Vrščaj B (2014b) Digitalna pedološka karta Občine Braslovče 1:10.000 [The soil map of Braslovče County 1:10,000]. Priprava kmetijskih strokovnih podlag Občine Braslovče

Ravbar M (1975) Karst erosion in the vicinity of Straža near Novo Mesto. Geogr Horiz 22:12–18

Remec-Rekar Š, Bat M (2003) Lakes. In: Uhan J, Bat M (eds) Richness of Slovenia's waters. Ministry of the Environment, Spatial Planning and Energy: Slovenian Environment Agency, Ljubljana

Renard KG (1997) Predicting soil erosion by water: a guide to conservation planning with the revised universal soil loss equation, Ex-Lib edn. United States Government Printing, Washington, D.C.

Repe B (2001) Field and laboratory methods for soil research, study materials. Geography Department, Faculty of Arts, University of Ljubljana, Ljubljana, Slovenia

Repe B (2002) Soil degradation threat to Slovenia's landscapes = Ogroženost slovenskih pokrajin zaradi degradacije prsti. Acta Geogr Slov 99–121

Repe B (2002) Soil degradation threat to Slovenia's landscapes. Acta Geogr Slov 99–121

Repe B (2004) Soils of Slovenia. In: Orožen Adamič M (ed) Slovenia: a geographical overview. Association of the Geographical Societies of Slovenia, Založba ZRC, Ljubljana, Slovenia, pp 51–56

Repe B (2005) Soil sets of the wrb classification for geography textbooks. Geogr Sch 15:71–76

Repe B (2005b) Applicability of a new FAO WRB soil classification for school geography. Presented at the natural disasters/14. Ilešičevi dnevi, 22–24 Sept 2005, Geography Department, Faculty of Arts, University of Ljubljana, Ljubljana, Slovenia, pp 15–16

Repe B (2006) Svetovna WRB klasifikacija prsti. Geogr Obz 53:9–22

Repe B (2006) World WRB soil classification. Geogr Horiz 53:9–22

Repe B (2006b) Geographical soil map and its use within geography. Doctoral thesis, Geography Department, Faculty of Arts, University of Ljubljana, Ljubljana, Slovenia

Repe B (2007a) Voda v prsti in ugotavljanje njenega razporejanja v odvisnosti od reliefa. Dela 91–106

Repe B (2007) Natural vegetation of Slovenia. Geogr Sch 16:21–28

Repe B (2007c) Geographical soil map—a different soil determination approach in Slovenia. Chang. Soils Chang. World Soils Tomorrow

Repe B (2008) Soil type extent determination using remote sensing classification methods. In: Perko D, Zorn M, Razpotnik N, Čeh M, Hladnik D, Krevs M, Podobnikar T, Repe B, Šumrada R (eds) Geographical information systems in Slovenia 2007–2008. Založba ZRC, pp 195–206

Repe B (2009) Soil loss in Prekmurje due to the permanent land use changes in the last 15 years. In: Kikec T (ed) Pomurje: sustainable regional development along the Mura River. Association of Slovenian Geographers; Pomurje Geographical Association, Ljubljana; Murska Sobota, pp 195–206

Repe B (2010) Prepoznavanje osnovnih prsti slovenske klasifikacije [Recognition of Slovenian soil classification soil types]. Geografski zbornik 2010:143–166

Repe B (2010b) Recognition of the Slovenian soil classification types. Dela 0:143. doi:10.4312/dela.34.8.143-166

Repe B, Blatnik M (2013) Plant altitude belts in the Gorenjska Region. In: Rogelj B, Mrak I, Potočnik Slavič I (eds) Gorenjska in the era of glocalization. Geography Department, Faculty of Arts, University of Ljubljana, Bled, pp 43–60

Repe B, Pristovšek A (2011) Geografski prispevek k poznavanju subakvalnih prsti ob Slovenski obali [A geographic contribution to subaquatic soils of Slovenian Cosdat]. In: Glasnik Univerze na Primorskem, Znanstveno-raziskovalnega središča Koper. Presented at the Četrti Melikovi dnevi: Geografski vidiki upravljanja z morjem in zaledjem, Koper, 10–11 junij 2011, Univerza na Primorskem, Znanstveno-raziskovalno središče Koper, Koper, pp 51–52

Republika Slovenija (1996) Uredba o mejnih, opozorilnih in kritičnih imisijskih vrednostih nevarnih snovi v tleh

Republika Slovenija (2011) Z A K O N O KMETIJSKIH ZEMLJIŠČIH uradno prečiščeno besedilo (ZKZ-UPB2), UL RS št. 71/2011

Resulović H, Čustović H (2002) Pedologija [Pedology]. Univerzitet u Sarajevu [University of Sarajevo], Sarajevo

Resulović H, Čustović H (2008) Sistematika tla/zemljišta (Soil taxonomy). Univerzitet u Sarajevu (University of Sarajevo), Sarajevo

Ribičič M (2002) Landslides, landslips and rockfalls. In: Ušeničnik B (ed) Disasters and protection against them. Administration for Civil Protection and Disaster Relief (Ministry of Defence), Ljubljana, pp 260–267

Robič D, Accetto M (2001) Overview of the syntaxonomical system of forest vegetation of Slovenia – University of Ljubljana, BF, Department for forestry and renewable resources, Ljubljana, manuscript

Rupreht J, Šporar M, Vidic N, Prus T, Vrščaj B (1985) Tla sekcije Gorica in Cerkno [Soils of sections Gorica and Cerkno]. Raziskovalno poročilo [Research report], Raziskovalna poročila CPVO. Univerza Edvarda Kardelja v Ljubljani, Biotehniška fakulteta, VTOZD za agronomijo [University of Ljubljana, Biotechnical Faculty, Agronomy Department], Ljubljana, Slovenija

SFS Report 2013 (2014) ZGS, Ljubljana, 131

Šifrer M (1965) New geomorphological discoveries in the coastal areas of Koper. Geogr Horiz

Šifrer M (1998) Quaternary surface. In: Fridl J (ed) Geographical atlas of Slovenia: the county in space and time. DZS, Ljubljana, pp 74–77

Simončič P, Robek R, Mali B, Železnik P, Kraigher H (2010) Vpliv težke mehanizacije na procese v gozdnih tleh. Kmečki glas, ISSN 0350-4093, vol. 67, priloga Gozdarstvo, p 13–15

Šinigoj J, Požar M, Krivic M, Vrščaj B, Godeša T (2008) Izdelava ArcGIS aplikacije iPREHOD; DS4 Izdelava GIS modela ocene prehodnosti terena za vozilo [Elaboration of iPREHOD ArcGIS application, WP4—elaboration of the land trafficability model]. Geološki zavod Slovenije [Slovenian Geological Survey], Ljubljana, Slovenia [Slovenia]

Škorić A (1986) Soil development, genesis and systematics. Sveučilište u Zagrebu, Zagreb

Škorić A (1986b) Postanak, razvoj i sistematika tla [Genesis, development ansd taxonomy of soils]. Sveučilište u Zagrebu [University of Zagreb], Zagreb

Škorić A, Filipovski Ǵ, Ćirić M (1973) Klasifikacija tala Jugoslavije [Yugoslav Soil Classification]. Sveučilište u Zagrebu [University of Zagreb], Zagreb

Škorić A, Filipovski Ǵ, Ćirić M (1985) Klasifikacija zemljišta Jugoslavije [Yugoslav soil classification]. ANU Bosne i Hercegovine [Academy of Science and Arts of Bosnia and Herzegovina], Sarajevo

Škvarč A (2012) Erosion in agriculture [WWW Document]. Fruit grow. Advices - Sunny Juicy Nasveti Za Goj. Sadja - Sončno Sočno. URL http://www.soncno-socno.si/za_sadjarje/nasveti_za_gojenje_sadja/. Accessed 1.4.15

Slovenian Environment Agency (2010) Presentation of Slovenia [WWW Document]. Present Slov. URL http://www.arso.gov.si/soer/predstavitev_Slovenije.html. Accessed 7.4.14

Stepančič D, Hočevar A, Gregorič V, Stritar A, Zupančič M, Kajfež-Bogataj L (1984a) Komentar k listu Murska Sobota. Osnovna Pedološka karta SFRJ, Pedološka karta Slovenije [Readings to the soil map of Slovenia—section Muska Sobota, soil map of Yugoslavia]. Raziskovalno poročilo [Research report], Raziskovalna poročila CPVO. Univerza Edvarda Kardelja v Ljubljani, Biotehniška fakulteta, VTOZD za agronomijo [University of Ljubljana, Biotechnical Faculty, Agronomy Department], Ljubljana, Slovenija

Stepančič D, Lobnik F, Prus T, Vidic N, Vrščaj B, Šrok D (1984b) Tla sekcije Kranj in Rogatec [Soils of sections Kranj and Rogatec]. Raziskovalno poročilo [Research report], Raziskovalna poročila CPVO. Univerza Edvarda Kardelja v Ljubljani, Biotehniška fakulteta, VTOZD za agronomijo [University of Ljubljana, Biotechnical Faculty, Agronomy Department], Ljubljana, Slovenia [Slovenia]

Stepančič D, Lobnik F, Vidic N, Stritar A, Seliškar A, Kajfež-Bogataj L, Marinček L (1985) Komentar k listu Ljubljana. Osnovna Pedološka karta SFRJ, Pedološka karta Slovenije [Readings to the soil map of Slovenia—section Ljubljana, soil map of Yugoslavia]. Raziskovalno poročilo [Research report], Raziskovalna poročila CPVO. Univerza Edvarda Kardelja v Ljubljani, Biotehniška fakulteta, VTOZD za agronomijo [University of Ljubljana, Biotechnical Faculty, Agronomy Department], Ljubljana, Slovenia [Slovenia]

Stepančič D, Lobnik F, Vidic N, Stritar A, Seliškar A, Kajfež-Bogataj L, Marinček L (1986a) Komentar k listu Ptuj. Osnovna Pedološka karta SFRJ, Pedološka karta Slovenije [Readings to the soil map of Slovenia—section Ptuj, soil map of Yugoslavia]. Raziskovalno poročilo [Research report], Raziskovalna poročila CPVO. Univerza Edvarda Kardelja v Ljubljani, Biotehniška fakulteta, VTOZD za agronomijo [University of Ljubljana, Biotechnical Faculty, Agronomy Department], Ljubljana, Slovenia [Slovenia]

Stepančič D, Rupreht J, Šporar M, Prus T, Vidic N, Kajfež-Bogataj L, Vrščaj B (1986b) Pedološke kartiranje SR Slovenije - Tolmin [Soil mapping in Slovenia—section Tolmin]. Raziskovalno poročilo [Research report], Raziskovalna poročila CPVO. Univerza Edvarda Kardelja v Ljubljani, Biotehniška fakulteta, VTOZD za agronomijo [University of Ljubljana, Biotechnical Faculty, Agronomy Department], Ljubljana, Slovenia [Slovenia]

Stepančič D, Rupreht J, Prus T, Šporar M, Vidic N, Kajfež-Bogataj L, Hodnik A, Vrščaj B, Zupan M, Hafner V (1987a) Pedološke kartiranje SR Slovenije - Kranjska gora - sever [Soil mapping in Slovenia—section Kranjska gora—North]. Raziskovalno poročilo [Research report], Raziskovalna poročila CPVO. Univerza Edvarda Kardelja v Ljubljani, Biotehniška fakulteta, VTOZD za agronomijo [University of Ljubljana, Biotechnical Faculty, Agronomy Department], Ljubljana, Slovenia [Slovenia]

Stepančič D, Rupreht J, Prus T, Šporar M, Vidic N, Kajfež-Bogataj L, Hodnik A, Vrščaj B, Zupan M, Hafner V (1987b) Pedološke kartiranje SR Slovenije - Postojna 4 [Soil mapping in Slovenia—section Postojna 4]. Raziskovalno poročilo [Research report], Raziskovalna poročila CPVO. Univerza Edvarda Kardelja v Ljubljani, Biotehniška fakulteta, VTOZD za agronomijo [University of Ljubljana, Biotechnical Faculty, Agronomy Department], Ljubljana, Slovenia [Slovenia]

Stritar A (1965) Typical sequences of soil types in Slovenia. Geogr Horiz 12

Stritar A (1990) Landscape and landscape Systems. Partizanska knjiga, Ljubljana, Slovenia

Stritar A (1990b) Krajina in krajinski sistemi [Landscape and landscape systems]. Partizanska knjiga, Ljubljana, Slovenia [Slovenia]

Stritar A (1991) Pedologija - kompendij [Pedology—compendium]. Univerza v Ljubljani, Biotehniška fakulteta, Agronomski oddelek [University of Ljubljana, Biotechnical Faculty, Agronomy Departments], Ljubljana, Slovenia [Slovenia]

Surina B (2005) Subalpine and alpine vegetation of the Krn Area in the Julian Alps. Scopolia 1–222

Sušin J (1964) Doprinos k poznavanju terre rosse v Slovenskem Primorju [Contribution to the knowledge of Terra Rossa in Slovenian Littoral Area]. Univerza Edvarda Kardelja v Ljubljani, Biotehniška fakulteta, VTOZD za agronomijo [University of Ljubljana, Biotechnical Faculty, Agronomy Department], Ljubljana, Slovenija [Slovenia]

Sušin J (1983) Gradivo za Pedološki slovar [Materials for dictionary of soil science], Kmetijski tehniški slovar. Nauk o tleh. Univerza Edvarda Kardelja v Ljubljani, Biotehniška fakulteta, VTOZD za agronomijo/University of Ljubljana, Biotechnical Faculty, Agronomy Department, Ljubljana, Slovenia [Slovenia]

Sušin J, Vrščaj B, Vernik T, Verbič F, Jeretina J, Zagorc B, Rednak M, Glad J, Bergant J, Čuden I (2009) Poročilo o izvedbi strokovnih nalog za Ministrstvo za okolje in prostor v letu 2008 in 2009: končno poročilo (KIS - Poročila o strokovnih nalogah No. 123). KIS - Poročila o strokovnih nalogah. Kmetijski inštitut Slovenije, Ljubljana, Slovenia

Sušnik A, Kurnik B (2003) Crop damage due to catastrophic drought in 2003. Ujma 54–60

Talni profili, digitalna baza podatkov (TP) [Soil profile; digital database (SP)] 2001

Tič I, Vrščaj B (2001a) MS ACCESS; tehnična dokumentacija o bazah podatkov na MS SQL strežniku PedonSQL. Center za pedologijo in varstvo okolja, Biotehiška fakulteta, Univerza v Ljubljani [Center for Soil and Environment Protection, Biotechnical Faculty, University of Ljubljana], Ljubljana, Slovenia [Slovenia]

Tič I, Vrščaj B (2001b) Talni Informacijski Sistem Slovenije - TIS. MS ACCESS - PEDOLOŠKI PROFILI (PP). Tehnična dokumentacija o bazi podatkov pedoloških profilov (PP) na MS SQL strežniku "PedonSQL" [Soil information system of Slovenia—SIS. MS access—soil profile data. Soil profile dataset technical documentation]. Center za pedologijo in varstvo okolja, Biotehiška fakulteta, Univerza v Ljubljani [Center for Soil and Environment Protection, Biotechnical Faculty, University of Ljubljana], Ljubljana, Slovenia [Slovenia]

Tič I, Vrščaj B (2001c) Talni Informacijski Sistem Slovenije - TIS. MS ACCESS - PKE. Tehnična dokumentacija o bazi podatkov pedokartografskih in pedosistematskih enot (PKE) na MS SQL strežniku "PedonSQL" [Soil information system of Slovenia—SIS. MS Access—SMU and STU data. Soil mapping units Technical documentation]. Center za pedologijo in varstvo okolja, Biotehiška fakulteta, Univerza v Ljubljani [Center for Soil and Environment Protection, Biotechnical Faculty, University of Ljubljana], Ljubljana, Slovenia [Slovenia]

Tič I, Vrščaj B (2001d) Talni Informacijski Sistem Slovenije - TIS. Priročnik za uporabo podatkov na MS SQL strežniku "PedonSQL" [Soil information system of Slovenia—SIS. Handbook on MS SQL soil dataset]. Center za pedologijo in varstvo okolja, Biotehiška fakulteta, Univerza v Ljubljani [Center for Soil and Environment Protection, Biotechnical Faculty, University of Ljubljana], Ljubljana, Slovenia [Slovenia]

Tič I, Vrščaj B (2001e) Tehnična dokumentacija o bazah podatkov analitskega laboratorija na MS SQL strežniku PedonSQL [Soil information system of Slovenia—SIS. MS access—soil profile data. Soil analytical database technical documentation]. Center za pedologijo in varstvo okolja, Biotehiška fakulteta, Univerza v Ljubljani [Center for Soil and Environment Protection, Biotechnical Faculty, University of Ljubljana], Ljubljana, Slovenia [Slovenia]

Tič I, Vrščaj B (2002) Talni Informacijski Sistem Slovenije - TIS. SQLtehnična dokumentacija. Tehnična dokumentacija o bazah podatkov na MS SQL strežniku "PedonSQL." Center za pedologijo in varstvo okolja, Biotehiška fakulteta, Univerza v Ljubljani, Ljubljana

Uhan J, Krajnc M (2003) Underground water. In: Uhan J, Bat M (eds) Richness of Slovenia's waters. Ministry of the Environment, Spatial Planning and Energy: Slovenian Environment Agency, Ljubljana

Urbančič M, Simončič P, Prus T, Kutnar L (2005) Atlas gozdnih tal [Atlas of forest soils]. Zveza gozdarskih društev Slovenije, Gozdarski vestnik in Gozdarski inštitut Slovenije [Union of Forestry Associations of Slovenia GDK and Slovenian Forestry Institute]

Verbič T (1998) Rocks. In: Fridl J (ed) Geographical atlas of Slovenia: the county in space and time. DZS, Ljubljana, pp 74–77

Verbič J, Jeretina J, Opara A, Vernik T, Vrščaj B (2011) Poročilo o izvedbi strokovnih nalog za Ministrstvo za okolje in prostor v letu 2010: končno poročilo (KIS - Poročila o strokovnih nalogah No. 123). KIS - Poročila o strokovnih nalogah. Kmetijski inštitut Slovenije, Ljubljana, Slovenia

VGI (2000) Slovenia's wetland inventory according to the Ramsar convention. Institute for Water of the Republic of Slovenia

Vidic NJ (1998) Soil-age relationships and correlations: comparison of chronosequences in the Ljubljana Basin, Slovenia and USA. CATENA 34:113–129. doi:10.1016/S0341-8162(98)00085-X

Vidic NJ, Lobnik F (1997) Rates of soil development of the chronosequence in the Ljubljana Basin, Slovenia. Geoderma 76:35–64. doi:10.1016/S0016-7061(96)00098-5

Vidic N, Pavich M, Lobnik F (1991) Statistical analyses of soil properties on a quaternary terrace sequence in the upper Sava River valley, Slovenia, Yugoslavia. Geoderma 51:189–211. doi:10.1016/0016-7061(91)90071-Z

Vovk Korže A (1995) Landscape-ecological units of the Northeastern Slovenia. Ph.D. dissertation, Oddelek za geografijo, Filozofska fakulteta, Univerza v Ljubljani, Ljubljana, Slovenia

Vovk Korže A (2003) New approach to the soils knowledge. Geogr Sch 12

Vovk Korže A (2007) The role of soils in the ecosystem. Dela 107–119

Vovk Korže A (2014) Soils and permaculture. Geogr Sch 23:21–27

Vovk Korže A, Lovrenčak F (2004) A manual for soil field research. Geography Department, Faculty of Arts, University of Ljubljana, Ljubljana; Maribor

Vrščaj B (1983) Navodila za delo s Commodore 64 [Instructions for work with commodore 64]. Ljubljana, Slovenia [Slovenia]

Vrščaj B (1987) Profili.exe računalniški pogram za obdelavo podatov talnih profilov [Profili.exe—a computer programe for soil profile data management]. Center za pedologijo in varstvo okolja, Biotehiška fakulteta, Univerza v Ljubljani [Center for Soil and Environment Protection, Biotechnical Faculty, University of Ljubljana], Ljubljana, Slovenia [Slovenia]

Vrščaj B (1992) Gnojenje.exe računalniški pogram za izračun gnojinih odmerkov. Ver 2.2 [Gnojenje.exe—a computer programme for soil fertility control. ver 2.2]. Center za pedologijo in varstvo okolja, Biotehiška fakulteta, Univerza v Ljubljani [Center for Soil and Environment Protection, Biotechnical Faculty, University of Ljubljana], Ljubljana, Slovenia [Slovenia]

Vrščaj B (1994) Gnojenje.exe računalniški pogram za izračun gnojinih odmerkov. Ver. 3.0b Gnojenje.exe računalniški pogram za izračun gnojinih odmerkov. Ver 3.0b [Gnojenje.exe—a computer programme for soil fertility control]. Katedra za pedologijo in varstvo okolja, Biotehniška fakulteta, Univerza v Ljubljani, Ljubljana

Vrščaj B (1996) Izdelava modela podatkovnega segmenta talnega informacijskega sistema. Thesis, Univerza v Ljubljani, Biotehniška fakulteta, Oddelek za agronomijo, Ljubljana

Vrščaj B (1999) Digitalni podatki tal Slovenije. Zasnova, vsebina in možnosti uporabe [Digital soil data of Slovenia. Design, content and potential aplications]. Zb. Bioteh. Fak. Univerze V Ljubljani 1999, p 23

Vrščaj B (2008) Strukturne spremembe kmetijskih zemljišč, njihova urbanizacija in kakovost v obdobju 2002–2007. Hmelj Bilt 15:73

Vrščaj B (2010) Krovno poročilo projekta Algoritmi digitalne kartografije in pretvorbene funkcije za potrebe nacionalnega talnega informacijskega sistema [Executive summary of the project algorithms of digital soil mapping and conversion tools for the national soil information system]. Raziskovalno poročilo [Research report], Raziskovalna poročila Kmetijskega inštituta Slovenije [Agricultural Institute of Slovenia Research Reports]. Kmetijski inštitut Slovenije, CL, Center za tla in okolje [Agricultural Institute of Slovenia, CL, Centre for Soil and Environment Research], Ljubljana, Slovenia [Slovenia]

Vrščaj B (2011) Primerjalna analiza izgub kmetijskih zemljišč na trasah hitre ceste 3. razvojne osi odseka Šentrupert - Velenje [Soil loss assessment on Šentrupert-Velenje Morotway]. Raziskovalno poročilo [Research report] No. 1. Kmetijski inštitut Slovenije, CL, Center za tla in okolje [Agricultural Institute of Slovenia, CL, Centre for Soil and Environment Research], Ljubljana, Slovenija [Slovenia]

Vrščaj B (2014a) Jerina - preliminarni izsledki morfologije vinogradniških tal na Krasu [Jerina—preliminary results of morphological examination of vineyard soil in the Karst]. Zb. Prispevkov Simp. Kraško Okolje Proj. Agrotur Proc. Symp. Karst Environ. Proj. Agrotur 2014, pp 7–22

Vrščaj B (2014b) Olive groves land suitability - SW Slovenia case study; Final report (research report No. 2014-06A). Kmetijski inštitut Slovenije, Oddelek za kmetijsko ekologijo in naravne vire, Center za tla in okolje [Agricultural Institute of Slovenia, Department for Agroecology and Natural Resources, Centre for Soil and Environment], Ljubljana, Slovenija [Slovenia]

Vrščaj B, Godeša T (2008) Zajem in obdelava pedoloških podatkov; DS2 Zajem in obdelava pedoloških podatkov [Soil data collection and processing, WP2]. Kmetijski inštitut Slovenije [Agricultural Institute of Slovenia], Ljubljana, Slovenia [Slovenia]

Vrščaj B, Kralj T (2014) Digitalna pedološka karta Slovenije po WRB 2014, 1:25.000 (PK25-WRB2014) [Digital soil map of Slovenia according to the WRB 2014, 1:25,000 (DSM25-WRB2014)]

Vrščaj B, Lobnik F (1997) Digitalni podatki tal Slovenije [Digital soil data of Slovenia], 16

Vrščaj B, Lobnik F (1999a) Establishement of the digital soil map of Slovenia in the scale 1:25.000. Res Rep Biotech Fac Univ Ljubl - Agric 73:287–300

Vrščaj B, Lobnik F (1999b) Talni informacijski sistem [Soil information system]. Zb Bioteh Fak Univerze V Ljubljani 1999, 23

Vrščaj B, Tič I (2001a) Tehnična dokumentacija o bazah podatkov na MS SQL strežniku PedonSQL [Soil information system of Slovenia —SIS. MS SQL server—PedonSQL technical documentation]. Center za pedologijo in varstvo okolja, Biotehiška fakulteta, Univerza v Ljubljani [Center for Soil and Environment Protection, Biotechnical Faculty, University of Ljubljana], Ljubljana, Slovenia [Slovenia]

Vrščaj B, Tič I (2001b) Tehnična dokumentacija o bazah podatkov talnega analitskega laboratorija na MS SQL strežniku PedonSQL. Center za pedologijo in varstvo okolja, Biotehiška fakulteta, Univerza v Ljubljani [Center for Soil and Environment Protection, Biotechnical Faculty, University of Ljubljana], Ljubljana

Vrščaj B, Zupan M, Lobnik F (1989) Računalniška obdelava podatkov v projektu tematska karta onesnaženosti zemljišč Celjske občine [Computer data processing in the thematic map on soil contamination of the Celje Municipality]. Zb Bioteh Fak Univerze V Ljubljani 1989:41–58

Vrščaj B, Prus T, Harts JJ, Arragon JV (1994a) Plotting directly on the topographic maps: map production technique. Presented at the EGIS/MARI '94, Utrecht: EGIS Foundation, Paris, pp 387–390

Vrščaj B, Prus T, Harts JJ, Arragon JV (1994b) Soil information system (SIS/TIS) in Slovenia. Presented at the EGIS/MARI '94, Utrecht: EGIS Foundation, Paris, pp 382–386

Vrščaj B, Prus T, Harts JJ, Arragon JV (1994c) Application of soil information system in the national project of irrigation in Slovenia. Presented at the EGIS/MARI '94, Utrecht: EGIS Foundation, Paris, pp 1457–1463

Vrščaj B, Prus T, Kralj T, Šporar M, Rupreht J, Tič I, Zupan M, Lobnik F, Štampar F, Korošec Z, Bernik R, Kastelec D, Knapič M (2003) Zasnova metodologije za vrednotenje zemljišč - končno poročilo. Univerza v Ljubljani, Oddelek za agronomijo, Center za pedologijo in varstvo okolja, Ljubljana

Vrščaj B, Prus T, Lobnik F (2005) Soil information and soil data use in Slovenia. Soil Resour Eur 2005:331–344

Vršič S, Valdhuber J, Pulko B, Stergar A (2000) Einfluss der Bodenpflege auf Erosion und Nährstoffbilanz. Begrün Im Weinbau Kolloqu Int Arbeitskreises 112–115

Wästerlund I, Hassan E A (1995) Forest harvesting systems friendly to the environment: p 30

Western AW, Grayson RB, Blöschl G, Willgoose GR, McMahon TA (1999) Observed spatial organization of soil moisture and its relation to terrain indices. Water Resour Res 35:797–810. doi:10.1029/1998WR900065

Wischmeier WH, Smith DD (1978) Predicting rainfall erosion losses: a guide to conservation planning. Science and Education Administration, U.S. Department of Agriculture

Wischmeier W, Johnson C, Cross B (1971) A soil erodibility nomograph for farmland and construction sites. J Soil Water Conserv 26:189–193

Wraber M (1969) Pflanzengeographische Stellung und Gliederung Sloweniens, Vegetatio, The Hague, 17(1–6):176–199

Železnik P, Mali B, Robek R, Simončič P, Kraigher H (2011) Vpliv težke mehanizacije na strojno sečnjo na tla in drobne korenine. In: KRČ, Janez (ed) Odzivi gozdne tehnike in gozdarstva na spremenjene razmere gospodarjenja : zbornik razširjenih izvlečkov. 1. izd. Ljubljana: BF, Odd. za gozdarstvo in obnovljive gozdne vire, p. 39–41

Zemljič M (1972) Erosion phenomena in Slovenia. For Bull 30:233–238

Zemljišni informacioni sistem Jugoslavije - kodni sistem [Soil information system of Yugoslavia—the coding system] (1978) Sarajevo

Žižek L, Planinšek Š, Skudnik M, Hladnik D, Kobal M, Okršlar G, Žlindra D, Simončič P (2013) Zaključno poročilo naloge Javna okoljska služba 2012. Ljubljana: Gozdarski inštitut Slovenije, p 286, http://eprints.gozdis.si/1746/

Zobec I (1930) Kmetijska kemija [Agricultural chemistry]. Jugoslovanska tiskarna v Ljubljani, Ljubljana, Slovenia [Slovenia]

Zorn M (2007) Recent geomorphic processes on fluvial-denudational relief on the example of Dragonja River Basin. Dissertation thesis, Geography Department, Faculty of Arts, University of Ljubljana, Ljubljana, Slovenia

Zorn M (2008) Erosion processes in Slovene Istria. Založba ZRC

Zorn M (2012) Soil erosion on flysch [WWW Document]. URL http://giam2.zrc-sazu.si/sites/default/files/zorn_erozija_prsti_flis.pdf. Accessed 1.4.15

Zorn M, Komac B (2002) Slope processes and the debris flow in Log Pod Mangartom. Geogr Bull 74:9–23

Zorn M, Komac B (2004) Recent mass movements in Slovenia. In: Orožen Adamič M (ed) Slovenia: a geographical overview. Association of the Geographical Societies of Slovenia, Založba ZRC, Ljubljana, Slovenia, pp 79–80

Zorn M, Komac B (2005a) Soil erosion on agricultural land in Slovenia. Ujma 19

Zorn M, Komac B (2005b) Soil erosion on agricultural land in slovenia —measurements of rill erosion in the Besnica Valley. Acta Geogr Slov 47

Zorn M, Komac B (2007) Probability modelling of landslide hazard. Acta Geogr Slov 47:139–169

Zorn M, Komac B (2008) Zemeljski plazovi v Sloveniji [Landslides in Slovenia], Georitem. Založba ZRC, Ljubljana, Slovenia [Slovenia]

Zorn M, Komac B (2008) Landslides in Slovenia, Georitem. Založba ZRC, Ljubljana

Zorn M, Petan S (2007) Measurement of interill soil erosion under different land use in Slovene Istria. In: Knapič M (ed) Thematic strategy for soil protection: conference at the World soil day, 5 Dec 2007. Soil Science Society of Slovenia, Ljubljana, pp 51–63

Zupan M, Grčman H, Prus T, Hodnik A, Vrščaj B (2002a) Praktikum iz pedologije [Lecture notes of soil science]. Univerza v Ljubljani, Biotehniška fakulteta, Oddelek za Agronomijo [University of Ljubljana, Biotechnical Faculty, Department of Agronomy], Ljubljana

Zupan M, Vrščaj B, Lobnik F (2002b) Soil pollution monitoring as a comprehensive system. Presented at the 17th world congress of soil science, 14–21 August 2002, Bangkok, Thailand, International society of soil science, Bangkok, pp 1410/1–1410/8

Zupan M, Grčman H, Lobnik F (2008) Raziskave onesnaženosti tal Slovenije [Soil contamination research in Slovenia]. Agencija RS za okolje; Center za pedologijo in varstvo okolja [Slovenian Environment Agency and Center for Soil Science and Environment], Ljubljana, Slovenia [Slovenia]

Zupančič M, Marinček L, Seliškar A, Puncer I (1987) Considerations on the phytogeographic division of Slovenia. Biogeographia - Biogeografia delle Alpi Sud-Orientali, 13:89–98

Zupančič M, Žagar V (1995) New views about the phytogeographic division of Slovenia. Razprave IV razreda, 26(1):3–30

Index

B. Vrščaj et al., *The Soils of Slovenia*, World Soils Book Series,
DOI 10.1007/978-94-017-8585-3

MIX
Papier aus verantwortungsvollen Quellen
Paper from responsible sources
FSC® C105338

If you have any concerns about our products,
you can contact us on
ProductSafety@springernature.com

In case Publisher is established outside the EU,
the EU authorized representative is:
Springer Nature Customer Service Center GmbH
Europaplatz 3, 69115 Heidelberg, Germany

Printed by Libri Plureos GmbH
in Hamburg, Germany